Magnetism of Molecular Conductors

Special Issue Editor
Manuel Almeida

MDPI • Basel • Beijing • Wuhan • Barcelona • Belgrade

MDPI

Special Issue Editor
Manuel Almeida
Universidade de Lisboa
Portugal

Editorial Office
MDPI
St. Alban-Anlage 66
Basel, Switzerland

This edition is a reprint of the Special Issue published online in the open access journal *Magnetochemistry* (ISSN 2312-7481) from 2016–2017 (available at: http://www.mdpi.com/journal/magnetochemistry/special_issues/molecular_conductors).

For citation purposes, cite each article independently as indicated on the article page online and as indicated below:

Lastname, F.M.; Lastname, F.M. Article title. *Journal Name* **Year**, *Article number*, page range.

First Edition 2018

ISBN 978-3-03842-931-9 (Pbk)
ISBN 978-3-03842-932-6 (PDF)

Table of contents

About the Special Issue Editor

Manuel Almeida, Senior Researcher and Professor, studied at the Technical University of Lisbon, where he obtained a Ph.D. in 1984 under the supervision of Prof. Luís Alcácer. After a post- doctoral stay with Professor Tobin J. Marks at the Northwestern University, USA, from 1985–1986, he joined the Portuguese National Laboratory LNETI to establish a research group on Solid State Physics and Chemistry, focused on the preparation and characterization of materials with unconventional electrical and magnetic properties. In 2013, his laboratory was integrated into the University of Lisbon, where he is leading research activities in materials chemistry and physics.

Preface to "Magnetism of Molecular Conductors"

Magnetism of Molecular Conductors

Manuel Almeida

C2TN-Centro de Ciências e Tecnologias Nucleares, Instituto Superior Técnico, Universidade de Lisboa, P-2695-066 Bobadela LRS, Portugal; malmeida@ctn.tecnico.ulisboa.pt; Tel.: +351-219-946-171

Received: 19 June 2017; Accepted: 21 June 2017; Published: 27 June 2017

The study of the magnetic properties of molecular conductors has experienced, during the last decades, a very significant evolution, comprising systems of increasing molecular complexity and moving towards multifunctional materials, namely by their incorporation in conducting networks of different paramagnetic centers. In this context, molecular magnetic conductors have emerged at the intersection between the fields of molecule-based conductors and molecule-based magnets as a very exciting class of multifunctional materials in which the interaction and synergy between conduction electrons and localized magnetic moments can lead to new phenomena, complex phase diagrams, and different ground states, with a large potential for technological applications, namely in electronic devices, sensors and in spintronics. Among these phenomena are unusual field-induced transitions, including magnetic field-induced superconductivity, very large magnetoresistance effects, conductors that are switchable by magnetic field, changes of magnetic ordering or spin state, etc.

This Special Issue of Magnetochemistry features a collection of research contributions illustrating recent achievements in different aspects of this topic concerning the development, study and understanding of the magnetic properties of molecular conductors and their applications. Quite different types of compounds are considered.

A contribution by Tamotsu Inabe et al. [1], reviews a series of compounds based on axially ligated phthalocyanines. Metal phthalocyanines are one of the first examples of compounds where, in addition to delocalized π-conduction electrons in the ligand, there can also be localized magnetic moments for some metals. The π–d interaction of these local moments embedded in the sea of conduction electrons has long since been identified as a source of possible interesting phenomena. In this contribution, the properties of $TPP[M(Pc)(CN)_2]_2$ compounds (TPP = tetraphenylphosphonium, Pc = phthalocyaninato), with M = Fe and Cr are reviewed, emphasizing carrier localization and charge disproportionation enhanced by the interaction between local magnetic moments and conduction π-electrons (π–d interaction), and the large negative magnetoresistance, reflecting the difference in the anisotropy of different d–d, π–d, and π–π interactions.

Two other contributions in this issue concern a family of compounds with two types of chains (conducting and magnetic), based on the organic perylene donor and inorganic $[M(mnt)_2]$ anions, which have been studied for more than 30 years, but are still unique among molecular materials. In a review by Jean-Paul Pouget et al. [2], the structural instabilities exhibited of these salts are reviewed and discussed in relation to the magnetic properties of the 1D spin-Peierls (SP) instability of the dithiolate stacks, showing, in particular, that α-$(Per)_2[M(mnt)_2]$ salts exhibit the physical properties expected of a two-chain Kondo lattice. In another contribution by Manuel Matos et al. [3], these compounds are also addressed, the properties of the solid solutions $(Per)_2[Pt_xAu_{(1-x)}(mnt)_2]$ being described, probing the incorporation of paramagnetic $[Pt(mnt)_2]$ impurities in diamagnetic chains, and the effect of breaking paramagnetic chains with diamagnetic centers.

Another contribution by Yugo Oshima et al. [4] concerns λ-$(BETS)_2FeCl_4$, (BETS = bis (ethylenedithio)tetraselenafulvalene), a very relevant compound in the context of the topic of this special issue, due to the magnetic field-induced superconducting state observed. Studies on the antiferromagnetic insulating phase of this compound are reviewed, and new ESR data on the solid solutions with diamagnetic anions, λ-(BETS)$_2$Fe$_x$Ga$_{1-x}$Cl$_4$, are provided, showing that there is no sign of paramagnetic Fe spins in the antiferromagnetic ground state, which has been a point of previous debate.

Another contribution by Maria Laura Mercuri [5] reviews Anilato-Based Molecular Materials, illustrating the large potential of anilato ligands, derivatives of the 2,5-dioxy-1,4-benzoquinone framework with various substituents (X = H, Cl, Br, I, CN, etc.) in different positions as molecular building blocks for the design of a rich variety of materials with peculiar magnetic and/or conducting properties.

Molecular radical units with unpaired electrons have been the basis of one of the earliest devised strategies for achieving organic conductors. In the contribution of Manuel Souto et al. [6], a new molecular dyad is reported, based on a monopyrrolo-tetrathiafulvalene electron donor linked by a π-conjugated bridge to a perchlorotriphenylmethyl radical, with interesting properties and the potential to give rise to new radical conductors in the solid state. The combination of radical units with electroactive donor networks is another strategy for preparing magnetic conductors, and in this issue Kazuki Horikiri and Hideki Fujiwara describe charge transfer salts of a new EDT-TTF (ethylenedithiotetrathiafulvalene) donor containing a radical through a π-conjugated vinylene spacer, with diamagnetic GaCl$_4$ and paramagnetic FeCl$_4$ anions with strong π–d interactions [7]. Another contribution by Hiroki Akutsu et al. describes two dmit-based salts with a stable organic radical-substituted ammonium cation, exhibiting magnetic contributions from segregated 2D anionic and cationic sub-lattices [8].

BEDT-TTF (Bis(ethylenedithio)tetrathiafulvalene or ET) is one of the most successful electron donors, being the basis of a very large number of charge transfer salts with metallic and superconducting properties. In the contribution of Samia Benmansour et al. [9], two novel paramagnetic conductors of this donor—as salts with oxalate anionic layers containing high spin Mn(III) (S = 2) and Mn(II) (S = 5/2) ions—are described. BEDT-TTF charge transfer salts are characterized by a very large structural diversity associated with different electrical and magnetic properties. Tadashi Kawamoto et al., in their contribution to this issue [10], describe structural, transport, and magnetic properties two δ-type polymorphs of the salt (BEDT-TTF)$_2$TaF$_6$ with Charge Ordering transitions.

The exploration of interaction effects between delocalized conduction electrons and paramagnetic centers by incorporation of spin-crossover units in conducting networks remains a challenge in molecular multifunctional conducting and magnetic materials. As an example of current efforts under such strategy, a contribution by Yuri N. Shvachko et al. [11] describes three conducting systems based on the electron acceptor TCNQ and spin cross-over Fe(III) cations, displaying strong interactions between local magnetic moments of Fe(III) ions and electron spins of the organic TCNQ network.

Finally, Royama Yamamoto et al. [12] explore the incorporation of photosensitive dyes to optically control and trigger conduction and magnetism in the photo-excited states of organic materials. In their contribution, these authors report a new type of salt based on [Ni(dmit)$_2$] (dmit = 1,3-dithiole-2-thione-4,5-dithiolate and 3,3′-Dihexyloxacarbocyanine monocation, exhibiting photoconductivity with photo-excited spins, demonstrating the possibility of preparing paramagnetic organic photo responsive semiconductors.

From the large diversity of compounds considered in this special issue, illustrating current strategies for the development of molecular magnetic conductors, it becomes clear that the magnetic conductors are a topic of increasing interest among molecular materials, where several significant developments and relevant applications are expected in the near future.

Conflicts of Interest: The authors declare no conflict of interest.

References

1. Inabe, T.; Hanasaki, N. Axially Ligated Phthalocyanine Conductors with Magnetic Moments. *Magnetochemistry* **2017**, *3*, 18. [CrossRef]
2. Pouget, J.-P.; Foury-Leylekian, P.; Almeida, M. Peierls and Spin-Peierls Instabilities in the Per$_2$[M(mnt)$_2$] Series of One-Dimensional Organic Conductors; Experimental Realization of a 1D Kondo Lattice for M = Pd, Ni and Pt. *Magnetochemistry* **2017**, *3*, 13. [CrossRef]
3. Matos, M.; Bonfait, G.; Santos, I.C.; Afonso, M.L.; Henriques, R.T.; Almeida, M. The solid solutions (Per)$_2$[Pt$_x$Au$_{(1-x)}$(mnt)$_2$]; Alloying para- and diamagnetic anions in two-chain compounds. *Magnetochemistry* **2017**, *3*, 22. [CrossRef]
4. Oshima, Y.; Cui, H.-B.; Kato, R. Antiferromagnetic Insulating Ground State of Molecular π–d System λ-(BETS)$_2$FeCl$_4$ (BETS = Bis(ethylenedithio)tetraselenafulvalene): A Theoretical and Experimental Review. *Magnetochemistry* **2017**, *3*, 10. [CrossRef]
5. Mercuri, M.L.; Congiu, F.; Concas, G.; Sahadevan, S.A. Recent Advances on Anilato-Based Molecular Materials with Magnetic and/or Conducting Properties. *Magnetochemistry* **2017**, *3*, 17. [CrossRef]
6. Souto, M.; Bendixen, D.; Jensen, M.; Díez-Cabanes, V.; Cornil, V.J.; Jeppesen, J.O.; Ratera, I.; Rovira, C.; Veciana, J. Synthesis and Characterization of Ethylenedithio-MPTTF-PTM Radical Dyad as a Potential Neutral Radical Conductor. *Magnetochemistry* **2016**, *2*, 46. [CrossRef]
7. Horikiri, K.; Fujiwara, H. New Ethylenedithio-TTF Containing a 2,2,5,5-Tetramethylpyrrolin-1-yloxyl Radical through a Vinylene Spacer and Its FeCl$_4^-$ Salt—Synthesis, Physical Properties and Crystal Structure Analyses. *Magnetochemistry* **2017**, *3*, 8. [CrossRef]
8. Akutsu, H.; Turner, S.S.; Nakazawa, Y. New Dmit-Based Organic Magnetic Conductors (PO-CONH-C$_2$H$_4$N(CH$_3$)$_3$)[M(dmit)$_2$]$_2$ (M = Ni, Pd) Including an Organic Cation Derived from a 2,2,5,5-Tetramethyl-3-pyrrolin-1-oxyl (PO) Radical. *Magnetochemistry* **2017**, *3*, 11. [CrossRef]
9. Benmansour, S.; Sánchez-Máñez, Y.; Gómez-García, C.J. Mn-Containing Paramagnetic Conductors with Bis(ethylenedithio)tetrathiafulvalene (BEDT-TTF). *Magnetochemistry* **2017**, *3*, 7. [CrossRef]
10. Kawamoto, T.; Kurata, K.; Mori, T.; Kumai, R. Charge Ordering Transitions of the New Organic Conductors δ$_m$- and δ$_o$-(BEDT-TTF)$_2$TaF$_6$. *Magnetochemistry* **2017**, *3*, 14. [CrossRef]
11. Shvachko, Y.N.; Starichenko, D.V.; Korolyov, A.V.; Kotov, A.I.; Buravov, L.I.; Zverev, V.N.; Simonov, S.V.; Zorina, L.V.; Yagubskii, E.B. The Highly Conducting Spin-Crossover Compound Combining Fe(III) Cation Complex with TCNQ in a Fractional Reduction State. Synthesis, Structure, Electric and Magnetic Properties. *Magnetochemistry* **2017**, *3*, 9. [CrossRef]
12. Yamamoto, R.; Yamamoto, T.; Ohara, K.; Naito, T. Dye-Sensitized Molecular Charge Transfer Complexes: Magnetic and Conduction Properties in the Photoexcited States of Ni(dmit)$_2$ Salts Containing Photosensitive Dyes. *Magnetochemistry* **2017**, *3*, 20. [CrossRef]

Manuel Almeida
Special Issue Editor

magnetochemistry

MDPI

Review

Axially Ligated Phthalocyanine Conductors with Magnetic Moments

Tamotsu Inabe [1],* and Noriaki Hanasaki [2],*

1   Department of Chemistry, Faculty of Science, Hokkaido University, Sapporo 060-0810, Japan
2   Department of Physics, Osaka University, Toyonaka, Osaka 560-0043, Japan
*   Correspondence: inabe@sci.hokudai.ac.jp (T.I.); hanasaki@phys.sci.osaka-u.ac.jp (N.H.);
     Tel.: +81-11-706-3511 (T.I.); Tel.: +81-6-6850-5751(N.H.)

Academic Editor: Manuel Almeida
Received: 15 March 2017; Accepted: 18 April 2017; Published: 23 April 2017

Abstract: This mini-review describes electrical conductivity, magnetic properties, and magnetotransport properties of one-dimensional partially oxidized salts composed of axially ligated phthalocyanines, $TPP[M(Pc)(CN)_2]_2$ (TPP = tetraphenylphosphonium, Pc = phthalocyaninato), with M of Fe (d^5, $S = 1/2$) and Cr (d^3, $S = 3/2$). These salts are isomorphous, and π–π interactions in the crystal, that becomes the origin of the charge carriers, are nearly the same. Both the Fe and Cr salts show carrier localization and charge disproportionation which is enhanced by the interaction between local magnetic moments and conduction π-electrons (π–d interaction). However, the magnetic properties are slightly different between them. M = Fe has been found to show unique anisotropic magnetic properties and antiferromagnetic short-range magnetic order between the d-spins. On the other hand, for M = Cr, its magnetic moment is isotropic. Temperature dependence of the magnetic susceptibility shows typical Curie–Weiss behavior with negative Weiss temperature, but the exchange interaction is complicated. Both M = Fe and M = Cr show large negative magnetoresistance, reflecting the difference in the anisotropy. The magnetoresistance ratio (MR) is larger in the Fe system than in the Cr system in the low magnetic field range, but MR in the Cr system exceeds that in the Fe system when the magnetic field becomes higher than 15 T. We discuss the mechanism of the giant negative magnetoresistance with reference to the d–d, π–d, and π–π interactions.

Keywords: phthalocyanine-based conductor; magnetic ion; π–d interaction; negative magnetoresistance

1. Introduction

As well as the planar phthalocyanines, M(Pc) (Figure 1a), axially ligated phthalocyanine anions, $[M(Pc)L_2]^-$ (Figure 1b), give electrically conducting partially oxidized salts by electrolysis [1]. So far, the anionic complexes with M of Co [2], Fe [3], Cr [4], Mn [5], and Ru [6] and L of CN, Cl, and Br [7] have been synthesized. The d orbitals in M show ligand-field splitting by the coordination of Pc^{2-} and L, and take low-spin d electron configuration. In this situation, Co^{III} (d^6) becomes non-magnetic, but the other metals have unpaired electrons, introducing local magnetic moments in the conduction paths. In this mini-review, we describe the magnetic, transport, and magnetotransport properties of the conductors of M = Fe (d^5, $S = 1/2$) and Cr (d^3, $S = 3/2$) with the cationic component of TPP (tetraphenylphosphonium) and L = CN. Before starting to describe the properties of these magnetic conductors, we briefly survey the structure and properties of the system with M = Co.

For M = Co and L = CN, a partially oxidized salt of $TPP[Co(Pc)(CN)_2]_2$ was obtained when the TPP salt ($TPP[Co(Pc)(CN)_2]$) was electrochemically oxidized. If the cationic part was exchanged, a series of conducting crystals with various dimensionality of the π–π interactions were obtained [8,9]. In this mini-review, we will focus on the TPP salts.

(a)

(b)

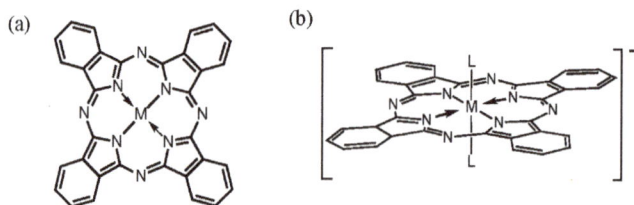

Figure 1. (a) Planar M(Pc) and (b) axially ligated $[M(Pc)L_2]^-$.

In TPP[Co(Pc)(CN)$_2$]$_2$, the Pc units form a one-dimensional (1D) π–π stacking chain with negligible interchain interactions (Figure 2a) [2]. The π–π stacking is uniform. Since each Pc ring is oxidized by 1/2e (TPP$^+$[Co^{3+}(Pc$^{1.5-}$)(CN$^-$)$_2$]$_2$), the HOMO (highest occupied molecular orbital of the Pc π-system) band becomes three-quarters-filled (metallic band). However, the temperature dependence of the resistivity showed thermally activated behavior (Figure 2b). The ^{59}Co-NQR experiments revealed that the ground state was a charge disproportionation phase (Figure 2c) [10]. This results from the fact that the π system is susceptible to the electron correlation effect due to the narrow conduction band. Indeed, the band width of the slipped π–π stacking system was estimated to be about 0.5 eV from the thermoelectric power measurements, which is much smaller compared with those of the typical face-to-face stacked systems, e.g., 0.88 eV for Ni(Pc)I [1].

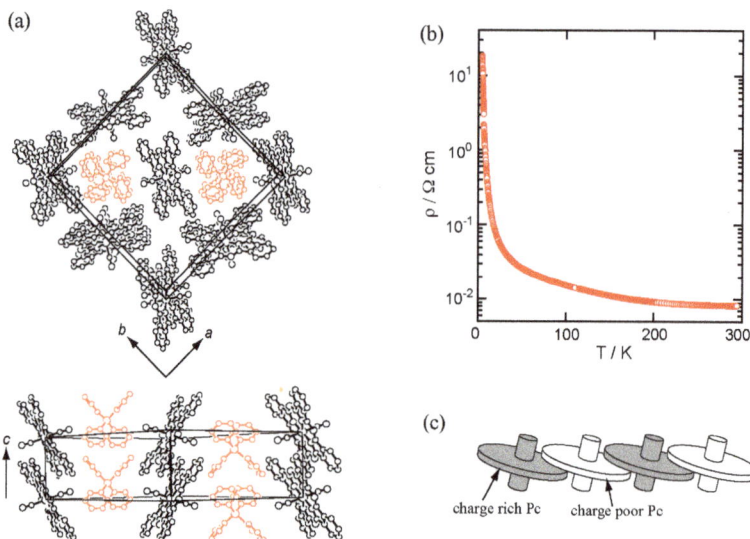

Figure 2. (a) Crystal structure of TPP[Co(Pc)(CN)$_2$]$_2$ (TPP = tetraphenylphosphonium); (b) temperature dependence of the single-crystal resistivity (along the *c*-axis) of TPP[Co(Pc)(CN)$_2$]$_2$; and (c) schematic picture of the charge disproportionation in TPP[Co(Pc)(CN)$_2$]$_2$.

We were interested in how the physical properties would be altered by exchanging non-magnetic Co^{III} (d^6, $S = 0$) with magnetic Fe^{III} (d^5, $S = 1/2$) in this TPP salt [3]. Fortunately, introduction of Fe^{III} negligibly affected the molecular structure of the Pc unit and the crystal structure in the TPP salt. This is because the ion size of Fe^{3+} is almost the same as that of Co^{3+} (ionic radius r(Fe^{3+}) = 0.64 Å and r(Co^{3+}) = 0.63 Å). Thus, the geometry of the Pc unit was not affected and the crystal of TPP[Fe(Pc)(CN)$_2$]$_2$ was isomorphous with TPP[Co(Pc)(CN)$_2$]$_2$ (Table 1). In addition, the difference in π–π interaction between these two crystals was found to be negligible (overlap integral between the Pc π-HOMOs; 0.0085 in the Co system and 0.0087 in the Fe system) [3]. Therefore, one can see that the Co compound can be a good reference of the pure π-system when one discusses the effect of the local magnetic moments on the physical properties in the Fe compound.

Table 1. Crystal Data of TPP[M(Pc)(CN)$_2$]$_2$.

Crystal Data of TPP[M(Pc)(CN)$_2$]$_2$			
	M = Co	**M = Fe**	**M = Cr**
Crystal system		Tetragonal	
Space group		$P4_2/n$	
a/Å	21.676 (8)	21.722 (2)	21.778 (2)
c/Å	7.474 (4)	7.448 (2)	7.4636 (6)
V/Å^3	3511 (3)	3514.4 (5)	3539.9 (5)
Z		2	

Similarly, replacement by Cr^{III} (d^3, $S = 3/2$) gave negligible effects on the molecular geometry, crystal structure (Table 1), and π–π interaction (overlap integral between the Pc π-HOMOs = 0.0091) because of almost the same ion size ($r(\mathrm{Cr}^{3+}) = 0.63$ Å) [4]. The Cr system is expected to have an isotropic magnetization because of the d^3 configuration under the axially deformed octahedral ligand field (Figure 3a). On the other hand, the Fe system is expected to have an anisotropic magnetization by the spin-orbit interaction, because the degenerate d$_{xz}$ and d$_{yz}$ orbitals accommodate the unpaired electron under the D_{4h} symmetry of low-spin d^5 (Figure 3b). Thus, the difference between the Fe and Cr systems lies in the presence of the anisotropy in addition to the absolute value of the magnetic moment.

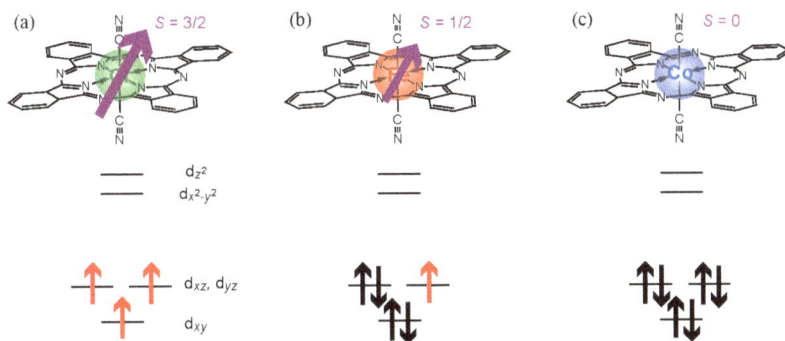

Figure 3. Schematic picture of the energy levels of d-orbitals in the (**a**) Cr^{III}(Pc)(CN)$_2$; (**b**) Fe^{III}(Pc)(CN)$_2$, and (**c**) Co^{III}(Pc)(CN)$_2$ units.

As described above, the replacement of M = Co by M = Fe and Cr in TPP[M(Pc)(CN)$_2$]$_2$ gives a good opportunity to discuss the π–d interactions by the various magnetic moments in the systems with common π–π interactions. Especially, we would like to emphasize that the Pc system is advantageous because the magnetic moment is introduced in the center of the π-ligand with fixed geometry. This situation is expected to yield larger π–d interactions compared with the other two-component systems with indirect π–d interactions between the individual π-conduction assemblies and counter ions with the local moment.

2. TPP[Fe(Pc)(CN)$_2$]$_2$ Magnetic Conductor

Since the discovery of the fascinating property of TPP[Fe(Pc)(CN)$_2$]$_2$, giant negative magnetoresistance [11], the magnetic exchange interactions between the π-spin and d-spin, including whether they are ferromagnetic or antiferromagnetic and how large they are, have been subjected to discussion. The experimental evidence was recently provided [12]. In conclusion, significantly strong ferromagnetic exchange interaction was found to exist between the Pc π-spin and Fe d-spin ($J_{\pi d}/k_B > 500$ K). This confirms the unique feature expected for the Pc system, namely, strong intramolecular π–d exchange interaction should work in the single Pc unit. It should be noted that the magnitude of $J_{\pi d}$ is extremely large compared with that estimated for the systems with π–d interactions between the individual π-conduction assemblies and counter ions with the local moment (order of ca. 10 K [13]).

With this feature in mind, the physical properties of TPP[Fe(Pc)(CN)$_2$]$_2$ are introduced. Firstly, its electrical conductivity is compared with that of the Co system [3]. The room-temperature resistivity (about 10^{-1} Ω cm) is one order of magnitude higher than that of the Co system, whereas the increase of the resistivity by lowering the temperature is rather steep. The value at 20 K is, thus, more than six orders of magnitude larger than that of the Co system (Figure 4a). This feature is considered to result from the development of charge disproportionation owing to the interaction between the local magnetic moment and π conduction electrons [14], and is consistent with the large $J_{\pi d}$.

Secondly, its magnetic properties are described. Figure 4b shows the temperature dependence of the magnetic susceptibility (χ_P) of the oriented crystals. Large anisotropy is a prominent feature [11]. In $B \perp c$, the susceptibility shows an anomaly at 20–25 K, indicating the occurrence of the antiferromagnetic magnetic order. However, the susceptibility for $B//c$ is significantly smaller and changes monotonically. The lower panel of Figure 4b shows a plot of $1/\chi_P$ vs. T. For both $B \perp c$ and $B//c$, the susceptibility reveals the Curie–Weiss-like behavior with similar Weiss temperature (-20 K $< \theta < -10$ K) in the temperature range of 120–300 K.

The magnetic anisotropy was found to result from the large anisotropy of the g-tensor (Figure 4c) from the angular dependence of the ESR spectra of [Fe(Pc)(CN)$_2$]$^-$ [15]. The origin of this anisotropy was considered to arise from the spin-orbit interaction under the situation in which the degenerate d_{xz} and d_{yz} orbitals accommodate the unpaired electron under the D_{4h} symmetry (Figure 3b) [15]. However, the quantum chemistry calculation revealed that the Jahn–Teller effect lowers the symmetry to D_{2h}, inducing splitting of d_{xz} and d_{yz} orbitals [7]. Even in this case, the energy difference between the ground state $(d_{xz})^2(d_{yz})^1$ and the excited state $(d_{xz})^1(d_{yz})^2$ configurations is very small (0.01–0.02 eV), and this situation suggests that the strong spin-orbit interaction plays an essential role in this system.

The numerical simulation based on the anisotropic Heisenberg model in one dimension was performed to explain the anomaly observed at 25 K for the susceptibility with $B \perp c$ and the large anisotropy [16]. The results indicate that the anomaly at 25 K is due to antiferromagnetic short range order formation of the d electrons and that the π-electrons fall into an antiferromagnetic state at the lower temperatures. By combining the simulation results for the other one-dimensional system (PTMA$_{0.5}$[Fe(Pc)(CN)$_2$]·CH$_3$CN, PTMA = phenyltrimethylammonium [17]), a model of charge-ordered ferrimagnetism was proposed for the spontaneous magnetization at low temperatures [18].

The most fascinating feature of this system is its magnetotransport properties, and the above-mentioned magnetic state is important to elucidate the origin of them. The negative magnetoresistance effect appears regardless of the direction of the magnetic field (Figure 5a). However, the magnitude of this effect precisely reflects the anisotropy of the magnetic susceptibility, namely, the effect is enhanced for $B \perp c$ while it is reduced for $B//c$ [11].

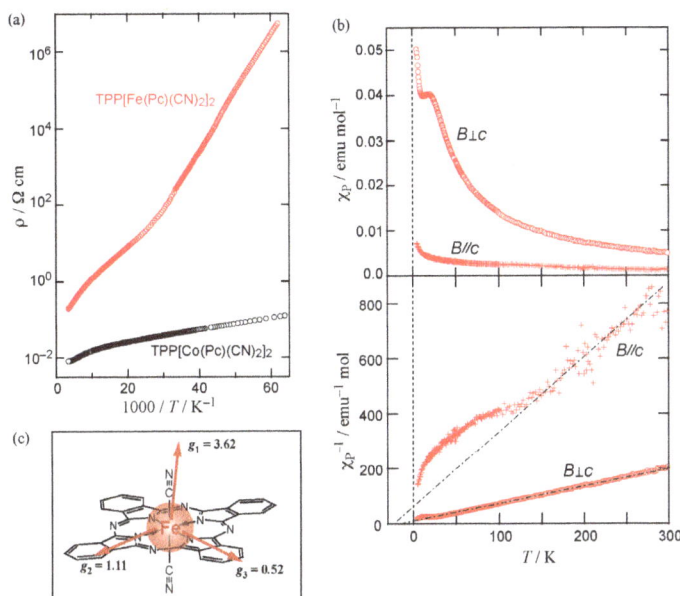

Figure 4. (**a**) Temperature dependence of the resistivity (along the *c* axis) of TPP[Fe(Pc)(CN)$_2$]$_2$ and TPP[Co(Pc)(CN)$_2$]$_2$; (**b**) [upper panel] χ_P vs. *T* plot of the magnetic susceptibility of TPP[Fe(Pc)(CN)$_2$]$_2$ measured in the magnetic field (1 T) perpendicular and parallel to the *c*-axis. [lower panel] $1/\chi_P$ vs. *T* plot for the same susceptibility data; (**c**) *g*-tensor anisotropy determined from the angular dependence of ESR of PNP[Fe(Pc)(CN)$_2$] single crystal (PNP = bis(triphenylphosphine)iminium).

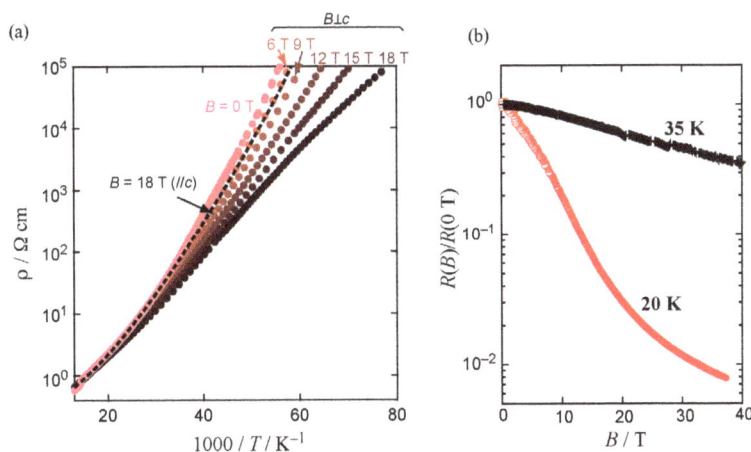

Figure 5. (**a**) Temperature dependence of the resistivity of TPP[Fe(Pc)(CN)$_2$]$_2$ measured along the *c* axis in magnetic fields (0–18 T) perpendicular to the *c* axis and that in a magnetic field of 18 T parallel to the *c* axis (dashed curve); (**b**) Field dependence ($B \perp c$) of the magnetoresistance at 20 and 35 K.

Figure 5b shows the field dependence of the magnetoresistance up to 37 T at 20 K [19] and at 35 K. The resistance decreases smoothly, indicating that the giant negative magnetoresistance effect is not due to a magnetic field-induced first order phase transition. Though the magnetoresistance is still decreasing in 37 T, even at this stage, the magnetoresistance ratio (MR = {[R(B) − R(0 T)]/R(0 T)} × 100 (%)) is as large as −99.5%. The MR at 35 K is relatively small, about −65%, indicating less development of the charge disproportionation at this temperature (vide infra).

Now, let us discuss the mechanism of the appearance of the giant negative magnetoresistance in TPP[Fe(Pc)(CN)$_2$]$_2$. At zero magnetic field, the charge disproportionation state in π-electrons is caused by the intersite Coulomb interaction V, as observed for the Co salt. At low temperatures, a short-range antiferromagnetic order between the localized d-spins appears ($|J_{dd}|/k_B$ ~32 K). The d–d interaction is assumed to be through a superexchange mechanism, since the interaction rapidly decreases when the d-spin concentration is diluted below 50% [20]. The π-electron interacts with the localized magnetic moment by the strong π–d interaction ($|J_{πd}|/k_B$ >500 K). As a result, the charge disproportionation is enhanced (charge-ordered ferrimagnetism: middle panel in Figure 6). If the π-electron hops to the neighboring site whilst keeping the spin state, the spin becomes antiparallel to the localized d-spin. This situation increases the energy of the electronic system by $J_{πd}$. Therefore, $J_{πd}$ effectively enhances the Coulomb effect of V.

Next, we consider the state in magnetic fields. At high temperature, there is an antiferromagnetic fluctuation enhancing the charge disproportionation. The resistivity may be increased only in this fluctuation region. Negative magnetoresistance is considered to be achieved by reducing this fluctuation region with applying magnetic fields. At low temperatures where weak ferromagnetism is observed, the antiferromagnetic order may grow considerably along the one-dimensional direction. As can be seen from the magnetization, high magnetic fields are required to destroy the short-range antiferromagnetic order of Fe d-spins. From the detailed magnetoresistance measurements of TPP[Fe$_x$Co$_{1-x}$(Pc)(CN)$_2$]$_2$, it was revealed that the charge disproportionation was rather developed in intermediate magnetic fields [20]. This is because the energy in the electronic system can decrease owing to the Zeeman effect when the π-electrons are localized. When the field becomes much higher, the short-range antiferromagnetic order of Fe d-spins is destroyed. At this stage, the local moment in the Pc unit becomes parallel to those in the neighboring units, and π-electrons can hop to the neighboring sites, leading to the reduction of the electrical resistance (lower panel in Figure 6) [21].

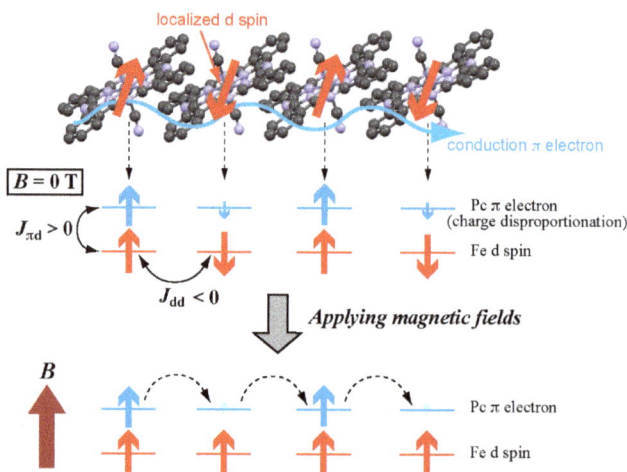

Figure 6. One-dimensional chain of [Fe(Pc)(CN)$_2$]$^{1/2-}$ (upper panel). Middle and lower panels show the schematic diagrams of the magnetic interactions at low temperature. At $B = 0$ T, Fe d-spins tend to form the antiferromagnetic short-range order that makes Pc π-electrons localized due to strong π–d interaction. In magnetic fields, breakage of the antiferromagnetic order of Fe d-spins allows Pc π-electrons to transfer to the neighboring site when the field is applied perpendicular to the *c* axis (1D chain direction).

3. TPP[Cr(Pc)(CN)$_2$]$_2$ Magnetic Conductor

As shown in Figure 3a, Cr^{III} in TPP[Cr(Pc)(CN)$_2$]$_2$ has $S = 3/2$, resulting in a larger magnetic moment than that of Fe^{III}. In addition, even under D_{4h} symmetry, all degenerate t$_{3g}$ orbitals accommodate an unpaired electron in the high spin state, thus the magnetization moment becomes almost isotropic, because the contribution from the spin-orbit momentum to the magnetism is quenched. Indeed, angular dependent ESR [4] and multifrequency ESR [22] experiments gave nearly isotropic g-value ($g = 1.995 \pm 0.005$). The temperature dependence of the static magnetic susceptibility was found to follow the Curie–Weiss law above 50 K (Figure 7a) with the Curie constant $C = 4.16$ emu K mol^{-1} (expected value for $S = 3/2$; 3.75 emu K mol^{-1} (the formula unit contains two Cr(Pc)(CN)$_2$ units)) and the Weiss temperature $\theta = -20$ K ($|J|/k_B = 8.2$ K from $\theta = zJS(S + 1)/3k_B$, where z (coordination number) = 2 for the 1-D system) [22]. Below 15 K, the observed susceptibility data is larger than that expected from the Curie–Weiss law. This may be due to contribution from the charge-order ferrimagnetism as observed for the Fe system. In the magnetization experiments under high magnetic fields, since the saturation of the magnetization was not observed even in 53 T within the measured temperature range, magnetic exchange interaction was estimated to be $|J_{dd}|/k_B > 11.9$ K [22].

Figure 7b shows the temperature dependence of the resistivity of TPP[Cr(Pc)(CN)$_2$]$_2$. Though the room-temperature resistivity is almost the same as the Fe system ($10^{-1} \sim 10^0$ Ω cm), the activation energy of the conduction is higher compared with the Fe system, suggesting that the charge disproportionation is more developed in the Cr system. Indeed, the current density–electric field plot of TPP[Cr(Pc)(CN)$_2$]$_2$ showed negative differential resistance (NDR) below 60 K [4]; the temperature at which NDR appeared was significantly higher than that in the Fe system (NDR appeared below 30 K [23]). These facts strongly support that the charge localization in the Cr system occurs at higher temperature compared with the Fe system.

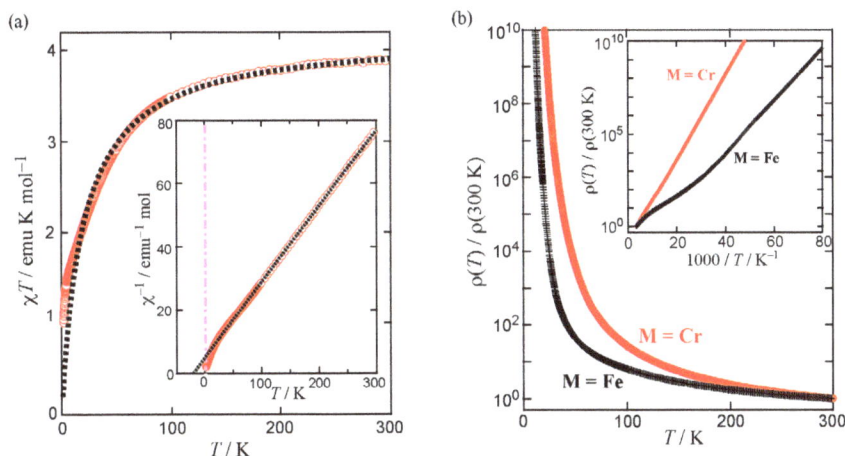

Figure 7. (**a**) Temperature dependence of the product of magnetic susceptibility and temperature (χT) in 1 T of TPP[Cr(Pc)(CN)$_2$]$_2$. The dotted line represents a best fit by the Curie–Weiss law for the data above 50 K. Inset: Temperature dependence of the inverse susceptibility; (**b**) Temperature dependence of the resistivity normalized by the value at 300 K ($\rho(T)/\rho(300$ K)) of TPP[M(Pc)(CN)$_2$]$_2$ with M = Fe and Cr. Inset: Arrhenius plot of the normalized resistivity.

Since the charge disproportionation was suggested to be more developed in the Cr system, it was expected to show larger magnetoresistance effects. The first report up to the field strength of 9 T at 20 K indicated that the effect is only 25% of that observed for the Fe system [4]. However, the measurements at higher magnetic fields indicated that the magnetoresistance effect of the Cr system at 35 K near 50 T was 1.3 times larger than that observed for the Fe system (Figure 8). It can be seen from Figure 8 that the non-magnetic Co system shows normal positive magnetoresistance.

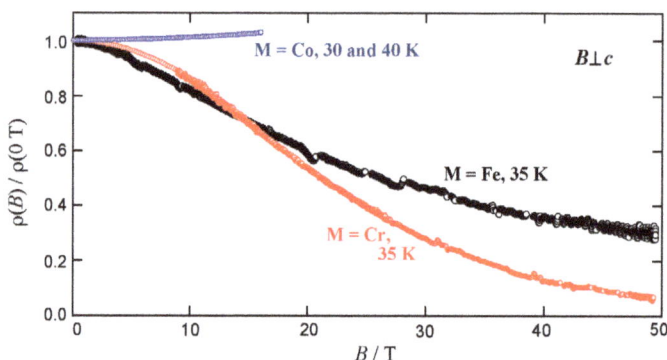

Figure 8. Magnetic-field dependence of normalized resistivity [$\rho(B)/\rho(0\ \text{T})$] of TPP[M(Pc)(CN)$_2$]$_2$ for M = Fe (35 K), Cr (35 K) and Co (30 and 40 K) in magnetic fields perpendicular to the *c* axis.

Now, let us consider the difference between the Fe and Cr systems. For the saturation of the magnetization, higher magnetic fields are required in the Cr system than in the Fe system. This suggests that the antiferromagnetic interaction between the d-spins is larger in the Cr system than in the Fe system. This stronger interaction in the Cr system makes the charge disproportionation of the π-electrons more developed. As a result, when the d-spins are aligned parallel by the external magnetic fields, the charge disproportionation is attenuated by a larger degree, making the magnetoresistance effect more pronounced in the Cr system. When the magnetic field strength is insufficient to align the d-spins parallel, the charge disproportionation is affected by a much smaller degree, resulting in a smaller magnetoresistance effect compared with the Fe system.

At the present stage, there is no quantitative evaluation of the field strength required to align the d local moments. For the Fe system, the field dependence of the magnetoresistance seems to approach the point of saturation at this temperature. However, it is suggested that much higher field strength is required for the saturation of the magnetoresistance as well as magnetization in the Cr system than in the Fe system.

So far, the magnetic moment of the π-electron has not been included in the discussion. In the system without local magnetic moment, TPP[Co(Pc)(CN)$_2$]$_2$, the susceptibility shows nearly constant small values in the high-temperature region (~5 × 10^{-4} emu mol^{-1}), because π-electrons can hop to the neighboring sites with the aid of finite *t* (transfer integral), resulting in the Pauli-like behavior [24]. However, at low temperatures, the susceptibility increases gradually where the line width of the ESR signal shows gradual broadening. This indicates that the π-electrons bear localized character along with the antiferromagnetic fluctuation [10], suggesting the growth of charge disproportionation.

This situation seems to be drastically changed by introducing a local magnetic moment (this accompanies large ferromagnetic π–d interaction). In the high-temperature region where the charge disproportionation is not fully developed, the susceptibility of the π-electrons is considered to become as small as that observed for the Co system (π-electrons can hop to the neighboring sites with certain probability due to the thermal fluctuation). At the present stage, we have not succeeded in estimating the contribution of the π-electrons from the observed susceptibility containing the large value of the localized magnetic moment. In the Fe system, there is a steep rise of the susceptibility in the low-temperature region. The magnetic torque experiments [16,18] suggested that the susceptibility of the π-electrons contributes to this steep rise. The π-electrons show an antiferromagnetic state below 13 K, and it is explained that this order leads to the charge-order ferrimagnetism (middle panel of Figure 6). In the Cr system, a steep rise of magnetization was also observed at low temperatures.

However, there are different aspects in the magnetic behavior, suggesting more complicated magnetic interactions in the Cr system.

Though the magnetic order state of spins originating from the π-electrons can be observable only at low temperatures, traces of the order state may remain up to higher temperatures, as is the case for the d-spins (in the temperature range in which the charge disproportionation is observable). As shown in Figure 8, the negative magnetoresistance is indeed observed at 35 K. At this temperature, the charge disproportionation is considered to develop, though the ordering of the π-spins was not observed. The spins originating from the π-electrons also respond to the magnetic field through the strong π–d interactions. The detailed magnetic structure of the π-electrons in the Cr system may be different from that in the Fe system. However, it can be assumed that, in both the systems at 35 K, the π-electrons are released from the localized state in the charge disproportionation by destroying small domains of antiferromagnetic order of d-spins by applying external magnetic fields.

4. Conclusions

This mini-review describes the series of isostructural Pc conductors of $TPP[M(Pc)(CN)_2]_2$ that shows giant negative magnetoresistance with M = Fe and Cr. After the discovery of giant negative magnetoresistance of the Fe system in 2000, many studies have been conducted. The detailed aspects of the charge disproportionation, π–d interactions, d–d interactions, and π–π interactions have been disclosed by comparison with the non-magnetic Co system, by theoretical studies, and by the elaborate magnetic and magnetotransport measurements. At zero magnetic field, short-range antiferromagnetic order of d-spins follows the development of charge disproportionation. From the above experimental results, a model of charge-ordered ferrimagnetism was proposed. In magnetic fields, the following mechanism was proposed. The antiferromagnetic order of the d-spins is destroyed by the magnetic field, which leads to destabilization of the charge disproportionation state of the π-electrons. This causes the decrease of the resistance from the value at zero field.

On the other hand, studies on the system with a larger magnetic moment, M = Cr, have recently been conducted. In contrast to the anisotropic moment of the Fe system, the Cr system showed an almost isotropic moment. The charge disproportionation was suggested to be more developed in the Cr system. The magnetic measurements at high fields have suggested that there are rather robust antiferromagnetic interactions in the Cr system, supporting more developed charge disproportionation. As a result, the field dependence of the magnetoresistance of the Cr system crosses that of the Fe system, resulting in a larger negative magnetoresistance effect of the Cr system at higher magnetic fields.

Magnetic properties at high fields suggested that the mechanism of the negative magnetoresistance in the Cr system might contain some different aspects compared with the Fe system. Nevertheless, there seems to be a common feature in these systems: breakage of magnetic interactions between the d-spins follows destabilization of the charge disproportionation of the π-electrons that leads to the reduction of the resistance.

The occurrence of π–d interactions is ensured when magnetic ions of M are introduced into $M(Pc)L_2$ units composing the molecular conductors. This is a valuable system that realizes unique magnetotransport properties. Though the $M(Pc)L_2$ unit is flexible toward substitution of the components, the structural framework is rather robust. Therefore, this is a useful building block for the design of functional π–d systems.

Acknowledgments: The authors acknowledge all the researchers who collaborated this study. The authors especially thank to the following co-workers for their great contribution: Hiroyuki Tajima at University of Hyogo, Masaki Matsuda at Kumamoto University, Hiroshi Murakawa and Mitsuo Ikeda at Osaka University.

Conflicts of Interest: The authors declare no conflict of interest.

References

1. Inabe, T.; Tajima, H. PhthalocyaninesVersatile components of molecular conductors. *Chem. Rev.* **2004**, *104*, 5503–5533. [CrossRef] [PubMed]

2. Hasegawa, H.; Naito, T.; Inabe, T.; Akutagawa, T.; Nakamura, T. A highly conducting partially oxidized salt of axially substituted phthalocyanine. Structure and physical properties of TPP[Co(Pc)(CN)$_2$]$_2$ {TPP = tetraphenylphosphonium, [Co(Pc)(CN)$_2$] = dicyano(phthalocyaninato)cobalt(III)}. *J. Mater. Chem.* **1998**, *8*, 1567–1570. [CrossRef]

3. Matsuda, M.; Naito, T.; Inabe, T.; Hanasaki, N.; Tajima, H.; Otsuka, T.; Awaga, K.; Narymbetov, B.; Kobayashi, H. A one-dimensional macrocyclic π-ligand conductor carrying a magnetic center. Structure and electrical, optical and magnetic properties of TPP[Fe(Pc)(CN)$_2$]$_2$ {TPP = tetraphenylphosphonium, [Fe(Pc)(CN)$_2$] = dicyano(phthalocyaninato)iron(III)}. *J. Mater. Chem.* **2000**, *10*, 631–636. [CrossRef]

4. Takita, Y.; Hasegawa, H.; Takahashi, Y.; Harada, J.; Kanda, A.; Hanasaki, N.; Inabe, T. One-dimensional phthalocyanine-based conductor with *S* = 3/2 isotropic magnetic centers. *J. Porphyr. Phthalocyanines* **2014**, *18*, 814–823. [CrossRef]

5. Matsuda, M.; Yamaura, J.; Tajima, H.; Inabe, T. Structure and magnetic properties of a low-spin manganese(III) phthalocyanine dycyanide complex. *Chem. Lett.* **2005**, *34*, 1524–1525. [CrossRef]

6. Yu, D.E.; Kikuchi, A.; Taketsugu, T.; Inabe, T. Crystal structure of ruthenium phthalocyanine with diaxial monoatomic ligand: Bis(triphenylphosphine)iminium dichloro(phthalocyaninato(2-))ruthenium(III). *J. Chem.* **2013**, *2013*, 486318. [CrossRef]

7. Yu, D.E.C.; Matsuda, M.; Tajima, H.; Kikuchi, A.; Taketsugu, T.; Hanasaki, N.; Naito, T.; Inabe, T. Variable magnetotransport properties in the TPP[Fe(Pc)L$_2$]$_2$ system (TPP = tetraphenylphosphonium, Pc = phthalocyaninato, L = CN, Cl, and Br). *J. Mater. Chem.* **2009**, *19*, 718–723. [CrossRef]

8. Asari, T.; Naito, T.; Inabe, T.; Matsuda, M.; Tajima, H. Novel phthalocyanine conductor containing two-dimensional Pc stacks, [PXX]$_2$[Co(Pc)(CN)$_2$] (PXX = *peri*-xanthenoxanthene, Co(Pc)(CN)$_2$ = dicyano(phthalocyaninato)cobalt(III)). *Chem. Lett.* **2004**, *33*, 128–129. [CrossRef]

9. Asari, T.; Ishikawa, M.; Naito, T.; Matsuda, M.; Tajima, H.; Inabe, T. Nearly isotropic two-dimensional sheets in a partially oxidized Co(Pc)(CN)$_2$ salt (Pc = phthalocyaninato). *Chem. Lett.* **2005**, *34*, 936–937. [CrossRef]

10. Hanasaki, N.; Masuda, K.; Kodama, K.; Matsuda, M.; Tajima, H.; Yamazaki, J.; Takigawa, M.; Yamaura, J.; Ohmichi, E.; Osada, T.; et al. Charge disproportionation in highly one-dimensional molecular conductor TPP[Co(Pc)(CN)$_2$]$_2$. *J. Phys. Soc. Jpn.* **2006**, *75*, 104713. [CrossRef]

11. Hanasaki, N.; Tajima, H.; Matsuda, M.; Naito, T.; Inabe, T. Giant negative magnetoresistance in quasi-one-dimensional conductor TPP[Fe(Pc)(CN)$_2$]$_2$: Interplay between local moments and one-dimensional conduction electrons. *Phys. Rev. B* **2000**, *62*, 5839–5842. [CrossRef]

12. Murakawa, K.; Kanda, A.; Ikeda, M.; Matsuda, M.; Hanasaki, N. Giant ferromagnetic π-*d* interaction in a phthalocyanine molecule. *Phys. Rev. B* **2015**, *92*, 054429. [CrossRef]

13. Mori, T.; Katsuhara, M. Estimation of πd-interactions in organic conductors including magnetic anions. *J. Phys. Soc. Jpn.* **2002**, *71*, 826–844. [CrossRef]

14. Hotta, C.; Ogata, M.; Fukuyama, H. Interaction of the ground state of quarter-filled one-dimensional strongly correlated electronic system with localized spins. *Phys. Rev. Lett.* **2005**, *95*, 216402. [CrossRef] [PubMed]

15. Hanasaki, N.; Matsuda, M.; Tajima, H.; Naito, T.; Inabe, T. Contribution of degenerate molecular orbitals to molecular orbital angular momentum in molecular magnet Fe(Pc)(CN)$_2$. *J. Phys. Soc. Jpn.* **2003**, *72*, 3226–3230. [CrossRef]

16. Tajima, H.; Yoshida, G.; Matsuda, M.; Nara, K.; Kajita, K.; Nishio, Y.; Hanasaki, N.; Naito, T.; Inabe, T. Magnetic torque and heat capacity measurements on TPP[Fe(Pc)(CN)$_2$]$_2$. *Phys. Rev. B* **2008**, *78*, 064424. [CrossRef]

17. Matsuda, M.; Naito, T.; Inabe, T.; Hanasaki, N.; Tajima, H. Structure and electrical and magnetic properties of (PTMA)$_x$[M(Pc)(CN)$_2$]·y(solvent) (PTMA = phenyltrimethylammonium and [M(Pc)(CN)$_2$] = dicyano(phthalocyaninato)M^{III} with M = Co and Fe). Partial oxidation by partial solvent occupation of the cationic site. *J. Mater. Chem.* **2001**, *11*, 2493–2497. [CrossRef]

18. Tajima, H.; Yoshida, G.; Matsuda, M.; Yamaura, J.; Hanasaki, N.; Naito, T.; Inabe, T. Magnetic torque and ac and dc magnetic susceptibility measurements on PTMA$_{0.5}$[Fe(Pc)(CN)$_2$]·CH$_3$CN: Origin of spontaneous magnetization in [Fe(Pc)(CN)$_2$] molecular conductors. *Phys. Rev. B* **2009**, *80*, 024424. [CrossRef]

19. Hanasaki, N.; Matsuda, M.; Tajima, H.; Ohmichi, E.; Osada, T.; Naito, T.; Inabe, T. Giant negative magnetoresistance reflecting molecular symmetry in dicyano(phthalocyaninato)iron compounds. *J. Phys. Soc. Jpn.* **2006**, *75*, 033703. [CrossRef]

20. Ikeda, M.; Kanda, A.; Murakawa, H.; Matsuda, M.; Inabe, T.; Tajima, H.; Hanasaki, N. Effect of localized spin concentration on giant magnetoresistance in molecular conductor TPP[Fe_xCo_{1-x}(Pc)(CN)$_2$]$_2$. *J. Phys. Soc. Jpn.* **2016**, *85*, 024713. [CrossRef]

21. Hanasaki, N.; Tateishi, T.; Tajima, H.; Kimata, M.; Tokunaga, M.; Matsuda, M.; Kanda, A.; Murakawa, H.; Naito, T.; Inabe, T. Metamagentic transition and its related magnetocapacitance effect in phthalocyanine-molecular conductor exhibiting giant magnetoresistance. *J. Phys. Soc. Jpn.* **2013**, *82*, 094713. [CrossRef]

22. Ikeda, M.; Kida, T.; Tahara, T.; Murakawa, H.; Nishi, M.; Matsuda, M.; Hagiwara, M.; Inabe, T.; Hanasaki, N. High magnetic field study on giant negative magnetoresistance in the molecular conductor TPP[Cr(Pc)(CN)$_2$]$_2$. *J. Phys. Soc. Jpn.* **2016**, *85*, 064713. [CrossRef]

23. Ishikawa, M.; Yamashita, S.; Naito, T.; Matsuda, M.; Tajima, H.; Hanasaki, N.; Akutagawa, T.; Nakamura, T.; Inabe, T. Nonlinear transport phenomena in highly one-dimsnsional M^{III}(Pc)(CN)$_2$ chains with π–d interaction (M = Co and Fe and Pc = phthalocyaninnato). *J. Phys. Soc. Jpn.* **2009**, *78*, 104709. [CrossRef]

24. Yamashita, S.; Naito, T.; Inabe, T. Purity effects on the charge-transport propertoes in one-dimensional TPP[Co^{III}(Pc)(CN)$_2$]$_2$ (TPP = tetraphenylphosphonium and Pc = phthalocyaninato) conductors. *Bull. Chem. Soc. Jpn.* **2009**, *82*, 692–694. [CrossRef]

magnetochemistry

MDPI

Article

Peierls and Spin-Peierls Instabilities in the Per$_2$[M(mnt)$_2$] Series of One-Dimensional Organic Conductors; Experimental Realization of a 1D Kondo Lattice for M = Pd, Ni and Pt

Jean-Paul Pouget [1,*], Pascale Foury-Leylekian [1] and Manuel Almeida [2]

[1]   Laboratoire de Physique des Solides, Université Paris-sud, CNRS UMR 8502, F91405 Orsay, France;
      pascale.foury@u-psud.fr
[2]   C2TN—Centro de Ciências e Tecnologias Nucleares, Instituto Superior Técnico, Universidade de Lisboa,
      P-2695-066 Bobadela LRS, Portugal; malmeida@ctn.tecnico.ulisboa.pt
*   Correspondence: jean-paul.pouget@u-psud.fr

Academic Editor: Carlos J. Gómez García
Received: 25 January 2017; Accepted: 17 February 2017; Published: 25 February 2017

Abstract: We consider structural instabilities exhibited by the one-dimensional (1D) (arene)$_2$X family of organic conductors in relation with their electronic and magnetic properties. With a charge transfer of one electron to each anion X, these salts exhibit a quarter-filled (hole) conduction band located on donor stacks. Compounds built with donors such as fluorenthene, perylene derivatives and anions X such as PF$_6$ or AsF$_6$ exhibit a high temperature (T_P ~170 K) conventional Peierls transition that is preceded by a sizeable regime of 1D $2k_F$ charge density wave fluctuations (k_F is the Fermi wave vector of the 1D electron gas located on Per stack). Surprisingly, and probably because of the presence of a multi-sheet warped Fermi surface, the critical temperature of the Peierls transition is considerably reduced in the perylene series α-(Per)$_2$[M(mnt)$_2$] where X is the dithiolate molecule with M = Au, Cu, Co and Fe. Special attention will be devoted to physical properties of α-(Per)$_2$[M(mnt)$_2$] salts with M = Pt, Pd and Ni which incorporate segregated S = 1/2 1D antiferromagnetic (AF) dithiolate stacks coexisting with 1D metallic Per stacks. We analyze conjointly the structural and magnetic properties of these salts in relation with the 1D spin-Peierls (SP) instability located on the dithiolate stacks. We show that the SP instability of Pd and Ni derivatives occurs in the classical (adiabatic) limit while the SP instability of the Pt derivative occurs in the quantum (anti-adiabatic) limit. Furthermore, we show that in Pd and Ni derivatives 1st neighbor direct and frustrated 2nd neighbor indirect (through a fine tuning with the mediated $2k_F$ RKKY coupling interaction on Per stacks) AF interactions add their contribution to the SP instability to stabilize a singlet-triplet gap. Our analysis of the data show unambiguously that magnetic α-(Per)$_2$[M(mnt)$_2$] salts exhibit the physics expected for a two chain Kondo lattice.

Keywords: One dimensional organic conductor; Peierls transition; spin-Peierls transition; Kondo lattice; frustrated antiferromagnetic chain

1. Introduction

Since the discovery of the so-called Peierls transition in the Krogmann salt K$_2$Pt(CN)$_4$Br$_{0.3}$-3H$_2$O (KCP) [1] in 1973, nearly than 45 years ago, many investigations have shown that most one-dimensional (1D) conductors are subject to a coupled electronic-structural instability transition at the $2k_F$ critical wave vector (k_F being the Fermi wave vector of the 1D electron gas). Due to the electron–phonon coupling the Peierls transition consists in a $2k_F$ modulated wave of bond distances, forming a so-called bond ordered wave (BOW), accompanied by a $2k_F$ modulation of the electronic density, forming a so-called charge density wave (CDW); these two waves being in quadrature (see [2]). The $2k_F$ modulation wave vector related to the 1D band filling is generally in incommensurate relation with the chain reciprocal wave vector, noted b* below ($2k_F$, expressed in b* reciprocal unit, amounts to half the number of conduction electrons, ρ, per repeat unit; the factor 2 is due to the spin degree of freedom). At the Peierls transition, T_P, the long range $2k_F$ modulation, which opens a gap at the Fermi level in the 1D band structure, drives a metal to insulator phase transition. However, because of the 1D nature of the underlying electronic instability, the Peierls transition is preceded by a very sizable regime of 1D $2k_F$ CDW/BOW fluctuations which extends up to 2–4 times T_P (the onset temperature of CDW/BOW fluctuations corresponds to about the mean-field Peierls temperature, T_P^{MF} defined more precisely below). Between T_P^{MF} and T_P, local fluctuations in direct space open a partial gap (i.e., a pseudo-gap) in the electronic structure. More explanations concerning these distinctive features can be found in a recent review [3].

$2k_F$ Peierls transitions are well observed in quasi-1D inorganic compounds such as the Krogmann salts, the blue bronzes, $K_{0.3}MoO_3$, and the transition metal tri-chalcogenides, $NbSe_3$ and TaS_3, built with chain of transition elements based inorganic polyhedron ($Pt(CN)_4$ square, MoO_6 octahedron and $NbSe_6/TaS_6$ anti-prism, respectively) between which there is a strong overlap of d wave functions [4]. Other significant examples can be found among organic conductors built with stacks of planar molecules between which there is a sizeable overlap of p_π molecular orbitals (MO). In organic salts, the metallic character is achieved either by a partial charge transfer ρ from stacks of donor (D) to stacks of acceptor (A), such as in TTF-TCNQ, or by a complete charge transfer from anion (X)/cation (Y) to D/A in 2:1 D_2X or A_2Y salts (TTF = tetrathia-fulvalene, TCNQ = tetracyano-quinodimethane). Only D_2X salts, with $\rho = 1/2$ hole charge transfer, will be considered in this paper (in these salts, the band structure built on the donor HOMO is quarter-filled in hole). An important characteristic of organic metals is that, with a stack built with tilted large planar molecules, the HOMO nodal structure leads to intra-stack transfer integral ($t_{//}$) often smaller or comparable to intra-molecular (U) or inter-molecular (V) Coulomb repulsion terms. Thus because of the relative importance of electron repulsions U and V with respect to $t_{//}$, organic conductors develop also another type of CDW instability at the critical $4k_F$ wave vector consisting in a Wigner (or Mott-Hubbard) type of charge localization [2]. Thus, as a non-degenerate HOMO level can be at most occupied by an hole whatever its spin, the critical wave vector of the charge localization mechanism, being associated to ρ, is $4k_F$ in 1D. In 2:1 organic salts, intensively studied in recent years [5], a spin-charge decoupling accompanies the charge localization phenomenon. When such a decoupling is achieved, the localized S = 1/2 degrees of freedom remain available to order in anti-ferromagnetic (AF) or non-magnetic singlet paired ground states. The singlet pairing is generally stabilized by a dimerization of the chain of localized spins. The transition which thus results opens a singlet-triplet gap in the AF magnetic excitation spectrum. Being analog to the metal-insulator Peierls transition, which opens a gap in a metallic excitation spectrum, the singlet pairing transition is called for this reason a spin-Peierls (SP) transition whose main characteristics are now well documented in the recent literature (see for example Reference [3,5]).

The perylene (Per) molecule, shown in Figure 1a, has played an important role in the development of the field of 1D organic conductors because the first molecular crystal exhibiting a metallic conductivity was found in 1954 when Per was exposed to Br [6]. Then many family of organic salts based on the Per donor and its derivatives were found to exhibit metallic properties [7]. Among them, 2:1 D_2X salts were found to be quasi-1D metals exhibiting a $2k_F$ BOW/CDW instability diverging into a Peierls metal-insulator transition (see Section 2). A very original physics is observed in the α phase of Per-dithiolate salts, named α-(Per)$_2$[M(mnt)$_2$] below, which mix 1D conducting and magnetic properties [7,8] (another β phase is semiconducting [7]). The structure of α-(Per)$_2$[M(mnt)$_2$] exhibits along the b direction (perpendicular to the plane of Figure 2) regular stacks of tilted and partially oxidized Per molecules which coexist with metal-bisdithiolene complex [M(mnt)$_2$]$^-$ stacks. [M(mnt)$_2$]$^-$, shown in Figure 1b, is a close shell molecule for M = Au and Cu, while [M(mnt)$_2$]$^-$ bears an unpaired spin 1/2 for M = Ni, Pd and Pt. Due to the charge transfer of one electron per dithiolate, leaving $\rho = 1/2$ hole per Per, α-(Per)$_2$[M(mnt)$_2$] forms a family of high conducting and anisotropic conductors (σ_b ~700 S/cm and $\sigma_b/\sigma_\perp$ ~10^3 at RT) [9] with a quarter-filled hole (or a three quarter filled electron) conduction band—see Figure 3a. With regular stacks of $M =$ Au and Cu close shell dithiolate molecules and dimerized stacks of $M =$ Fe and Co dithiolate molecules, only the Per stack is electro-active. These systems exhibit a Peierls instability that was reported more than 25 years ago [8]. Its salient features will be summarized in Section 2 and compared to those shown by other D$_2$X salts of arene donors. Magnetic α-(Per)$_2$[M(mnt)$_2$] salts incorporating dithiolate stacks with $M =$ Ni, Pd and Pt are very original systems since dithiolate stacks, where each [M(mnt)$_2$]$^-$ bears a spin S $= 1/2$, coexist with conducting Per stacks subject to a Peierls instability. In addition, these regular dithiolate stacks forming S $= 1/2$ AF chains are subject to a SP instability [9]. The SP instability of the dithiolate stacks was however poorly studied. Thus, a complete analysis of their SP instability will be object of Section 3. Section 3 will also consider the unprecedented coupling between the SP and Peierls instabilities from which a new physics emerges.

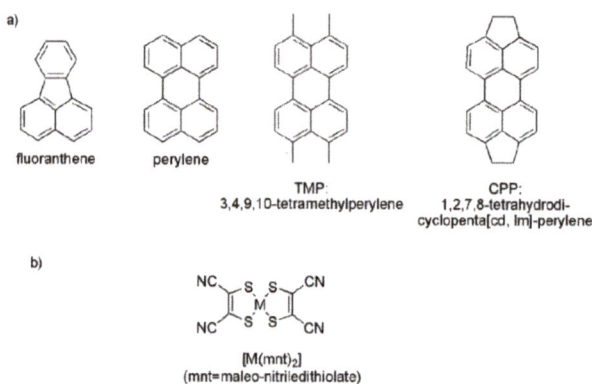

Figure 1. Chemical structure of: (**a**) fluoranthene (FA), perylene donors and its derivatives TMP and CPP; and (**b**) the dithiolate acceptor [M(mnt)$_2$].

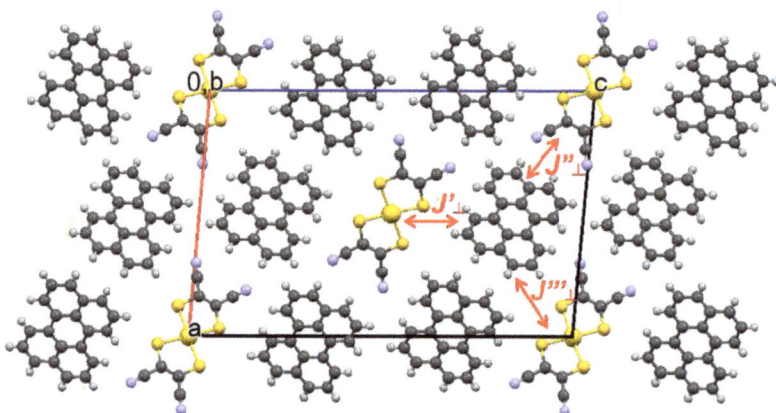

Figure 2. Crystal structure of α-Per$_2$[M(mnt)$_2$] projected along the stack direction b, which shows the presence of segregated Per and dithiolate stacks. In this structure, each [M(mnt)$_2$] stack fills tunnel delimited by six Per stacks, and there is one Per stack inside each triangular array of first neighbors [M(mnt)$_2$] stacks. First neighbor inter-stack [M(mnt)$_2$]—Per AF exchange coupling $J_\perp$ are schematically indicated (note that there are three different types of interactions per Per).

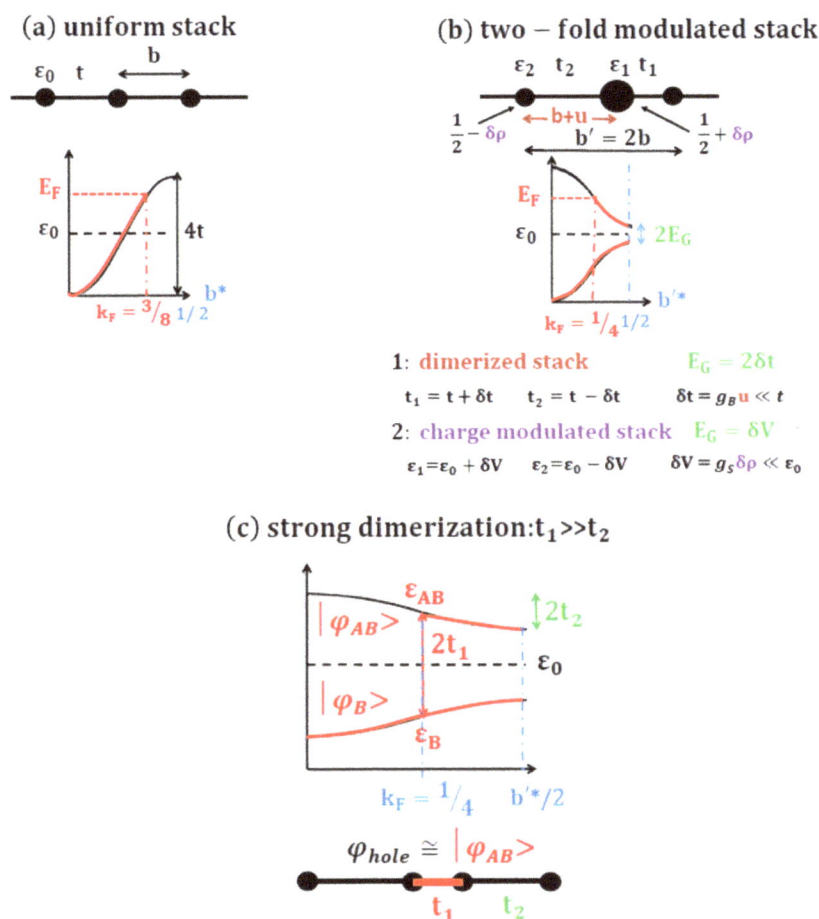

Figure 3. Three-quarter electron (one-quarter hole) filled band structure of D_2X arene salts for: a uniform donor stack (**a**); and a two-fold modulated donor stack (**b**). The positive band dispersion is due to the graphitic type of overlap of donor molecules. (**b**) considers the two extreme situations: (1) of a dimerized stack, and (2) of a charge modulated stack. The expression of the band gap $2E_G$ at the Brillouin zone boundary $b'*/2$ is given in (1), for a small modulation of transfer integrals t, and in (2) for a small modulation of site energies ε. The case of a strongly dimerized stack is considered in (**c**). In this situation the wave functions/energies are basically that of well decoupled bonding $/\Psi_B>/\varepsilon_B$ and anti-bonding $/\Psi_{AB}>/\varepsilon_{AB}$ states of the dimer. For a one quarter hole band filling, the system can be considered as having an half-filled AB band where the hole wave function is basically localized in the anti-bonding state of the dimer.

2. Peierls Instability in Per$_2$X Salts and Its Per Substituted Derivatives

In this section, we summarize the Peierls instabilities exhibited by various D_2X arene cation radical salts whose main characteristics are given in Table 1. This table shows that depending of the salt the donor stack can be either uniform or two-fold modulated. In the second case, two-fold bond or site modulations of the donor stack periodicity are generally the consequence of specific interactions with intrinsically dimerized anion stacks or with chemically alternated anion-solvent columns. Such differentiated interactions will either modulate inter-donor bond distances and/or modulate the external potential on donor sites with the consequence to modulate electronic parameters such as the intra-stack transfer integral t or the one-electron site energy ε (see Figure 3b). In all these cases, there is a band folding in b*/4 or b'*/2 together with the opening of a band gap $2E_G$, whose expression is given in Figure 3b. Furthermore, if there is a lateral disorder between columns of anions surrounding a given donor stack, the modulation of the stack will be also disordered.

1D CDW systems are very sensitive to disorder. Disorder either limits the lifetime of electron–hole pairs of wave vector $2k_F$, which are the building blocks of the CDW, or pins the phase of the $2k_F$ BOW/CDW modulation. Electron backscattering on impurities reverses the wave vector of one constituent of the electron–hole pair (let say from $+k_F$ to $-k_F$), which destroy the CDW pairing. Consequently, the backscattering process gives a finite lifetime at the electron–hole pair. This lifetime effect depresses the Peierls transition, as observed in the solid solution $Per_2[Au_xPt_{1-x}(mnt)_2]$ [10]. Local pinning of the phase of the CDW on a random distribution of lattice defects limits the longitudinal and transverse spatial coherence of the 3D CDW order (for more details see [3]).

Table 1. Characteristics of arene cation radical salts exhibiting a Peierls instability. The table gives the structure of the donor stack. The Peierls transition temperature (T_P) and the temperature of minimum of resistivity (T_ρ) are taken from conductivity measurements of Ref. [7,8]. The mean field Peierls temperature (T_P^{MF}) is calculated from the electrical gap $2\Delta_0$ using Equation (2). The dimension and temperature range of $2k_F^D$ BOW pre-transitional fluctuations are also indicated. [a] Dithiolate sacks composed of strongly paired $[Fe(mnt)_2]$ or $[Co(mnt)_2]$ units should induce a doubling of the Per stack periodicity. The expected Per two-fold stack deformation due to its dimerized surrounding has not been determined in the Co compound [11], but it has been recently found to be small in the Fe compound [12]. [b] The question mark leaves open a possible doubling of CPP stack periodicity induced by its surrounding of alternated anion–solvent columns [13]. [c] A doubling of TMP stack periodicity is expected from the detection of a charge order (i.e., $4k_F$ CDW) on donor stack by ^{13}C-CPAS-NMR [13]. However, local NMR measurement is unable to probe the spatial extend of this order.

Compound	Donor Stacking	T_P (K)	$2k_F^D$ BOW Pre-Transitional Fluctuations	T_ρ (K)	T_P^{MF} (K)
$Per_2[Cu(mnt)_2]$	Uniform	32	not observed	40	66
$Per_2[Au(mnt)_2]$	Uniform	12	not observed	16	11.5
$Per_2[Fe(mnt)_2]$	Dimerized [a]	73	3D $\leq$ 80 K	180	165
$Per_2[Co(mnt)_2]$	Dimerized [a]	58	3D $\leq$ 65 K	160	200
$(CPP)_2PF_6 + CH_2Cl_2$	Uniform? [b]	158	1D above RT	>300	?
$(CPP)_2AsF_6 + CH_2Cl_2$	Uniform? [b]	170	1D above RT	?	?
$(TMP)_2PF_6 + CH_2Cl_2$	$4k_F$ site CDW [c]	<20	1D $\leq$ 210 K	>300	?
$(TMP)_2AsF_6 + CH_2Cl_2$	$4k_F$ site CDW [c]	<20	1D $\leq$ 200 K	?	?
$(FA)_2 PF_6$	Dimerized	180	1D above RT	>300	400–600

2.1. Uniform Stack

For ρ = 1/2 hole per arene, the 1D donor band structure is quarter-filled in holes ($\frac{3}{4}$ filled in electrons)—see Figure 3a. For an uniform stack of periodicity b, the critical Peierls wave vector is $2k_F^D = b^*/4$ for the hole filling ($2k_F^D = 3b^*/4$ for the electron filling). In the mean field approximation (which neglects 1D pre-transitional $2k_F^D$ fluctuations), the Peierls transition occurs for:

$$T_P^{MF} \approx C\,E_F e^{-1/\lambda_{2k_F}} \tag{1}$$

In Equation (1), E_F is the hole Fermi energy and, since $2k_F$ phonons pair a $-k_F$ hole to a $+k_F$ electron, λ_{2k_F} is the reduced $2k_F$ electron–phonon coupling [3]. C is a constant, of few units, which depends on the shape of the band dispersion near E_F (for a free electron dispersion $C \approx 2.25$). The Peierls transition opens a gap 2Δ at $\pm k_F$ in the 1D band structure. At 0 K, the gap $2\Delta_0$ is related to T_P^{MF} by the BCS-type correspondence law:

$$2\Delta_0 = 3.56 T_P^{MF} \tag{2}$$

Table 1 reports $T_P{}^{MF}$ deduced, via Equation (2), from the activation energy Δ_0 of the conductivity measured below T_P.

Note that with $2k_F^D = b^*/4$, the Peierls superstructure stabilizes a 4b periodicity in stack direction. Since $4 \times 2k_F^D = b^*$, the $2k_F$ BOW/CDW modulation wave length is in fourth-fold commensurate relation with the chain periodicity b, so that the phase of the CDW modulation should be pinned in the structure by the four-fold lattice potential.

2.2. Two-Fold Modulated Stack

Table 1 shows that in many salts the donor stack periodicity is doubled. In that case with a stack periodicity b' = 2b, the critical Peierls wave vector should be $2k_F^D = b'^*/2$ (see the folded band structure shown in Figure 3b). Note also that the new reciprocal periodicity b'* of the two-fold modulated structure amounts to $4k_F^D$ which is the critical wave vector associated to a charge localization phenomena (see Section 1). In such a modulated structure, the band folding opens a band gap $2E_G$ at $\pm b'^*/2$ due either to a dimerization of the stacks or to the presence of non-equivalent donor sites. Generally the bond dimerization due to successive molecular shifts $\pm u/2$ (leading to a differentiation of transfer integral by $\delta t = g_B u$) or to a charge unbalance on each donor $\pm \delta \rho$ (leading to a difference of HOMO potential energy $\delta V = g_S \delta \rho$) are small quantities, so that with $E_G = 2\delta t$ or $E_G = \delta V$ one has $E_G \ll t$ (in these last expressions g_i is the bond (B) or site (S) electron–phonon coupling and t is the transfer integral of the uniform stack). Note that in these situations the underlying dimerization or charge modulation effect can be viewed as achieved by the presence of a static $4k_F$ BOW or $4k_F$ CDW, of amplitude $u/2$ or $\delta \rho$, respectively, on the Per stack. For a weak modulation and for a weakly interacting electron gas, these effects do not change appreciably the physics of the Peierls instability with respect to the one of a quarter-filled band. In particular, the $2k_F$ BOW/CDW modulation remains mainly pinned on the structure by a fourth-order lattice potential as for the uniform stack (the two-fold pinning lattice potential proportional to u or $\delta \rho$ being much smaller). This statement is not true for a sizeable band gap $2E_G$ in presence of strong electron repulsions where the $4k_F$ electron-electron scattering Umklapp term induces a $4k_F$ BOW or $4k_F$ CDW charge localization as observed for example in Fabre salts (TMTTF)$_2$X [14]; TMTTF = tetramethyl-tetratia-fulvalene.

In the case of a strong dimerization shown in Figure 3c, where $t_{intra} \gg t_{inter}$, bonding and anti-bonding states of the dimer are well decoupled in energy. Thus one hole tends to be localized on the anti-bonding state of each dimer, and the anti-bonding band can be considered as half-filled. In this situation equivalent to an half-filled 1D system, each dimer can be considered as a rigid unit and the $2k_F$ BOW/CDW of wave length 2b' will basically modulate the inter-dimers distances. There is thus strong pinning of the $2k_F$ CDW on each dimer by a two-fold lattice potential.

In the case of a two-fold modulated stack, there is an additional process entering in the $2k_F$ electron–phonon coupling Peierls mechanism. Since one has $4k_F^D = b'^*$, umklapp scattering processes should also contribute to the $2k_F$ electron–phonon coupling mechanism. The associated reduced electron–phonon coupling constant, λ_{um}, which is proportional to the amplitude of the $4k_F$ BOW or of the $4k_F$ CDW, adds to λ_{2k_F} in Equation (1). One simply gets, in the mean-field approximation, the relation:

$$T_P^{MF} \approx C\, E_F e^{-1/(\lambda_{2k_F} + \lambda_{um})} \tag{3}$$

In the limit of a strongly dimerized stack the contribution of umklapp processes could be of the same magnitude as the normal $2k_F$ electron–phonon coupling process. In that case with $2\lambda_{2k_F} \approx \lambda_{2k_F} + \lambda_{um}$, one obtains an enhanced mean-field Peierls transition temperature given by:

$$T_P^{MF} \approx C E_F e^{-1/2\lambda_{2k_F}} \tag{4}$$

2.3. Estimation of the Electron-Phonon Coupling

In a 3D solid made by a collection of weakly coupled chains, the true 3D Peierls transition temperature, T_P, is generally depressed by a sizeable fraction of $T_P{}^{MF}$ (see Table 1). This is due the presence of an important regime of 1D structural fluctuations which destroy the mean-field 1D order between $\sim T_P^{MF}$ and T_P [3]. In that temperature range, local 1D $2k_F$ lattice fluctuations form a pseudo-gap in the electronic density of states, which corresponds to a local formation of the Peierls gap. As the pseudo-gap reduces progressively the effective number of carriers in the vicinity of the Fermi level it is generally accompanied by an upturn at T_ρ in the thermal dependence of the electrical conductivity ($T_\rho \sim T_P^{MF} > T_P$, see Table 1) and by its decreases below T_ρ. Note that in a pure 1D system 1D CDW and BOW thermal fluctuations completely suppress the Peierls transition. The non-zero value of T_P reported in Table 1 is due to inter-chain coupling (several relevant mechanisms are described in Ref. [3]).

Using Equation (1) or (4) for uniform or strongly two fold modulated stacks, it is possible to estimate λ from the knowledge of $T_P{}^{MF}$ and E_F. In α-(Per)$_2$[M(mnt)$_2$] the linear thermal dependence of the hole-like thermo-power leads to a bandwidth of $4t_{//} = 0.6$eV [7,8] which gives for a quarter filled 1D hole band $E_F = \sqrt{2}t_{//} \approx 0.2$eV. Using a free hole dispersion in the vicinity of E_F, as assessed by the band structure calculation of Ref. [15], one gets with $T_P{}^{MF} \approx T_\rho$ in Equation (1):

$$\lambda_{2k_F} \sim 0.2 \text{ for the M = Au salt} \tag{5}$$

$$\lambda_{2k_F} \sim 0.25 \text{ for the M = Cu salt} \tag{6}$$

Assuming a sizable Per stack dimerization due to the chemical bonding of M(mnt)$_2$ units into dimers, one gets using Equation (4):

$$\lambda_{2k_F} \sim 0.2 \text{ for the M = Fe and Co salts} \tag{7}$$

These values are comparable to λ_{2k_F} calculated with Equation (1) in KCP (0.2) and in the blue bronze K$_{0.3}$MoO$_3$ (0.25).

If one assumes the same band width in (FA)$_2$ PF$_6$ as in α-(Per)$_2$[M(mnt)$_2$] one gets, with Equation (1) which neglects the stack dimerization (see below), $\lambda_{2k_F} = 0.6$. A similar reduced sizable electron-coupling $\lambda_{2k_F} = 0.6$ is obtained for the trans-polyacetylene, (CH)$_x$.

2.4. Peierls Instability in (Arene)$_2$ PF$_6$ and AsF$_6$ Salts

Table 1 shows that D$_2$X salts where D is fluoranthene (FA) or perylene derivatives substituted with four methyl groups (TMP) or two cyclopentanes (CPP) (Figure 1a) exhibit, in presence of monovalent anions such as $X = $ PF$_6$ and AsF$_6$, a sizable regime of 1D $2k_F$ fluctuations pre-transitional to the Peierls transition. Such 1D $2k_F$ BOW fluctuation regime, associated to a significant electron–phonon coupling, have been detected in the FA, CPP and TMP salts by X-ray diffuse scattering measurements [13,16–18]. In (CPP)$_2$AsF$_6$ + CH$_2$Cl$_2$ the susceptibility associated to the 1D $2k_F$ BOW instability, $\chi_{BOW}(2k_F)$, follows a Curie–Weiss dependence (which corresponds to a regime of Gaussian fluctuations of the amplitude of the order parameter) which diverges at a 2nd order Peierls transition at $T_P = 170$ K (Figure 4) [18]. Similar results are obtained in (FA)$_2$PF$_6$ [16]. $2k_F$ BOW/CDW fluctuations strongly affect the electron density of states at the Fermi level by forming a pseudo-gap, precursor to the Peierls gap. This progressively reduces the effective number of carriers available for the charge transport so that the electrical conductivity measured in these systems decreases in the temperature range of existence of 1D fluctuations below $T_\rho \sim T_P^{MF}$ [17–20]. All these features are those of a conventional Peierls instability.

In most of the salts, $2k_F$ fluctuations are 3D coupled at the Peierls transition at T_P which thus stabilizes a 3D long range order (LRO) of $2k_F$ BOW/CDW modulations. This 3D Peierls transition is characterized by the appearance below T_P of $2k_F$ superstructure reflections whose intensity $I_{sat}(q_P)$

is proportional to the square of the amplitude of the BOW/CDW modulation (see Figure 4 for $(CPP)_2AsF_6 + CH_2Cl_2$ [13,18]).

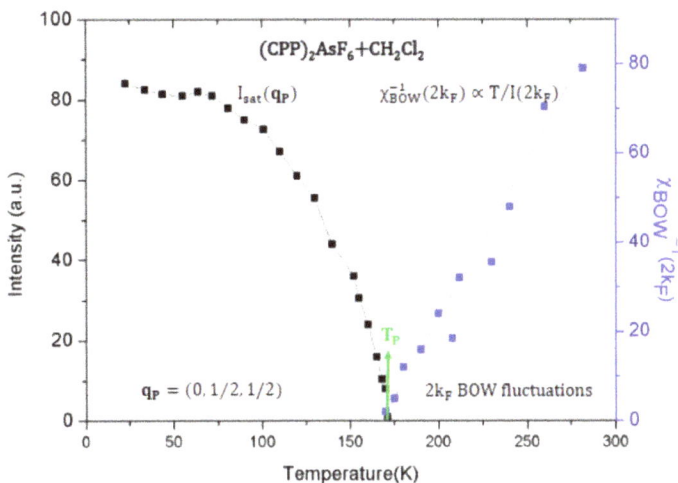

Figure 4. Temperature (T) dependence of the inverse of the $2k_F$ BOW susceptibility $\chi_{BOW}^{-1}(2k_F)$ above T_P and of the $2k_F$ Peierls satellite intensity $I_{sat}(q_P)$ below T_P in $(CPP)_2AsF_6 + CH_2Cl_2$ (adapted from [13,18]). $\chi_{BOW}^{-1}(2k_F)$ is proportional to $T/I(2k_F)$, where $I(2k_F)$ is the intensity of the X-ray diffuse lines at $2k_F$.

In the Peierls ground state of $(FA)_2PF_6$ it has been observed that the $2k_F$ BOW/CDW modulation could collective slide under the action of an external electric field exceeding a threshold field value of $E_T \sim 0.2$ V/cm [19,20]. As the threshold field depends on the pinning energy of the CDW on impurity and on the order of commensurability of the lattice potential, the finding of a similar E_T in dimerized $(FA)_2PF_6$ and in α-$(Per)_2Au(mnt)_2$ ($E_T \sim 0.5$ V/cm [21]) exhibiting a regular Per stack means that the pinning is mainly achieved by the fourth-order lattice potential in $(FA)_2PF_6$. Thus, the $2k_F$ CDW modulation is basically that of a quarter filled band in $(FA)_2PF_6$.

Table 1 shows that TMP and CPP salts incorporate one solvent molecule ($S = CH_2Cl_2$) per anion, X. This leads to the formation of mixed X-S columns in stack direction with a short range lateral order between neighboring X-S columns [13]. In CPP, inter-columnar disorder does not prevent the occurrence of a high temperature Peierls transition at $T_P = 158$–170 K (see Table 1 and Figure 4). $(TMP)_2AsF_6 + CH_2Cl_2$ exhibits also a sizeable regime of $2k_F$ fluctuations below ~200 K, but at the difference of its CPP analog it exhibits only a short range lateral BOW order at 20 K [18]. The absence of a 3D long range BOW order could be due a strong pinning of the $2k_F$ modulation on X-S disorder. In addition, and by analogy with earlier findings in the $[(TMTSF)_{1-x}(TMTTF)_x]_2ReO_4$ solid solution [22], the divergence of the $2k_F$ BOW instability could be inhibited by the existence of a $4k_F$ CDW order on the TMP stack already at RT.

2.5. Peierls Instability in Per Stack of α-$(Per)_2[M(mnt)_2]$

α-$(Per)_2[M(mnt)_2]$ salts with $M = Au$, Cu, Fe and Co exhibit a quite small Peierls modulation which is assessed by the detection below T_P of extremely weak superlattice reflections at the $2k_F^D = 1/4b^*$ reciprocal position in the Cu salt [23] and somewhat stronger reflections at the $2k_F^D = 1/2b'^*$ reciprocal position in dimerized Fe and Co salts [24]. No superlattice reflections have been detected up to now in the Au salt (however one expects from the relative magnitude of Peierls gaps superlattice reflections one order of magnitude smaller in the Au salt than in the Cu salt). Probably because of the weakness of the Peierls instability in all these salts the pre-transitional regime of 1D $2k_F$-BOW fluctuations could not be detected. Their characteristics temperature T_P, T_ρ and T_P^{MF} are reported in Table 1.

The $2k_F^D = 1/4b^*$ CDW stabilized below T_P in the Au compound collectively slides under the action of an external electric field exceeding E_T ~0.5 V/cm and exhibits the basic features of non-linear conductivity phenomena observed in CDW inorganic systems such as the blue bronze and the transition metal tri-chalcogenides [21,25]. Note, however, that the detection of a much larger threshold field E_T ~9 V/cm in the Pt salt [25] could be explained by the presence of a two-fold lattice pinning potential brought by the SP dimerization of the [Pt(mnt)$_2$] stacks (see below in Section 3).

As expected for a standard Peierls system, the critical temperature T_P of the Au compound decreases as H^2 for modest magnetic fields [26]. More interestingly, the CDW ground state is found to be unstable for large magnetic fields H_c ~33–37 T exceeding the so-called Pauli limit $H_P \approx 22.5$ T [27,28]. In addition, unexpected features observed near H_c suggest the occurrence of field induced CDW states due to orbital effects related to the presence of a warped Fermi surface (FS) [28,29].

Because of the presence of two slightly shifted sets of double quasi-1D warped FS [15] combined with the opening of a Peierls gap only slightly larger than inter-stack transfer integrals ($t_\perp$ ~2 meV), a destabilization of the Peierls instability of the Au compound is expected when pressure enhances the nesting breaking components of the FS. More precisely, it has been calculated [30] for a system exhibiting a warped FS incompletely nested by the $2k_F$ wave vector that the Peierls ground state should vanish when Δ_0 becomes smaller than the typical energy of hole and electron pockets remaining after the incomplete FS nesting process. Indeed it has been observed a vanishing of the CDW ground state of the Au compound above $P_c \approx 5$ kbar. More interestingly it is found, when the low temperature metallic state is restored under pressure, that α-(Per)$_2$[Au(mnt)$_2$] becomes a superconductor below T_S ~0.3 K [31].

Finally, it is interesting to remark that the weak Peierls gap, $2\Delta_0 \approx 3.5$ meV, of the Au salt [7,8] is smaller than a typical acoustic phonon frequency (Ω_c ~4.5–7 meV in TTF-TCNQ for $2k_F = 3/4b^*$ [32]). Thus, with $2\Delta_0 < \hbar\Omega_c$, the Peierls transition of the Au salt should occur in the non-adiabatic regime [33]. The situation should be contrasted to that of the Cu salt where with $2\Delta_0 \approx 20$ meV $> \hbar\Omega_c$ [7,8] the Peierls transition should occur in the adiabatic regime, as for the (arene)$_2$ PF$_6$ and AsF$_6$ salts previously considered.

Note that in the non-adiabatic regime the 0 K amplitude of the Peierls gap should be reduced by quantum fluctuations from its mean-field value given by the BCS relation-ship (2). In addition, the non-adiabaticity of the Peierls mechanism should increase under pressure because of the hardening of the frequency of phonon modes. This effect should reduce the Peierls gap, and the vanishing of the CDW ground state in the Au salt under pressure should occur at a quantum critical point.

2.6. Comparison between Peierls Instabilities in (Arene)$_2$PF$_6$/AsF$_6$ and in α-(Per)$_2$[M(mnt)$_2$] Salts

While the donor stack of all these different families are made with molecules of similar chemical characteristics, there is a sizeable difference between Peierls instabilities in FA and CPP salts and those in Per salts. There is in particular a difference of more than a factor two in the Peierls transition temperature T_P and in the temperature range [T_ρ-T_P] of 1D CDW fluctuations for these two series of salts. As seen in Table 1, T_ρ occurs well above RT for FA and CPP salts while T_ρ is around 160–180 K (~2 times T_P) for the M = Co and Fe derivative and very close to T_P for the Cu and Au derivatives. Table 2 shows that, in the M = Ni, Pd and Pt derivatives, T_P and T_ρ are comparable to those of the Au and Cu derivatives.

A possible explanation relies on the number of donor conduction bands crossing the Fermi level (and also on the importance of their warping): one band for CPP and TMP salts, two bands for FA salt and four bands for Per salts. In the case of α-(Per)$_2$[M(mnt)$_2$], the total FS includes four sheets composed of two slightly shifted sets of warped open FS [15]. With such a band structure, the best nesting wave vector of the global FS should be poorly defined. The poor FS nesting should smoothen the thermal divergence of the $2k_F$ electron–hole response function [30] and thus reduces T_P^{MF} and T_P.

A somewhat similar situation is found in the quarter-filled (TSeT)$_2$Cl organic salt which, with a poorly nested FS composed of four warped sheets, exhibits a modest Peierls transition at T_P = 26 K [34] (TSeT = tetraseleno-tetracene).

3. Peierls and Spin-Peierls Instabilities in α-(Per)$_2$M(mnt)$_2$] with M = Ni, Pd and Pt

A special attention must be devoted to Per salts incorporating [M(mnt)$_2$] acceptors with M = Ni, Pd and Pt, because the charge transfer from Per to dithiolate complexes leads to the formation of spin $\frac{1}{2}$ magnetic [M(mnt)$_2$]$^-$ species. One thus obtains two types of electro-active donor and acceptor stacks. However, at variance with TTF-TCNQ where the incommensurate charge transfer gives rise to two types of conducting stacks, each subject to its own CDW instability, the [M(mnt)$_2$]$^-$ stack is not conducting because there is a Mott–Hubbard localization of one electron per [M(mnt)$_2$]. With such a charge localization each [M(mnt)$_2$]$^-$ molecule bears an unpaired spin $\frac{1}{2}$. One thus obtains S = 1/2 AF coupled dithiolate stacks which coexist with metallic Per stacks in the same structure (Figure 2). As for the M = Au and Cu derivatives, Per stacks of the Pt, Pd and Ni derivatives are subject to a Peierls instability whose main characteristics are given in Table 2.

At the difference of the α-Per$_2$[M(mnt)$_2$] series, it is interesting to remark that with an incommensurate charge transfer of 0.82 electron from Li to [Pt(mnt)$_2$], the 1D conductor Li$_{0.82}$[Pt(mnt)$_2$] (H$_2$O)$_2$, whose stack structure resembles to some extend to that of KCP, undergoes a conventional Peierls transition at T_P = 215 K which is preceded by a sizeable regime of $2k_F$ BOW fluctuations [35]. These data show that the Peierls instability of the acceptor dithiolate stacks is achieved by a sizeable electron–phonon coupling as found for the donor stack in (arene)$_2$ PF$_6$ and AsF$_6$ salts. The presence of critical phonon modes coupled to electronic degrees of freedom at first order in molecular displacement should persist in S = 1/2 AF electron localized [M(mnt)$_2$] chains. Via a magneto-elastic spin–phonon coupling, these modes should modulate the AF exchange coupling between dithiolate molecules. Such a modulation is the basic ingredient to obtain a stack dimerization allowing to pair spins $\frac{1}{2}$ into a SP ground state.

3.1. Spin-Peierls Structural Fluctuations

The key role of the lattice counterpart of the SP instability of the M = Ni, Pd and Pt dithiolate stacks is evidenced by the detection of a sizeable regime of 1D structural fluctuations appearing as diffuse lines of strong intensity located in b*/2 on X-ray diffuse patterns. Such diffuse lines have been observed in Pt ([9]—see also Figure 5a), Ni [23] and Pd [9] derivatives. The reduced critical wave vector of the fluctuations q_{SP} = b*/2 corresponds to an incipient instability towards a doubling (dimerization) of the dithiolate stack periodicity. This structural modulation should drive the pairing of neighboring S = 1/2 spins into magnetic singlets.

The above quoted diffuse lines have been detected below 30 K, 100 K and 100 K in the Pt, Ni and Pd salts, respectively. This onset temperature, given in Table 2, is taken as the SP mean-field temperature (T_{SP}^{MF}) of the dithiolate stacks. T_{SP}^{MF} is twice larger than the temperature at which Peierls fluctuations begin to manifest on Per stacks (i.e., T_ρ, in Table 2). This provides clear evidence that the SP instability on dithiolate stack starts before the Peierls instability on Per stack.

The SP structural fluctuations of Ni, Pd and Pt derivatives can be quantitatively analyzed from the thermal dependence of the intensity and profile (along the stack direction b) of the diffuse lines [9], namely from:

- The q_{SP} peak intensity $I(q_{SP})$, the $T/I(q_{SP})$ ratio gives a quantity proportional to the inverse SP structural susceptibility $\chi_{SP}^{-1}(q_{SP})$ (Figure 6).
- The half-width at half-maximum (corrected by the experimental resolution) directly gives the inverse coherence length of the SP fluctuation in stack direction ξ_b^{-1} (Figure 7).

$\chi_{SP}(q_{SP})$ exhibits a Curie–Weiss type divergence for the Pd and Pt salts as predicted for the high temperature fluctuations of the amplitude of the order parameter (i.e., regime of Gaussian fluctuations) [36]. The linear thermal dependence of the inverse susceptibility fitting the data, shown in Figure 6:

$$\chi_{SP}^{-1}(q_{SP}) \propto (T - T_{SP}) \tag{8}$$

allows defining T_{SP} at about 28 K and 7.5 K for the Pd and Pt salts. For the Ni salt, $\chi_{SP}^{-1}(q_{SP})$ changes slope upon cooling. Figure 6 shows that the high temperature data extrapolate to a "T_{SP}" of ~25 K.

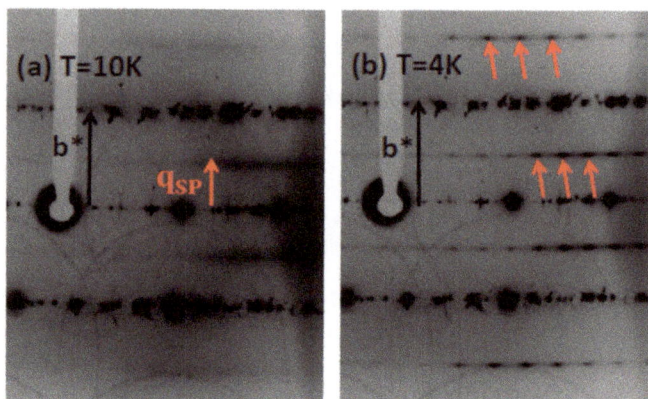

Figure 5. X-ray diffuse scattering patterns from α-Per$_2$[Pt(mnt)$_2$] at: 10 K (**a**); and 4 K (**b**). In (**a**), 1D SP critical lattice fluctuations gives rise to intense diffuse lines located at half distance (i.e., at the reduced q_{SP} = b*/2 wave vector) between horizontal layers of main Bragg reflections. In (**b**), red arrows show that these lines have condensed into broad diffuse spots, corresponding to the establishment of a 3D SP SRO.

Table 2. Characteristics of α-Per$_2$[M(mnt)$_2$] salts for M = Pt, Ni and Pd which exhibit both a Peierls and a spin-Peierls instability on the Per and dithiolate stacks respectively. The metal-insulator Peierls transition temperature (T_P) and the temperature of minimum of resistivity (T_ρ) are taken from electrical measurements of References [7,8]. The SP critical temperature and the nature of the short range (SRO) or long range (LRO) SP order detected below T_{SP} are indicated as well as the temperature range of observation of 1D SP fluctuations. "T_{SP}" is the temperature at which $\chi_{SP}^{-1}(q_{SP})$ and ξ_b^{-1} extrapolate to zero in the Ni and Pt derivatives. In the Pd derivative, T_{SP} is the true 3D SP transition at which $\chi_{SP}^{-1}(q_{SP})$ and ξ_b^{-1} vanish.

Per$_2$[M(mnt)$_2$]	T_P/T_{SP} (K)	b*/4 Peierls Modulation	b*/2 Spin-Peierls Modulation	T_ρ (K)	1D SP Fluctuations
M = Pt	8.2/"7.5"	not observed	SRO in all directions	18	≤30 K
M = Ni	25/"25–45"	LRO	SRO in all directions	50	≤100 K
M = Pd	28/28	not observed	LRO	50–80	≤100 K

The correlation length increases in an inverse square root law as predicted in the regime of Gaussian fluctuations of the amplitude of the order parameter [36]. The inverse correlation length, shown in Figure 7, is fitted by this square root thermal dependence:

$$\xi_b^{-1} \propto \sqrt{(T - T_{SP})} \tag{9}$$

The extrapolation at zero of Equation (9) leads to the same T_{SP} as does Equation (8) for the Pd and Pt salts.

For the Pd salt, $\chi_{SP}^{-1}(q_{SP})$ and ξ_b^{-1} vanish at T_{SP} = 28 K. Below T_{SP}, SP superstructure reflections are detected at the reciprocal wave vector component q_{SP} = b*/2 [9]. There is thus below T_{SP} = 28 K a long range SP order. The Pd salts thus exhibits all the structural characteristics of a well-defined 2nd order SP transition.

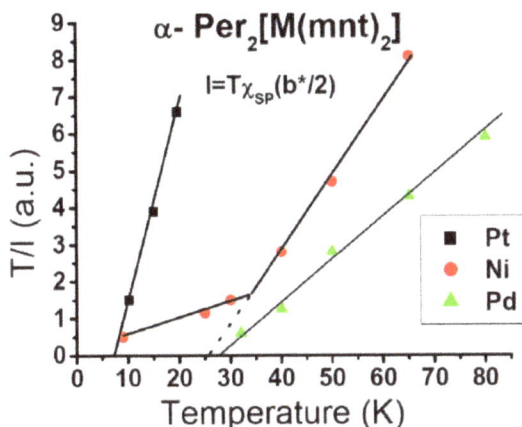

Figure 6. Temperature (*T*) dependence of the inverse spin-Peierls susceptibility $\chi_{SP}^{-1}(q_{SP})$ for the Ni, Pd and Pt derivatives. $\chi_{SP}^{-1}(q_{SP})$ follows basically a Curie–Weiss law (Equation (8)). For the Ni salt, the dashed line extrapolates the thermal dependence of the high temperature data of $\chi_{SP}^{-1}(q_{SP})$ towards "T_{SP}".

In the Pt salt, $\chi_{SP}^{-1}(q_{SP})$ and ξ_b^{-1} tend to vanish at about 7.5 K. However, the X-ray pattern taken at 4 K (Figure 5b) shows that only a SP SRO is achieved at low temperature. There is thus a pseudo-SP transition at "T_{SP} " ≈ 7.5 K as indicated in Table 2.

In the Ni salt, there is no vanishing of ξ_b^{-1} and $\chi_{SP}^{-1}(q_{SP}).\xi_b^{-1}$ saturates below 50 K while the high temperature data extrapolate to zero around 45 K. $\chi_{SP}^{-1}(q_{SP})$ changes slope below 35 K while the high temperature data linearly extrapolate to zero around 45 K. These two "T_{SP}" extrapolation values are indicated in Table 2.

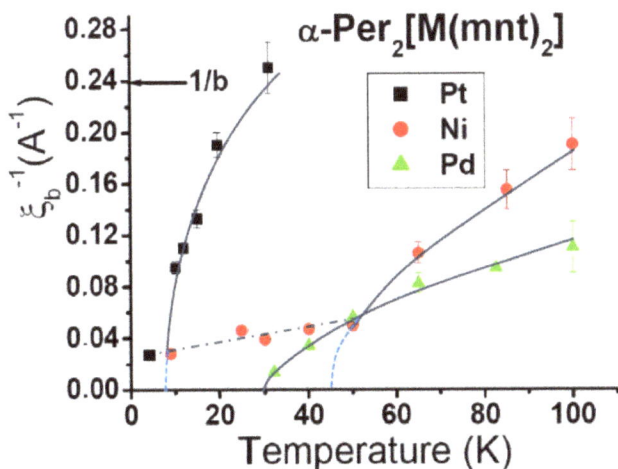

Figure 7. Temperature dependence of the inverse correlation length ξ_b^{-1} in stack direction for the Ni, Pd and Pt derivatives. For the Ni and Pt salts, the dashed lines extrapolate, using Equation (9), the thermal dependence of the high temperature data of ξ_b^{-1} towards "T_{SP}". The dotted dashed lines gives the low temperature saturation values of ξ_b^{-1} for the Ni and Pt salts.

3.2. Spin-Peierls and Peierls Orders

Table 2 gives the Peierls critical temperature (on Per stack) deduced from the metal to insulator transition detected by conductivity measurements [7,8]. Very weak $2k_F^D = b^*/4$ super-lattice reflections have been detected below T_P in the Ni salt [23]. They have an intensity comparable to those found below T_P in the Cu salt. No $b^*/4$ super-lattice reflections have been detected in the Pd and Pt salts. They could be too weak to be detected.

T_{SP} and "T_{SP}" values reported in Table 2 are the temperature at which 1D SP fluctuations diverges or tends to diverge as discussed above. Sharp SP superstructure reflections, of longitudinal component $q_{SP} = b^*/2$, are observed below $T_{SP} = 28$ K in the Pd salt [9].

"T_{SP}" is a pseudo SP transition temperature for the Ni salt since the low temperature saturation of $\chi_{SP}^{-1}(q_{SP})$ and ξ_b^{-1} reveals a short range dimerization order in all the directions [23]. At 9 K, the SP dimerization extends on 35 Å (i.e., 8b) in stack direction and a well-defined phasing between neighboring dimerization extend only on 14 Å in transverse directions. Fourteen Angstroms correspond to about the distance between first neighboring dithiolate stacks. Figure 2 shows that the transverse interaction between first neighbor dithiolate stacks should be mediated through a perylene stack. For this reason such a coupling should be weak, which explains why dimerization of only first neighbor dithiolate stacks are correlated in the SP ground state. Note that in the Ni derivative such a SP SRO cannot be due to disorder in the material because sharp $b^*/4$ superstructure reflections are also observed. Since a similar SP SRO is observed in the Pt derivative (see below), the SRO has certainly an intrinsic origin.

A somewhat similar situation is observed in the Pt derivative. A specific heat anomaly indicates the occurrence of a thermodynamic phase transition around 8 K [37]. However, Table 2 indicates a T_P (8.2 K corresponding to the metal-insulator transition detected from transport measurements) distinct from "T_{SP}" (7.5 K temperature at which 1D SP fluctuations tend to diverge and at which the EPR signal drop [38]). The finding of a T_{SP} distinct from T_P seems to be sustained by the extrapolation to 0 T of the magnetic phase diagram of the Pt salt [39].

The establishment of a low temperature spin-singlet ground state on [Pt(mnt)$_2$] chains is evident from EPR [38], ^{1}H NMR spectra and spin relaxation $(1/T_1)$ rate [39,40] measurements. However, local magnetic measurements cannot probe the spatial extend of the SP order. Thus, a possible explanation of the decoupling between T_P and "T_{SP}" could rely on the fact that the SP dimerization order remains incomplete. The X-ray pattern shown in Figure 5b proves that there is not a long range SP dimerization at 4 K: the SP dimerization extends on 36 Å (i.e., 9b) in stack direction and 16 Å (i.e., the first neighbor dithiolate distance) in transverse direction. Note that these SP coherence lengths are comparable to those found in the Ni derivative. In addition, an incomplete SP pairing means that unpaired spins 1/2 should remain present. A minority of such spins has been recently identified by low temperature NMR measurements [41]. Note that the presence of low temperature unpaired spins agrees with the observation of a finite spin susceptibility in the SP ground state of the Pt salt [40,41].

The simultaneous presence of incomplete transverse and longitudinal SP orders and of unpaired spins $\frac{1}{2}$ can be rationalized using a previous interpretation of similar findings in doped CuGeO$_3$ SP systems [42]. In the SP ground state, the minimum of inter-chain coupling energy generally imposes an out of phase transverse phasing between dimerization on first neigbouring chains (this corresponds to a minimum of Coulomb coupling between dimerized charged chains). This minimum of energy implies that dimerization should be shifted by b between first neighboring chains. In presence of such a SP pattern, the staggered transverse order can be easily broken by keeping the dimerization of two first neighboring chains in phase. This linear defect, which consists of an absence of relative shift between dimerization located on two first neighbor chains, can be viewed as adding a local transverse phase shift of π in the SP pattern. This defect costs a maximum of inter-chain coupling energy. In that situation, the best way to limit this cost is to reduce the spatial extent of the linear defect by restoring the natural out of phase SP inter-chain order. This can be achieved by limiting the extent of the defect by ending the linear defect by two unpaired spins $\frac{1}{2}$. Each unpaired spin corresponds to a defect of dimerization which changes the phase of the intra-chain SP dimerization by $\pm\pi$. The creation of pairs of $\pm\pi$ dimerization defects thus limits the spatial extend of the longitudinal SP order (the average distance between the two dimerization defects being of the order of ξ_b). As each dimerization defect bears an unpaired spin $\frac{1}{2}$, the limitation of the SP order in chain direction implies a presence of S = 1/2 free spins. Such magnetic defects have been detected in the SP ground state of the Pt derivative. In our description, the limitation of the longitudinal order is caused by the break of the transverse order, but the reverse is also true. Generally, unpaired spins are created by any type of defect interrupting the spatial coherence of the longitudinal SP order.

3.3. Mechanism of the SP Instability in α-(Per)$_2$[M(mnt)$_2$]

Table 2 gives the temperature at which 1D SP fluctuations begin to be detected. This temperature is taken as the mean-field temperature of the SP transition T_{SP}^{MF} [36]. One thus has $T_{SP}^{MF} \approx 30$ K for the Pt salts and 100 K for the Pd and Ni salts. T_{SP}^{MF} is related to the mean field SP gap, Δ^{MF}, by the mean-field correspondence relationship, which can be expressed for the Heisenberg chain by [43]:

$$\Delta^{MF} \approx 2.47 k_B T_{SP}^{MF} \tag{10}$$

With the above quoted T_{SP}^{MF} values, one gets $\Delta^{MF} \approx 75$ K for the Pt salts and $\Delta^{MF} \approx 250$ K for the Pd and Ni salts.

In this framework, the mechanism driving the SP transition depends on the relative value of the mean-field gap Δ^{MF} with respect to the q_{SP} critical phonon energy: $\hbar\Omega_C \approx 50$–100 K for TA-LA phonon frequencies in organics [44]. For the Heisenberg chain, the SP transition occurs in the classical (adiabatic) limit when $\hbar\Omega_C \leq \Delta^{MF}/2$ [45]. In the opposite limit of weak Δ^{MF}, $\hbar\Omega_C \geq \Delta^{MF}/2$, the SP transition occurs in the quantum (anti-adiabatic) limit. Note that if Δ^{MF} is too small, such that $\Delta^{MF} \leq 0.7 \hbar\Omega_C$ [46], the zero point phonon quantum fluctuations kills the SP dimerization. In that case, the SP gap vanishes exponentially at a quantum critical point beyond which a spin liquid state is stabilized. The 0 K phase diagram of the SP ground state of the AF Heisenberg chain is shown in Figure 8. This figure shows that the SP transition of Pd and Ni salts occurs in the adiabatic limit while the SP transition of the Pt salt occurs in the non-adiabatic limit.

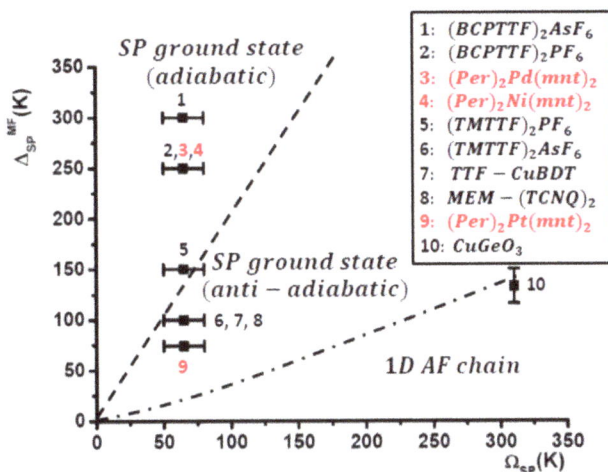

Figure 8. Nature of the SP ground state as a function of the mean-field gap, Δ_{MF} defined by Equation (7), and of the critical phonon frequency Ω_c for the SP Heisenberg chain (from [45]), together with the location of typical SP compounds (taken from Reference [3]). The α-Per$_2$[M(mnt)$_2$] salts with M = Ni, Pd and Pt are indicated in red.

In the adiabatic limit, when the dynamics of SP structural fluctuations is slow compared to those of magnetic degrees of freedom, a local SP pairing, which develops below $T_{SP}{}^{MF}$ on ξ_b progressively forms a pseudo-gap in the density of states of magnetic excitations. This manifests by a depression in the thermal dependence of the spin susceptibility. In the opposite anti-adiabatic limit, the dynamics of SP structural fluctuations is so rapid that the spin susceptibility remains unaffected by the SP fluctuations until T_{SP}. A typical example of an adiabatic SP transition is the (BCP-TTF)$_2$X series and of an anti-adiabatic SP transition is MEM-TCNQ [2,3,36,47] (BCP-TTF = benzocyclo-pentyl-tetrathiafulvalene; MEM = methyl-ethyl-morpholinium). Additional examples are given in Figure 8.

In order to analyze the influence of the SP structural instability on the spin degrees of freedom let us first consider the spin susceptibility, $\chi_S(T)$, of the Pd, Ni and Pt salts whose thermal dependence is shown in Figure 9a [8]. In a first approximation, the dithiolate sublattice can be described by a collection of isolated S = 1/2 AF chains. Then, each magnetic chain can be modeled by the simplest Heisenberg Hamiltonian:

$$H_{dith.} = J_1 \sum_j S_j S_{j+1} \tag{11}$$

where J_1 is the first neighbor exchange interaction. With such a Hamiltonian, the thermal dependence and the magnitude of the spin susceptibility $\chi_S(T)$ can be exactly calculated [48]. With this description Figure 9b gives at each temperature the effective first neighbor exchange interaction $J_{1\ eff}(T)$ taken from the absolute value of the spin susceptibility $\chi_S(T)$ for the Pd, Ni and Pt salts [49]. This figure shows in particular that:

- In the Pd salt, $J_{1\ eff}$ saturates at 260 K above 100 K.
- In the Pt salt, $J_{1\ eff}$ saturates at 35 K below 90 K.
- In the Ni salt, $J_{1\ eff}$ increases linearly between 300 and 100 K.

Figure 9b shows also deviations at these simple dependences. This means that Hamiltonian (Expression (9)) must be completed by additional contributions. Below we consider coupling of this Hamiltonian with SP structural fluctuations (as already considered in Reference [36]) and exchange coupling with conduction electron spins on Per stacks (as already considered in Reference [40]).

Let us start with the SP instability of the Pd salt which occurs in the adiabatic limit. For this salt relevant energies of its SP instability ($J_1 \approx 260$ K and $T_{SP}{}^{MF} \approx 100$ K) are comparable to those already reported in the (BCP-TTF)$_2$X series ($J_1 \approx 270$ K, $T_{SP}{}^{MF} \approx 120$ K for X = AsF$_6$ and $J_1 \approx 330$ K, $T_{SP}{}^{MF} \approx 100$ K for X = PF$_6$ [47]). All these organic compounds have also similar critical phonon frequencies, Ω_C. Figure 9a shows that the spin susceptibility of the Pd salt behaves above 100 K as $\chi_S(T)$ of an S = 1/2 AF chain with $J_1 \approx 260$ K, which is also the case for (BCP-TTF)$_2$X. Below $T_{SP}{}^{MF} \approx 100$ K SP lattice fluctuations develop on the correlation length ξ_b (whose thermal dependence of its inverse is given in Figure 7) a local spin singlet S = 0 state. This local non-magnetic order induces below 100 K a pseudo-gap in the magnetic excitation spectrum which manifests (see Figure 9a) by a decrease of the spin susceptibility with respect to $\chi_S(T)$ of the uniform 1D AF chain when ξ_b increases. This feature corresponds to a net increase of $J_{1\ eff}$ below 100 K (see Figure 9b). This thermal behavior, which resembles to the one measured and calculated in (BCP-TTF)$_2$X salts, corresponds to the coupling of AF fluctuations to SP critical lattice fluctuations [36]. The drop of $\chi_S(T)$ is the same in the Pd derivative and in (BCP-TTF)$_2$X between 100 K and 50 K. Below 50 K $\chi_S(T)$ of the Pd salt drops abruptly to nearly vanishes at $T_{SP} = 28$ K (Figure 9a), while $\chi_S(T)$ of (BCP-TTF)$_2$X SP continues to decrease monotonously and remains finite at its SP transition [36,47]. Thus the vanishing of $\chi_S(T)$ below 50 K in the Pd derivative cannot be due to standard SP structural fluctuations which diverge continuously in temperature (see Figures 6 and 7) as those of (BCP-TTF)$_2$X [36,47].

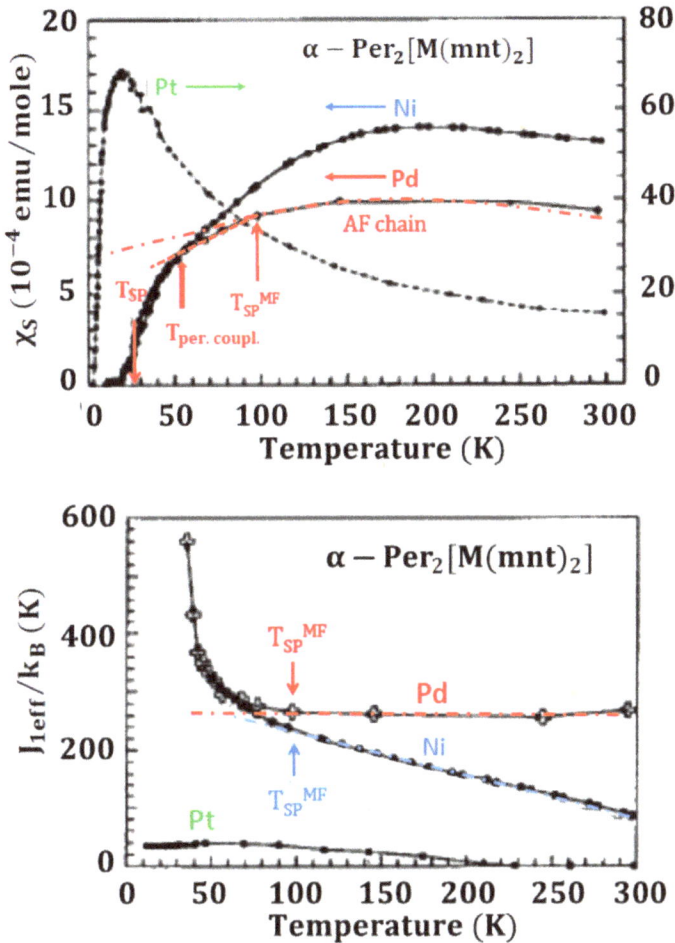

Figure 9. (**a**) Temperature dependence of the spin susceptibility, $\chi_S(T)$ of the Ni, Pd and Pt derivatives (**b**) Thermal dependence of the effective first neighbor AF exchange interaction $J_{1\,eff}(T)$ deduced from the amplitude of $\chi_S(T)$. In (**a**), $\chi_S(T)$ of the Pd derivative can be analyzed in the following way: (i) above $T_{SP}^{MF} \approx 100$ K $\chi_S(T)$ follows the thermal behavior of the 1D AF chain, (ii) between T_{SP}^{MF} and $T_{per.\,coupl.} \approx 50$ K the slight drop of $\chi_S(T)$ is due to the development of a pseudo-gap caused by 1D SP fluctuations, and (iii) between $T_{per.\,coupl}$ and $T_{SP} \approx 28$ K the large drop of $\chi_S(T)$ is due to an enhancement of the singlet-triplet gap due to frustration of AF interactions. A 3D SP transition occurs at T_{SP}. In (**b**), note that the rapid increase of $J_{1\,eff}(T)$ below 100 K (T_{SP}^{MF}) occurs in the temperature range where SP fluctuations are present. (Part (a) is completed from data of reference [8] and part (b) is modified from reference [49]).

In order to account for the extra vanishing of $\chi_S(T)$ observed below 50 K in the Pd derivative, one has to consider some thermal dependent modification of magnetic exchange interactions on dithiolate stacks. Below we propose that such a modification arises from the AF exchange coupling between the localized spins S_j on dithiolate molecule "*j*" and the spin density $s_l(x)$ on neighboring "*l*" metallic Per chain. Since the net decrease of $\chi_S(T)$ coincides with the development of a $2k_F$ density wave instability on the Per stack (which should grow below about $T_\rho \approx 50$–80 K—see Table 2) the new interaction appears to be driven by the onset of a collective response on the Per stacks below a quite well defined temperature, noted $T_{per.\,coupl}$ in Figure 9a, which is close to T_ρ. Below $T_{per.\,coupl}$, the $2k_F$ electron–hole instability on the Per stack, $\chi_{eh}(2k_F^D, T)$, submitted to an "external" AF field originating from the localized spins on dithiolate stacks responds by setting a $2k_F$ spin density wave (SDW) $s_l(x)$, as schematically illustrated by Figure 10.

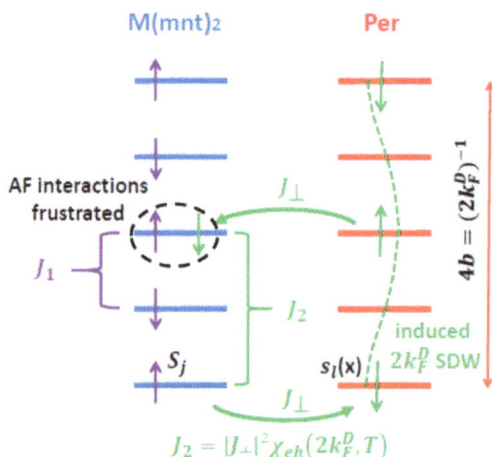

Figure 10. Schematic representation of competing AF S = 1/2 exchange couplings in α-Per$_2[M(mnt)_2]$ derivatives for M = Ni, Pd and Pt. J_1 is the first neighbor direct exchange coupling on dithiolate stack. J_2 is the second neighbor indirect RKKY exchange coupling mediated by the induced $2k_F^D$ SDW on the Per stack.

Following the notations of Ref. [40], this additional interaction can be modeled by adding to the direct exchange Hamiltonian (Equation (11)) an inter-chain exchange coupling Hamiltonian involving the spins density $s_l(x)$ on neighboring Per sacks:

$$H_{coupl.} = \sum_{l,\,per.} \int \left[H_l^{1D\;cond}(x) + J_\perp \sum_{j,\,dith.} S_j s_l(x) \right] dx \tag{12}$$

The first term of the right member of Equation (12) $H_l^{1D\;cond}(x)$ is the Hamiltonian of the 1D conduction electron gas located on the Per stack "l"; x being the stack direction. In the second right member of Equation (12) $J_\perp$ is the transverse exchange coupling between nearest dithiolate and per spins, respectively S_j and $s_l(x)$. Note, as indicated in Figure 2, that there are three different types of $J_\perp$ interactions per Per (the strongest one should occur in the direction of maximum overlap of dithiolate and Per MOs). Experimental evidence for a sizeable transverse exchange coupling $J_\perp$, (or equivalently for a fast inter-chain spin exchange regime) is provided by the observation for Pd [50] and Pt [40] derivatives of a single EPR line at a g value intermediate between those of $[M(mnt)_2]$ and of Per molecules.

Due to the quasi-1D nature of its electron gas, the Per stack exhibits a divergent electron–hole response $\chi_{eh}(2k_F^D, T)$, which sizably grows below T_ρ for $2k_F^D = b^*/4$. In this regime, a spin S° located on a dithiolate molecule placed at the origin should polarize, through the AF exchange coupling $J_\perp$, the electronic spin density $s_l(x)$ located on neighboring per stacks, as shown in Figure 10. This induces a SDW $s(x)$ on Per stacks whose thermal and spatial dependences are given by the 1D Fourier transform of $\chi_{eh}(2k_F^D, T)$ (assumed here to be that of a free electron gas):

$$s(x) \sim \chi_{eh}\left(2k_F^D, T\right) \frac{\cos(2k_F^D x)}{x} e^{-x/\xi_T} \qquad (13)$$

In Expression (13), the SDW which oscillates with the period 4b is also damped by the thermal electronic length ξ_T issued from the thermal broadening of the FS. At a distance $x = mb$ from the origin the oscillating spin density $s(x)$ exhibits, through $J_\perp$, a magnetic coupling with the spin S^m located on the near neighbor dithiolate molecule "m". Through the oscillating spin polarization of Per stack, this induces a spatially dependent effective exchange coupling between spins S^o and S^m distant of mb on the same dithiolate stack. Such an indirect oscillating interaction is known in the literature as the Rudernann–Kittel–Kasuya–Yoshida (RKKY) interaction. It behaves basically in 1D as

$$J_{RKKY}^m\left(2k_F^D, T\right) \sim -|J_\perp|\chi_{eh}\left(2k_F^D, T\right) \frac{\cos(m\pi/2)}{m} e^{-mb/\xi_T} \qquad (14)$$

Note that the spatial variation of the RKKY interaction is exactly given in 1D by $Si(2k_F x) - \pi/2$ where $Si(x)$ is the sine integral function (see Ref. [51]).

The mediated interaction (Expression (14)) provides in particular an indirect AF coupling between spins located on every second molecules ($m = 2$) on the dithiolate stack (note in Expression (14) that $J_{RKKY}^m(2k_F^D, T)$ does not depends on the sign of $J_\perp$). As shown in Figure 10 this indirect AF coupling competes with the intra-stack second neighbor effective ferromagnetic coupling due to the succession of two first-neighbor direct AF interaction J_1. Furthermore because of the 1D nature of the electron gas on Per stack $\chi_{eh}(2k_F^D, T)$ should diverge upon cooling (with a logarithmic thermal divergence for a 1D free electron gas). Thus the amplitude of the second neighbor AF interaction $J_{RKKY}^{m=2}(2k_F^D, T)$ is enhanced when T decreases and tends to overcome J_1.

1D magnetic chain with first, J_1, and second, J_2, neighbor frustrated AF interactions presents a very unusual phase diagram at 0 K (see for example [52]). When the ratio of exchange coupling $\alpha = J_2/J_1$ is smaller than $\alpha_c \approx 0.24$, the ground state is a gapless spin fluid state as for a simple Heisenberg chain, with a quasi-long range AF order. The nature of the ground state changes when $\alpha \geq \alpha_c$, through a quantum critical point located at α_c, into a spin gapped state with a long range dimer order. When the ratio of exchange coupling α is larger than α_c a singlet-triplet gap opens in absence of any coupling with the phonon field. In the Pd salt, the rapid vanishing of the spin susceptibility observed below 50 K (Figure 9a) could be caused by the rapid growth of the ratio of exchange coupling α above α_c due to the thermal increase of $J_{RKKY}^{m=2}(2k_F^D, T)$ below T_ρ. However, as SP critical fluctuations continue to diverge upon cooling the total singlet-triplet gap should superimpose the effects of the SP lattice dimerization and of the frustration of AF coupling. Such combined effects have been considered in the literature [53,54].

The SP instability of the Ni salt also occurs in the adiabatic limit. There are however some differences between Ni and Pd salts. A noticeable difference is that $J_{1\,eff}(T)$ increases linearly when T decreases between 300 and 100 K (Figure 9b), while $J_{1\,eff}(T)$ is constant in the same temperature range for the Pd salt. It is thus possible that the increase of $J_{1\,eff}(T)$ results from an enhancement of the intra-molecular overlap between dithiolate MO due a continuous sliding of neighboring Ni(dmit)$_2$ molecules upon cooling. However in spite of this effect there is a net deviation at the linear increases of $J_{1\,eff}(T)$ below 100 K when SP critical fluctuations develops. This could be due, as for the Pd salt, to the growth of a pseudo-gap in the AF magnetic excitation spectrum. In addition, similarly to the Pd salt, $\chi_S(T)$ abruptly drops below 50 K and vanishes around 20 K (Figure 9a). Note however that if below 50 K $\chi_S(T)$ exhibits the same temperature dependence for the two salts, Figures 6 and 7 show that in the same temperature range SP critical structural fluctuations behave differently for the Ni and Pd derivatives. While SP fluctuations diverge at $T_{SP} = 28$ K in the Pd derivative, SP fluctuations reduce their divergence below ~ 50 K (for ξ_b) and ~ 35 K (for χ_{SP}) in the Ni derivative. Thus, the rapid decrease of $\chi_S(T)$ in the Ni derivative cannot be due to the critical growth of SP fluctuations. As $\chi_S(T)$ considerably decreases below $T_\rho \approx 50$ K when $\chi_{eh}(2k_F^D, T)$ sizably increases, the spin gap opening is more likely due to a frustration effect between second neighbor indirect $J_{RKKY}^{m=2}(2k_F^D, T)$ and first neighbor direct J_1 AF interactions on the Ni dithiolate stack as shown in Figure 10.

Figure 9a shows that the thermal dependence of $\chi_S(T)$ below ~40 K resembles that of a thermally activated excitation process. For an activated process through a gap Δ, the spin susceptibility of a classical assembly of spin $\frac{1}{2}$ behave as:

$$\chi_s(T) \propto \frac{1}{T}e^{-\Delta/T} \tag{15}$$

Data of Figure 9a give Δ ~130 K, which amount to about $J_1/2$. We thus propose that the gap which develops below 50 K is a singlet triplet gap mainly set by frustration effects between first and second neighbor AF interactions when $\alpha = J_1/J_2 \geq \alpha_c \approx 0.24$. Note that the opening of a gap Δ~$J_1/2$ in absence of sizeable lattice dimerization requires a sizeable AF frustration ratio with α ~0.5 [54]. $\alpha = 0.5$ corresponds to the so-called Majumdar–Ghosh point where the ground state of the AF chain, which corresponds to two possible dimerization patterns formed by a succession of disconnected singlet dimers, is twofold degenerate.

The magnetic properties of Pt salt have been already considered in Ref. [40] first suggesting the presence of mediated RKKY interactions through the Per stack. However, the physics of Pt salt differs on many aspects from the one exhibited by Pd and Ni salts. Firstly, the SP instability of the Pt salt occurs in the non-adiabatic limit. Thus, the development of critical structural SP fluctuations below 30 K (T_{SP}^{MF}) should not open a pseudo-gap in the spin excitations. Accordingly, the thermal dependence of $\chi_S(T)$ should not deviate appreciably between T_{SP}^{MF} and T_{SP} from the extrapolated high temperature dependence of $\chi_S(T)$ (more precisely Figure 9b shows that J_1_{eff} does not change appreciably below 30 K). A similar behavior is shown by $\chi_S(T)$ in the non-adiabatic SP compound MEM-(TCNQ)$_2$ [2,3,47]. Secondly, SP fluctuations start below 30 K in the temperature range where the AF correlations are not developed (AF correlations develop when $\chi_S(T)$ begins to decrease below ≈ 20 K). The nature of the driving force of the SP instability in the Pt salt must be questioned because the SP instability should be triggered, in presence of AF correlations, by quantum fluctuations of the AF chain (i.e., in the temperature range below the maximum of $\chi_S(T)$).

All these features require a clarification of the various types of effective spin-spin interactions occurring on the Pt dithiolate stack. In particular and in addition to the first neighbor AF interaction J_1 (estimated at ~35 K from the fit of $\chi_S(T)$ shown in Figure 9b) additional AF interactions which competes with J_1 should be considered. Firstly, one expects to have below $T_\rho \approx 18$ K, when the electron–hole response of the Per stack develops a second neighbor mediated AF interaction $J_{RKKY}^{m=2}(2k_F^D, T)$ as for the Pd and Ni derivatives. Secondly in the non-adiabatic regime, the SP magneto-elastic coupling leads to a renormalization of J_1 and induces a second neighbor AF interaction J_2 [46]. For these two reasons the spin-spin Hamiltonian of the dithiolate stack should include first and second neighbor AF competing interactions which modify the thermal dependence of $\chi_S(T)$ when only J_1 is considered. This requires a more complete analysis of the thermal dependence of $\chi_S(T)$. However, as $\chi_S(T)$ does not decrease drastically on approaching T_{SP}, the frustration ratio $\alpha = J_1/J_2$ should be probably less than α_c, at the difference of Pd and Ni derivatives.

3.4. Nature of the Ground State

In the previous section, we have shown, especially for the Pd and Ni derivatives, that Per$_2$[M(mnt)$_2$] salts exhibit magnetic properties on dithiolate stacks coupled to a $2k_F$ density wave instability of the conducting Per stacks. Such features place Per$_2$[M(mnt)$_2$] among Kondo lattices where localized spins are coupled to itinerant spins of the conduction electron gas. Theory of 3D Kondo lattice based on Hamiltonians similar to the one given by (12) shows a competition between Kondo effect on each magnetic site and magnetic ordering arising from the RKKY interaction (see for example [55]). The Kondo effect, discovered in 1964 [56] by considering the interaction between a single magnetic impurity and conduction electrons, tends to stabilize a ground state characterized by the formation of a local singlet between the spin of conduction electrons and the spin of the impurity. In the Per-dithiolate series the 1D anisotropy of both the transfer integrals in the metallic subsystem and of the J_1 AF exchange coupling in the magnetic subsystem place $Per_2[M(mnt)_2]$ among the 1D Kondo lattices. Note also since, as shown in Figure 2, magnetic chains are spatially decoupled from the conducting chains that α-$Per_2[M(mnt)_2]$ belongs to the category of two-chain 1D Kondo lattices. In this respect α-$Per_2[M(mnt)_2]$ differs from standard one chain Kondo lattices where, because of the presence of several electron species, spin localized and conducting electrons are located on the same chain such as in metal-phthalo-cyanine-iodine $Cu(pc)I$ [57] or $BaVS_3$ [58] There is an abundant literature on 1D Kondo lattices based on elaborated theoretical considerations [51,59]. However, experimental clear-cut evidence of 1D Kondo lattice effects are quite sparse in the literature. Following our present analysis $Per_2[M(mnt)_2]$ organic salts with M = Ni and Pd should be a good realization of 1D Kondo lattice physics.

In this framework, it has been previously proposed that spin dimerization observed in $Per_2[M(mnt)_2]$ salts could be explained using a 1D Kondo lattice model at quarter filling with some kind of RKKY interaction between localized moments [60]. However the realization of such a ground state is not obvious because it has been numerically shown that a frustrated spin-1/2 Heisenberg chain coupled to adiabatic phonons can exhibit a tetramerized phase for a large enough frustration ratio α and a large spin-lattice coupling [61]. These different theoretical finding show that the physics of $Per_2[M(mnt)_2]$ should be quite subtle because there are two competing periodicity in the system: $2k_F^D = b^*/4$ for the Peierls instability on the Per stack and $2k_F^{SP} = b^*/2$ for the SP instability on the dithiolate stack.

In presence of non-magnetic dithiolate stacks the Peierls transition stabilizes the $2k_F^D$ modulation as shown by experimental studies of M = Cu, Co and Fe compounds (Table 1). Complications arise in salts where with M = Ni, Pd and Pt the dithiolate stack is magnetic and where, with a sizeable magneto-elastic coupling, a SP instability develops at $2k_F^{SP}$, which is two times $2k_F^D$. In Ni, Pd and Pt salts there is no experimental evidence that both $2k_F^D$ and $2k_F^{SP}$ LRO are simultaneously stabilized. In the Ni derivative where frustration effects are more apparent there is below 25 K a $2k_F^D$ LRO and a $2k_F^{SP}$ SRO. It is thus possible that the $b^*/2$ SP divergence on the dithiolate stacks stops around 50–35 K when the sizably frustrated spin system (α ~0.5) becomes quite strongly coupled to the adiabatic phonon field. In that case, the system should prefer to be tetramerized as predicted in Ref. [61]. In $Per_2[Ni(mnt)_2]$ the $2k_F^D$ LRO was initially attributed to the Peierls modulation on the Per stack. However one cannot exclude that a component of this modulation should originate from a distortion of the $Ni(dmit)_2$ stacks. In the Pd derivative the situation is different because a $2k_F^{SP}$ LRO is detected without any evidence of a $2k_F^D$ Peierls LRO; a feature which remains to be explained. In the Pt derivative there is a short range SP order in all the directions (of spatial extend comparable to the one of the Ni derivative) but Figure 5b does not provide any evidence of a $2k_F^D = b^*/4$ Peierls modulation. However, such a Peierls modulation could be too weak to be detected because the Peierls gap is quite small (one expects from the relative magnitude of the Peierls gaps super-lattice reflections 3 times less intense in the Pt salt than those detected in the Ni salt). Note however that the existence of a CDW modulation in the Pt derivative is assessed by the observations of non-linear conductivity effect due to the sliding of CDWs under electric field [25].

A key parameter of control of the phase diagram of $Per_2[M(mnt)_2]$ relies on the presence of sizeable inter-stack coupling. A close inspection of the structure shown in Figure 2 shows that if there are many direct interactions between Per stacks [15] the interaction between dithiolate stacks should be mediated through Per stacks. Thus, if Per stacks are the source of RKKY mediated interactions between localized spins on dithiolate stacks, one expects induced $2k_F^D$ SDW fluctuations on Per stacks (Figure 10). Such SDW fluctuations should compete with $2k_F^D$ BOW/CDW fluctuations at the origin of the Peierls instability of the Per stack. Up to now, there is no evidence of such a magnetic instability on the Per stack. However, it is possible that the modulation of Per stacks below T_P should be a mixed $2k_F^D$ SDW-CDW as found in the magnetic ground state of (TMTSF)$_2$PF$_6$ [62,63]. This could arise in particular in the Pd derivative where standard $2k_F^D$ Peierls superstructure reflections have not been detected.

In addition, in order to establish the RKKY mediated interaction the setting of a $2k_F^D$ electron–hole response function on the Per stack is of fundamental importance because it tunes the magnitude of the indirect AF coupling interaction via its thermal divergence below T_ρ. As the Peierls instability starts on Per stacks below T_ρ, which is lower than T_{SP}^{MF}, one observes on Ni and Pd dithiolate stacks firstly a SP instability below T_{SP}^{MF}, then secondly below T_ρ a vanishing of the singlet-triplet gap due to the growth of frustrated 2nd neighbor AF interactions. If the $2k_F^D$ electron–hole instability develops at higher temperature, as found in the (arene)$_2$ PF$_6$/AsF$_6$ salts, AF frustration will be set before the start of the SP instability. Thus, with the opening of a high temperature singlet-triplet gap due to AF frustration, the driving force for the SP instability could not be activated. These features show that the magnetic properties of dithiolate stacks result from a fine tuning with the Peierls instability on Per stacks.

4. Conclusions

In this paper, we have extensively reviewed the structural properties of the α-Per$_2$[M(mnt)$_2$] series of organic conductors. When the dithiolate stack is diamagnetic for M = Au or Cu or strongly dimerized for M = Co and Fe, the Per stack undergoes a $2k_F^D$ = b*/4 Peierls instability. However, the Peierls transition occurs at temperatures, T_P ~12–73 K, more than twice smaller than those T_P ~160–180 K found in other quarter-filled D$_2$X arene cation radical salts where D is either FA or substituted Per and X is a monovalent anion such as PF$_6$ and AsF$_6$. In the case of 1D S = 1/2 AF dithiolate stack for M = Ni, Pd and Pt, a SP instability develops at $2k_F^{SP}$ = b*/2. As this last wave vector is twice larger than $2k_F^D$, a very rich ground states is observed. The SP instability of Ni and Pd derivatives occurs in the classical limit with the formation of a pseudo-gap, in the AF magnetic excitations spectrum, driven by the growth of structural SP fluctuations below 100 K. Surprisingly, the spin susceptibility of these two salts drops below 50 K to finally vanish around 20 K. We attribute this unexpected behavior to the development of a singlet-triplet gap caused by frustration of S = 1/2 AF interactions on dithiolate stacks. Frustration is attributed to the presence of a second neighbor indirect RKKY S = 1/2 AF interaction mediated by a fine tuning with the $2k_F^D$ electron–hole instability of the Per stack. This subtle coupling between magnetic and conducting chains shows that the family of α-Per$_2$[M(mnt)$_2$] compounds provides for their Ni and Pd derivatives a remarkable realization of 1D Kondo lattices. A somewhat different magnetic behavior, with no clear-cut manifestation of the 1D Kondo lattice effects, is observed in the Pt derivative whose SP instability occurs in the quantum limit.

Acknowledgments: Earlier structural studies reported in this review have been performed by Vasco da Gama, Rui Henriques, Vita Ilakovac and Sylvain Ravy. One of us (JPP) recognizes very fruitful discussions with Claude Bourbonnais.

Conflicts of Interest: The authors declare no conflict of interest.

References

1. Comès, R.; Lambert, M.; Launois, H.; Zeller, H.R. Evidence for a Peierls Distortion or a Kohn Anomaly in One-Dimensional Conductors of the Type K$_2$Pt(CN)$_4$Br$_{0.30}$.xH$_2$O. *Phys. Rev. B* **1973**, *8*, 571–575. [CrossRef]
2. Pouget, J.-P. Bond and charge ordering in low-dimensional organic conductors. *Phys. B* **2012**, *407*, 1762–1770. [CrossRef]
3. Pouget, J.-P. The Peierls instability and charge density wave in one-dimensional electronic conductors. *C. R. Phys.* **2016**, *17*, 332–356. [CrossRef]
4. Schlenker, C.; Dumas, J.; Greenblatt, M.; van Smaalen, S. (Eds.) Physics and Chemistry of Low-Dimensional Inorganic Conductors. In *Nato ASI Ser. B Phys*; Plenum: New York, NY, USA, 1996; Volume 354.
5. Pouget, J.-P. Interplay between electronic and structural degrees of freedom in quarter-filled low dimensional conductors. *Phys. B* **2015**, *460*, 45–52. [CrossRef]

6. Akamatsu, A.; Inokuchi, H.; Matsunaga, Y. Electrical Conductivity of the Perylene-Bromine Complex. *Nature* **1954**, *173*, 168–169. [CrossRef]
7. Almeida, M.; Henriques, R.T. Perylene Based Conductors. Chapter 2. In *Handbook of Organic Conductive Molecules and Polymers Volume 1 "Charge Transfer Salts, Fullerenes and Photoconductors"*; Nalva, H.S., Ed.; John Wiley & Sons Ltd.: Chichester, UK, 1997; pp. 87–149.
8. Gama, V.; Henriques, R.T.; Bonfait, G.; Almeida, M.; Ravy, S.; Pouget, J.P.; Alcacer, L. The interplay between conduction electrons and chains of localized spins in the molecular metals (Per)$_2$M(mnt)$_2$, M = Au, Pt, Pd, Ni, Cu, Co and Fe. *Mol. Cryst. Liq. Cryst.* **1993**, *234*, 171–178. [CrossRef]
9. Henriques, R.T.; Alcacer, L.; Pouget, J.P.; Jérome, D. Electrical conductivity and x-ray diffuse scattering study of the family of organic conductors (perylene)$_2$M(mnt)$_2$, (M = Pt, Pd, Au). *J. Phys. C Solid State Phys.* **1984**, *17*, 5197–5208. [CrossRef]
10. Monchi, K.; Poirier, M.; Bourbonnais, C.; Matos, M.J.; Henriques, R.T. The Peierls transition in Per$_2$[Au$_x$Pt$_{1-x}$(mnt)$_2$]: Pair-breaking field effects. *Synth. Met.* **1999**, *103*, 2228–2231. [CrossRef]
11. Almeida, M.; Gama, V.; Santos, I.C.; Graf, D.; Brooks, J.S. Counterion dimerisation effects in the two-chain compound (Per)$_2$[Co(mnt)$_2$]: Structure and anomalous pressure dependence of the electrical transport properties. *CrystEngComm* **2009**, *11*, 1103–1108. [CrossRef]
12. Santos, I.C.; Gama, V.; Silva, R.A.L.; Almeda, M. to be submitted (2017).
13. Ilakovac, V.; Ravy, S.; Moradpour, A.; Firlej, L.; Bernier, P. Disorder and electronic properties of substituted perylene radical-cation salts. *Phys. Rev. B* **1995**, *52*, 4108–4122. [CrossRef]
14. Emery, V.J.; Bruinsma, R.; Barisic, S. Electron-Electron Umklapp Scattering in Organic Superconductors. *Phys. Rev. Lett.* **1982**, *48*, 1039–1043. [CrossRef]
15. Canadell, E.; Almeida, M.; Brook, J. Electronic band structure of α-(Per)$_2$M(mnt)$_2$ compounds. *Eur. Phys. J. B* **2004**, *42*, R453. [CrossRef]
16. Ilakovac, V.; Ravy, S.; Pouget, J.P.; Riess, W.; Brütting, W.; Schwoerer, M. CDW instability in the 2/1 organic conductor (FA)$_2$PF$_6$. *J. Phys. IV Fr.* **1993**, *3*, C2-137–C2-140. [CrossRef]
17. Peven, P.; Jérome, D.; Ravy, S.; Albouy, P.A.; Batail, P. Physical properties of the quasi-one dimensional substituted perylene cation radical salt. *Synth. Met.* **1988**, *17*, B405–B410. [CrossRef]
18. Ilakovac-Casses, V. Etude de l'influence du désordre sur les instabilités et les propriétés physiques des conducteurs et supraconducteurs organiques. Thesis, Université Paris-Sud, Orsay, France, 1994.
19. Reiss, W.; Schmid, W.; Gmeiner, J.; Schwoerer, M. Observation of charge density wave transport phenomena in the organic conductor (FA)$_2$PF$_6$. *Synth. Met.* **1991**, *41–43*, 2261–2267.
20. Reiss, W.; Brütting, W.; Schwoerer, M. Charge transport in the quasi-one-dimensional organic charge density wave conductor (Fluorenthene)$_2$PF$_6$. *Synth. Met.* **1993**, *55–57*, 2664–2669. [CrossRef]
21. Lopes, E.B.; Matos, M.J.; Henriques, R.T.; Almeida, M.; Dumas, J. Charge density wave non-linear transport in the molecular conductor (Perylene)$_2$Au(mnt)$_2$ (mnt = maleonitriledithiolate). *Europhys. Lett.* **1994**, *27*, 241–246. [CrossRef]
22. Ilakovac, V.; Ravy, S.; Pouget, J.P.; Lenoir, C.; Boubekeur, K.; Batail, P.; Dolanski Babic, S.; Biskup, N.; Korin-Hamzic, B.; Tomic, S.; Bourbonnais, C. Enhanced charge localization in the organic alloys [(TMTSF)$_{1-x}$(TMTTF)$_x$]$_2$ReO$_4$. *Phys. Rev. B* **1994**, *50*, 7136–7139. [CrossRef]
23. Gama, V.; Henriques, R.T.; Almeida, M.; Pouget, J.-P. Diffuse X-ray scattering evidence for Peierls and "spin-Peierls" like transitions in the organic conductors (Perylene)$_2$M(mnt)$_2$ [M = Cu, Ni, Co and Fe]. *Synth. Met.* **1993**, *55–57*, 1677–1682. [CrossRef]
24. Gama, V.; Henriques, R.T.; Almeida, M.; Bourbonnais, C.; Pouget, J.-P.; Jérome, D.; Auban-Senzier, P.; Gotschy, B. Structual and magnetic investigations of the Peierls transition of α-(Per)$_2$M(mnt)$_2$ with M = Fe and Co. *J. Phys. I Fr.* **1993**, *3*, 1235–1244. [CrossRef]
25. Lopes, E.B.; Matos, M.J.; Henriques, R.T.; Almeida, M.; Dumas, J. Charge Density Wave Dynamics in Quasi-One Dimensional Molecular Conductors: a Comparative Study of (Per)$_2$M(mnt)$_2$ with M = Au, Pt. *J. Phys. I Fr.* **1996**, *6*, 2141–2149. [CrossRef]
26. Bonfait, G.; Matos, M.J.; Henriques, R.T.; Almeida, M. The Peierls transition under high magnetic field. *Physics B* **1995**, *211*, 297–299. [CrossRef]
27. Graf, D.; Brooks, J.S.; Choi, E.S.; Uji, S.; Dias, J.C.; Almeida, M.; Matos, M. Suppression of a charge-density-wave ground state in high magnetic fields: Spin and orbital mechanisms. *Phys. Rev. B* **2004**, *69*, 125113. [CrossRef]

28. Brooks, J.S.; Graf, D.; Choi, E.S.; Almeida, M.; Dias, J.C.; Henriques, R.T.; Matos, M. Magnetic field dependence of CDW phases in $Per_2M(mnt)2$ (M = Au, Pt). *J. Low Temp. Phys.* **2006**, *142*, 787–803. [CrossRef]

29. Brooks, J.S.; Graf, D.; Choi, E.S.; Almeida, M.; Dias, J.C.; Henriques, R.T.; Matos, M. Magnetic field dependent behavior of the CDW ground state in $Per_2M(mnt)2$ (M = Au, Pt). *Curr. Appl. Phys.* **2006**, *6*, 913–918. [CrossRef]

30. Hasegawa, Y.; Fukuyama, H. A theory of phase transition in quasi-one-dimensional electrons. *J. Phys. Soc. Jpn.* **1986**, *55*, 3978–3990. [CrossRef]

31. Graf, D.; Brooks, J.S.; Almeida, M.; Dias, J.C.; Uji, S.; Terashima and Kimata, M. Evolution of superconductivity from a charge-density-wave ground state in pressurized $(Per)_2$ $[Au(mnt)_2]$. *Europhys. Lett.* **2009**, *85*, 27009. [CrossRef]

32. Shirane, G.; Shapiro, S.M.; Comès, R.; Garito, A.F.; Heeger, A.J. Phonon dispersion and Kohn anomaly in tetratiafulvalene-tetracyanoquinodimethane (TTF-TCNQ). *Phys. Rev. B* **1976**, *14*, 2325–2334. [CrossRef]

33. Caron, L.G.; Bourbonnais, C. Two-cutoff renormalization and quantum versus classical aspects for the one-dimensional electron-phonon system. *Phys. Rev. B* **1984**, *29*, 4230–4241. [CrossRef]

34. Goze, F.; Audouard, A.; Brossard, L.; Laukhin, V.N.; Ulmet, J.P.; Doublet, M.L.; Canadell, E.; Pouget, J.P.; Zavodnik, V.E.; Shibaeva, R.P.; et al. Magnetoresistance in pulsed fields, band structure calculations and charge density wave instability in $(TSeT)_2Cl$. *Synth. Met.* **1995**, *70*, 1279–1280. [CrossRef]

35. Ahmad, M.M.; Turner, D.J.; Underhill, A.E.; Jacobsen, C.S.; Mortensen, K.; Carneiro, K. Physical properties and the Peierls instability of $Li_{0.82}[Pt(S_2C_2(CN)_2)_2]$. $2H_2O$. *Phys. Rev. B* **1984**, *29*, 4796–4799. [CrossRef]

36. Dumoulin, B.; Bourbonnais, C.; Ravy, S.; Pouget, J.P.; Coulon, C. Fluctuation effects in low-dimensional spin-Peierls systems: Theory and experiment. *Phys. Rev. Lett.* **1996**, *76*, 1360–1363. [CrossRef] [PubMed]

37. Bonfait, G.; Matos, M.J.; Henriques, R.T.; Almeida, M. Spin-Peierls instability in $Per_2[M(mnt)_2]$ compounds probed by specific heat. *J. Phys. IV Colloq.* **1993**, *3*, 251–254. [CrossRef]

38. Henriques, R.T.; Alcacer, L.; Almeida, M.; Tomic, S. Transport and magnetic properties on the family of perylene-dithiolate conductors. *Mol. Cryst. Liq. Cryst.* **1985**, *120*, 237–241. [CrossRef]

39. Green, E.L.; Brooks, J.S.; Kuhns, P.L.; Reyes, A.P.; Lumata, L.L.; Almeida, M.; Matos, M.J.; Henriques, R.T.; Wright, J.A.; Brown, S.E. Interaction of magnetic field-dependent Peierls and spin-Peierls ground states in $(Per)_2[Pt(mnt)_2]$. *Phys. Rev. B* **2011**, *84*, 121101(R). [CrossRef]

40. Bourbonnais, C.; Henriques, R.T.; Wzietek, P.; Kongeter, D.; Voiron, J.; Jerome, D. Nuclear and electronic resonance approaches to magnetic and lattice fluctuations in the two-chain family of organic compounds $(perylene)_2$ $[M(S_2C_2(CN)_2)_2]$ (M = Pt, Au). *Phys. Rev. B* **1991**, *44*, 641–651. [CrossRef]

41. Green, E.L.; Lumata, L.L.; Brooks, J.S.; Kuhns, P.; Reyes, A.; Brown, S.E.; Almeida, M. ^1H and ^{195}Pt NMR Study of the Parallel Two-Chain Compound $Per_2[Pt(mnt)_2]$. *Crystals* **2012**, *2*, 1116–1135. [CrossRef]

42. Pouget, J.-P.; Ravy, S.; Schoeffel, J.P.; Dhalenne, G.; Revcolevschi, A. Spin-Peierls lattice fluctuations and disorders in $CuGeO_3$ and its solid solutions. *Eur. Phys. J. B* **2004**, *38*, 581–598. [CrossRef]

43. Orignac, E.; Chitra, R. Mean-field theory of the spin-Peierls transition. *Phys. Rev. B* **2004**, *70*, 214436. [CrossRef]

44. Pouget, J.P. Microscopic interactions in CuGeO3 and organic Spin-Peierls systems deduced from their pretransitional lattice fluctuations. *Eur. Phys. J. B* **2001**, *20*, 321–333, and 2001, *24*, 415. [CrossRef]

45. Citro, R.; Orignac, E.; Giamarchi, T. Adiabatic-antiadiabatic crossover in a spin-Peierls chain. *Phys. Rev. B* **2005**, *72*, 024434. [CrossRef]

46. Weiße, A.; Hager, G.; Bishop, A.R.; Fehske, H. Phase diagram of the spin-Peierls chain with local coupling: Density-matrix renormalization-group calculations and unitary transformations. *Phys. Rev. B* **2006**, *74*, 214426.

47. Liu, Q.; Ravy, S.; Pouget, J.P.; Coulon, C.; Bourbonnais, C. Structural fluctuations and spin-Peierls transitions revisited. *Synth. Met.* **1993**, *55–57*, 1840–1845. [CrossRef]

48. Eggert, S.; Affleck, I.; Takahashi, M. Susceptibility of the Spin 1/2 Heisenberg Antiferromagnetic Chain. *Phys. Rev. Lett.* **1994**, *73*, 332–335. [CrossRef] [PubMed]

49. Gama, V. O Papel das Cadeias Conductoras e das Cadeias Magnéticas nos Compostos da Familia $Per_x[M(mnt)2]$ (x = 2, M = Cu, Ni, Co e Fe; x = 1, M = Co). Thesis, Universidade Técnica de Lisboa, Lisbon, Portugal, 1993.

50. Alcacer, L.; Maki, A.H. Magnetic Properties of Some Electrically Conducting Perylene-Metal Dithiolate Complexes. *J. Phys. Chem.* **1976**, *80*, 1912–1916. [CrossRef]

51. Gulácsi, M. The one-dimensional Kondo lattice model at partial band filling. *Adv. Phys.* **2004**, *53*, 769–937. [CrossRef]

52. Nomura, K.; Okamoto, K. Critical properties of $S = 1/2$ antiferromagnetic XXZ chain with next-nearest-neighbour-interactions. *J. Phys. A Math. Gen.* **1994**, *27*, 5773–5788. [CrossRef]

53. Augier, D.; Poilblanc, D. Dynamical properties of low-dimensional $CuGeO_3$ and NaV_2O_5 spin-Peierls systems. *Eur. Phys. J. B* **1998**, *1*, 19–28. [CrossRef]

54. Watanabe, S.; Yokoyama, H. Transition from Haldane Phase to Spin Liquid and Incommensurate Correlation in Spin-1/2 Heisenberg chains. *J. Phys. Soc. Jpn.* **1999**, *68*, 2073–2097. [CrossRef]

55. Coqblin, B.; Núñez-Reigeiro, M.D.; Theumann, A.; Iglesias, J.R.; Magalhães, S.G. Theory of the Kondo lattice: competition between Kondo effect and magnetic order. *Philos. Mag.* **2006**, *86*, 2567–2580. [CrossRef]

56. Kondo, J. Resistance Minimum in Dilute Magnetic Alloys. *Prog. Theor. Phys.* **1964**, *32*, 37–49. [CrossRef]

57. Quirion, G.; Poirier, M.; Liou, K.K.; Ogawa, M.; Hoffman, B.M. Possibility of a one-dimensional Kondo system in the alloys of Cu_xNi_{1-x} (Phtalocyaninato)I. *J. Phys. (Paris) Colloq.* **1988**, *49*, C8-1475–C8-1476. [CrossRef]

58. Foury-Leylekian, P.; Leininger, P.; Ilakovac, V.; Joly, Y.; Bernu, S.; Fagot, S.; Pouget, J.P. Ground state of the quasi-1D correlated correlated electronic system $BaVS_3$. *Physics B* **2012**, *407*, 1692–1695. [CrossRef]

59. Tsunetsugu, H.; Sigrist, M.; Ueda, K. The ground-state phase diagram of the one-dimensional Kondo lattice model. *Rev. Mod. Phys.* **1997**, *69*, 809–863. [CrossRef]

60. Xavier, J.C.; Pereira, R.G.; Miranda, E.; Affleck, I. Dimerization Induced by the RKKY Interaction. *Phys. Rev. Lett.* **2003**, *90*, 247204. [CrossRef] [PubMed]

61. Becca, F.; Mila, F.; Poilblanc, D. Teramerization of a frustrated spin-1/2 chain. *Phys. Rev. Lett.* **2003**, *91*, 067202. [CrossRef] [PubMed]

62. Pouget, J.P.; Ravy, S. Structural Aspects of the Bechgaard Salts and Related Compounds. *J. Phys. I Fr.* **1996**, *6*, 1501–1525. [CrossRef]

63. Pouget, J.P.; Ravy, S. X-ray evidence of charge density wave modulations in the magnetic phases of $(TMTSF)_2PF_6$ and $(TMTTF)_2Br$. *Synth. Met.* **1997**, *85*, 1523–1528. [CrossRef]

magnetochemistry

MDPI

Article

The Solid Solutions (Per)$_2$[Pt$_x$Au$_{(1-x)}$(mnt)$_2$]; Alloying Para- and Diamagnetic Anions in Two-Chain Compounds

Manuel Matos [1,2], Gregoire Bonfait [3,4], Isabel C. Santos [3], Mónica L. Afonso [3], Rui T. Henriques [1] and Manuel Almeida [3,*]

[1] Instituto de Telecomunicações, P-1049-001 Lisboa, Portugal; mmatos@deq.isel.pt (M.M.); rui.henriques@lx.it.pt (R.T.H.)
[2] Chemical Engineering Department, ISEL, P-1959-007 Lisboa, Portugal
[3] C²TN-Centro de Ciências e Tecnologias Nucleares, Instituto Superior Técnico, Universidade de Lisboa, P-2695-066 Bobadela LRS, Portugal; gb@fct.unl.pt (G.B.); icsantos@ctn.tecnico.ulisboa.pt (I.C.S.); mochaves@sapo.pt (M.L.A.)
[4] Physics Department, Faculty of Sciences and Technologies, Universidade Nova de Lisboa, P-2829-516 Caparica, Portugal
* Correspondence: malmeida@ctn.tecnico.ulisboa.pt; Tel.: +351-219-946-171

Academic Editor: Carlos J. Gómez García
Received: 28 April 2017; Accepted: 26 May 2017; Published: 13 June 2017

Abstract: The α-(Per)$_2$[M(mnt)$_2$] compounds with M = Pt and Au are isostructural two-chain solids that in addition to partially oxidized conducting perylene chains also contain anionic chains that can be either paramagnetic in the case of M = Pt or diamagnetic for M = Au. The electrical transport and magnetic properties of the solid solutions (Per)$_2$[Pt$_x$-Au$_{(1-x)}$(mnt)$_2$] were investigated. The incorporation of paramagnetic [Pt(mnt)$_2$] impurities in the diamagnetic chains, and the effect of breaking the paramagnetic chains with diamagnetic centers for the low and high Pt range of concentrations were respectively probed. In the low Pt concentration range, there is a fast decrease of the metal-to-insulator transition from 12.4 K in the pure Au compound to 9.7 K for $x = 0.1$ comparable to the 8.1 K in the pure Pt compound. In the range $x = 0.50-0.95$, only β-phase crystals could be obtained. The spin-Peierls transition of the pure Pt compound, simultaneous with metal-to-insulator (Peierls) transition is still present for 2% of diamagnetic impurities ($x = 0.98$) with transition temperature barely affected. Single crystal X-ray diffraction data obtained a high-quality structural refinement of the α-phase of the Au and Pt compounds. The β-phase structure was found to be composed of ordered layers with segregated donors and anion stacks, which alternate with disordered layers. The semiconducting properties of the β-phase are due to the disorder localization effects.

Keywords: organic conductors; perylene; metal-dithiolate; organic alloys; conductivity; thermoelectric power; magnetic susceptibility; spin-Peierls transition; Peierls transition

1. Introduction

The charge transfer salts from the organic aromatic electron donor perylene (Per) and monoanionic transition metal bis-maleonitriledithiolate complexes [M(mnt)$_2$] (M = Au, Pt, Pd Cu, Ni Fe and Co), (Scheme 1) with general formula (Per)$_2$[M(mnt)$_2$], constitute a large family of compounds that since the first seminal paper in 1974 [1,2], have been the topic of a large number of successive studies and the subject of a few reviews [3–5], including a recent one focused on structural Peierls and spin-Peierls inabilities in this issue [6]. These studies soon provided evidence that these compounds can be obtained in two polymorphs, α- and β- [7]. The α-phases, with highly conducting quasi-1D metallic properties at higher temperatures, are particularly interesting since after more than 40 years of developments in the field of molecular conductors, they remain a unique family of compounds with two types of chains (magnetic and conducting). The β-phases present a less interesting semiconducting behavior, which is associated with a so far poorly resolved disordered structure. In spite of many efforts and several studies, the conditions that favor the formation of any of these two polymorphic phases have remained uncontrolled, but often electrocrystallization batches are entirely composed of only one of the polymorphs.

$[M(mnt)_2]$
(mnt=maleo-nitriledithiolate)

(Per)=perylene

Scheme 1. Molecular structure of perylene and $[M(mnt)_2]$.

The main feature of the α-phases is that conducting and magnetic chains can exist side by side in the same structure and in mutual interaction. The conducting chains in these compounds are provided by regular stacks of partially oxidized perylene molecules $(Per)^{1/2+}$ which behave as a $3/4$-filled 1D electron band system. The square planar anionic complexes $[M(mnt)_2]^-$ are also stacked alongside and in between the donor stacks (see Section 2.1). For some metals (e.g., M = Ni, Pd, Pt), the anions have a localized $S = 1/2$ magnetic moment, while in other metals such as Au and Cu the anions are diamagnetic. The interaction between these two types of chains and the competition between the instabilities of these 1D magnetic and conducting systems is a unique feature of this family of compounds that has been the central point of discussion in different studies.

Among the different members of this family, the pair of better-characterized compounds with neighbor transition metals in the periodic table, Au and Pt, where the anionic chains are respectively diamagnetic and paramagnetic, are special members which through comparison can establish evidence for the role of the paramagnetic chains in the properties of these compounds. Both compounds present a metal-to-insulator transition at 8 K and 12 K respectively that has been ascribed to a Peierls distortion (tetramerisation) of the perylene chains, which in a first approach behave as a $3/4$-filled band system. The Pt compound presents, in addition to the Peierls transition involving the conducting perylene chains, a spin-Peierls transition at the same critical temperature that corresponds to a dimerization of the spin-carrying units $[Pt(mnt)_2]^-$. The structural aspects of these transitions were the subject of a detailed review in this Special Issue of *Magnetochemistry* [6].

In this paper, we present results of a detailed study of the electrical transport and magnetic properties of the solid solutions $(Per)_2[Pt_x-Au_{(1-x)}(mnt)_2]$, for which some preliminary results were previously reported [8,9]. This study illustrated for the α-phases both the effect of incorporating paramagnetic impurities in the diamagnetic anionic chains and the breaking effect of diamagnetic centers in the paramagnetic anionic chains for the low and high Pt range of concentrations, respectively. In the range $x = 0.50-0.95$, only β-phase crystals could be grown and a partial description of the crystal structure of this disordered phase was obtained.

2. Results and Discussion

Single crystals with the composition $(Per)_2[Pt_x-Au_{(1-x)}(mnt)_2]$ in the entire composition range $x = 0-1$ were obtained by electrocrystallization from dichloromethane solutions of perylene and variable relative amounts of the gold and platinum complexes, under general galvanostatic conditions, as previously described for the pure compounds [10,11]. The quality and shape of the crystals was found; however, it was variable and dependent on the composition x. For x in the range 0.0–0.4 and 0.98–1.0, larger needle or elongated plate-shaped crystals could be obtained, with the largest ones reaching $10 \times 0.5 \times 0.08$ mm^3 for the pure Pt and Au compounds. However, in the range $x = 0.5-0.95$, only very small needle-shaped crystals could be obtained, which as described below, were all found to pertain to the β-phase polymorph.

The Scanning Electron Microscopy–Energy Dispersive Spectroscopy (SEM–EDS) microanalysis of the different crystals obtained in all composition ranges (Figure 1) revealed that the relative Pt and Au concentrations in the crystals closely follow those of the solution, denoting the formation of Au–Pt solid solutions in the full range of compositions, as expected in view of the similar size of the Pt and Au atoms and close similarities of the two isostructural compounds with identical lattice parameters [10,12]. However, as seen in Figure 1, there is a small deviation of the crystal composition from that of the solution, in the sense that the crystals tend to slightly enrich in Pt.

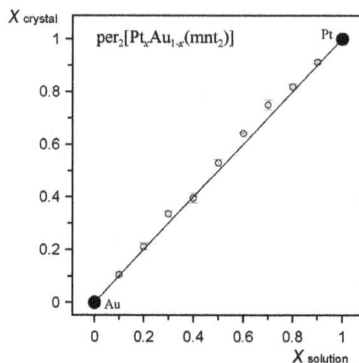

Figure 1. Pt and Au concentration of crystals (Per)$_2$[Pt$_x$-Au$_{(1-x)}$(mnt)$_2$] versus growth solution concentration. The line represents a linear relation with equal solid and liquid concentrations.

The isostructural nature of the Au and Pt compounds, space group P2$_1$/n, and the close similarity of lattice and structural parameters, are further confirmed by single crystal X-ray diffraction of pure Au and Pt compounds performed in this work, which enabled a crystal structure refinement with significantly higher quality than the original data published more than 30 years ago. Crystal data are summarized in Table 1. However, as further denoted by the electrical transport properties described below, for the range of composition from x = 0.45 to 0.95, all preparations lead only to β-phase crystals and no crystals of the α-phase could be obtained. The structure of the β-polymorph in spite of its disorder nature could be further enlightened by X-ray diffraction in very small single crystals using synchrotron radiation.

Table 1. Crystal and structural refinement data for (Per)$_2$[M(mnt)$_2$] compounds.

Compound	β-(Per)$_2$[Pt(mnt)$_2$]	α-(Per)$_2$[Pt(mnt)$_2$]	α-(Per)$_2$[Au(mnt)$_2$]
Sp.Gr.	P-1	P2$_1$/n	P2$_1$/n
a/Å	4.0105(3)	16.4258(3)	16.5632(9)
b/Å	15.2221(11)	4.1733(1)	4.1342(2)
c/Å	20.3352(13)	26.6028(4)	26.5226(12)
α/°	78.481(4)	90.0	90.0
β/°	88.200(5)	95.053(1)	94.990(3)
γ/°	86.334(5)	90.0	90.0
V/Å^3	1213.71(17)	1816.53(6)	1809.27(16)
Z, D$_{calc}$ (Mg/m^3)	1, 1.253 [1]	2, 1.792	2, 1.802
R$_1$/R$_w$ [I > 2σ(I)]	0.0775/0.1834	0.0191/0.0416	0.0694/0.1135
T/K	150(2)	150(2)	150(2)
CCDC [2]	1545581	1545582	1545583

[1] Taking into account disordered molecules in the structure for Z = 1.333, D$_{calc}$ = 1.783. [2] Cambridge Crystallographic Data Centre reference number.

2.1. Crystal Structure

The structural refinements confirmed that compounds α-(Per)$_2$[Au(mnt)$_2$] and α-(Per)$_2$[Pt(mnt)$_2$] are isostructural, as previously reported [10,12], belonging to the $P2_1/n$ group with very similar lattice parameters (Table 1). The main peculiarity of these compounds is the existence of two segregated perylene and [M(mnt)$_2$] stacks along the *b*-axis, as shown in Figure 2 for the Pt compound. There are only very small differences in the crystalline structure of these compounds with anions of identical dimensions, therefore making possible the growth of the solid solutions in the entire range of composition without any major structural changes expected.

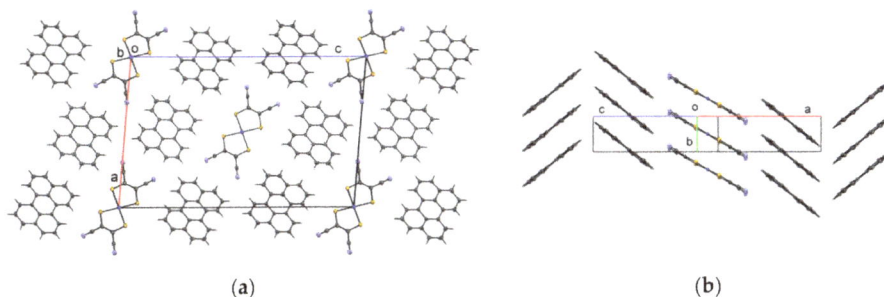

(a)   (b)

Figure 2. Crystal structure of α-(Per)$_2$[Pt(mnt)$_2$]: (**a**) view along the *b*-axis, where segregated Per and [Pt(mnt)$_2$] stacks can be seen. Each [Pt(mnt)$_2$] stack is surrounded by six Per stacks, and each donor stack has three Per and three [Pt(mnt)$_2$] stacks as first neighbors; (**b**) Partial view perpendicular to the stacking *b*-axis of parallel donor and acceptor stacks.

Besides the α-type structure summarized above, the single crystals studied by X-ray diffraction revealed also the occurrence of β-polymorphs, as first reported in the M = Ni and Cu compounds [7], and later observed in other compounds of this family with undimerised anions, namely for M = Pd [13,14] and Pt [15]. This β-polymorph is characterized by entirely different lattice parameters (Table 1) with a smaller cell volume ~1220 Å^3. This volume is too small to accommodate all the molecular units expected for Z = 2 as in the α-phases, which have cell volumes of the order of ~1840 Å^3, and it rather corresponds to Z = 1.33 assuming a similar density, denoting a disordered structure where at least one of the cell parameters should be a multiple. The disordered nature of the structure is also indicated by the presence of strong and temperature-independent diffuse scattering planes at $\pm na^*/6$, with n = 1, 2, and 3, as shown in Figure 3d) obtained with synchrotron radiation, indicating that the repeat unit along *a* is six-fold (~25 Å), as previously observed in the Ni and Cu compounds [7].

A full/clear structural refinement of the disordered β-phases could not be obtained. However, using diffraction data collected from a small Pt single crystal, it was possible to devise a structure containing ordered segregated donors and anion stacks along *a*, which alternated along *c*, with disordered layers. The disorder of these layers could not be satisfactorily modeled, but seemed to consist mainly in perylene molecules in two possible positions (50% occupation factors), regularly stacked along *a*.

The ordered stacks presented perylene and [Pt(mnt)$_2$] molecules tilted towards the stacking axis in the same orientation at variance with the alternated tilting of the α-phase, and with respective interplane distances of 3.270(6) Å and 3.521(4) Å. The overlap mode was virtually identical to that observed in the α-phase where these distances are 3.287(3) and 3.609(2) Å, respectively. The disordered perylene molecules had a tilting angle and overlap mode identical to the ordered ones, and the disordered layer contained additional electron density (in voids of V = 23,723 Å^3, corresponding to 19.5% of unit cell volume), corresponding to additional perylene and [Pt(mnt)$_2$] units, the positions of which could not be modelized. It is worth mentioning that the six-folding of the lattice parameter *a* may correspond to two [Pt(mnt)$_2$] units and four perylene units in a packing pattern that repeats every 6a but with no coherence in the b,c plane.

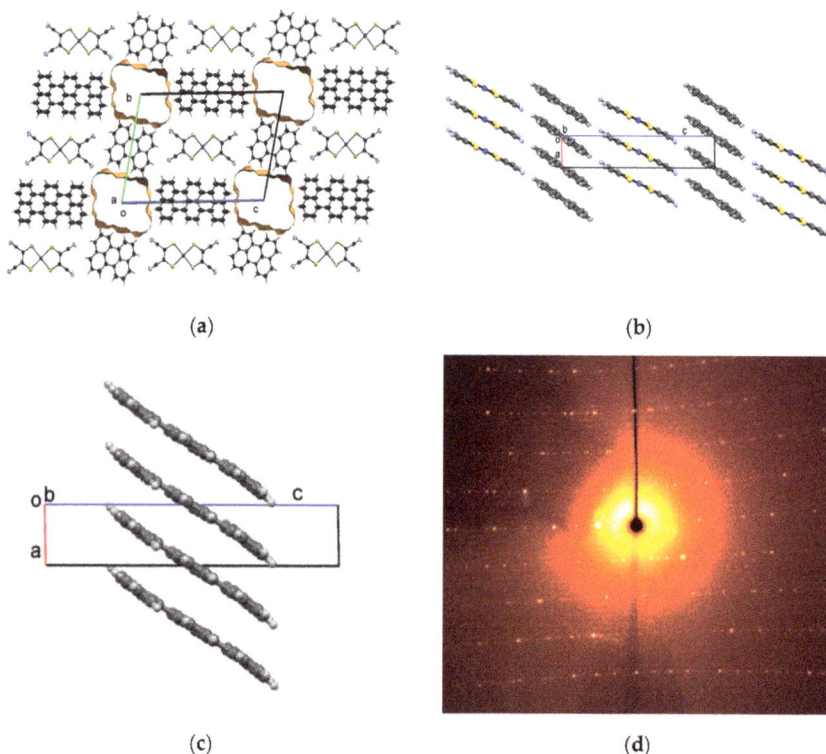

(a)

(b)

(c)

(d)

Figure 3. (a) Crystal structure of β-(Per)$_2$[Pt(mnt)$_2$] viewed along the stacking a-axis, with ordered layers of segregated perylene and [Pt(mnt)$_2$] stacks, alternating with disordered layers, in which only perylene molecules in two possible positions (50% occupation factors) could be devised; (b) Partial view along b of an ordered layer with parallel donor and acceptor stacks; (c) Partial view along b of a disordered stack of perylene molecules; (d) Diffraction pattern of β-(Per)$_2$[Pt(mnt)$_2$] crystal (mochromatic synchrotron radiation, fixed film fixed crystal) showing diffuse scattering planes at ±na*/6 (vertical axis).

2.2. Electrical Transport Properties—Conductivity and Thermoelectric Power

The results of the electrical transport properties measurements of (Per)$_2$[Pt$_x$-Au$_{(1-x)}$(mnt)$_2$] crystal are shown in Figure 4. The electrical resistivity ρ of the pure Pt and Au compounds closely followed the previously published results for the α-phases [16,17], with room temperature values of circa 900 S/cm (700–1200), and a clear metallic regime down to metal insulator (Peierls) transitions at 8.1 and 12.4 K respectively, which are seen by well-defined maxima in the derivative dln ρ/d(1/T). The crystals with increasing concentration in Pt, up to x = 0.40 showed a behavior comparable to that of the pure Au compound (Figure 4). There was a small sample dependence of both the sharpness of the transition and its temperature, typically ±0.1 K, which is ascribed to sample quality variations. In spite of these variations, it is clear that upon increasing Pt concentration there is first a sensible decrease of the metal-to-insulator transition from 12.2 K for x = 0 to 9.7 K for x = 0.1, followed by a much slower decrease for higher concentrations (Table 2). For x = 0.5 to 0.95, an entirely different regime of the electrical conductivity is observed; the room temperature conductivities are much smaller

(~40–60 S/cm) with a thermally-activated regime in the entire temperature range, identical to that previously described in β-phases [7,13,15]. For $x = 0.98$, a α-type behavior is recovered in samples from some batches, while other batches still present the β-type behavior.

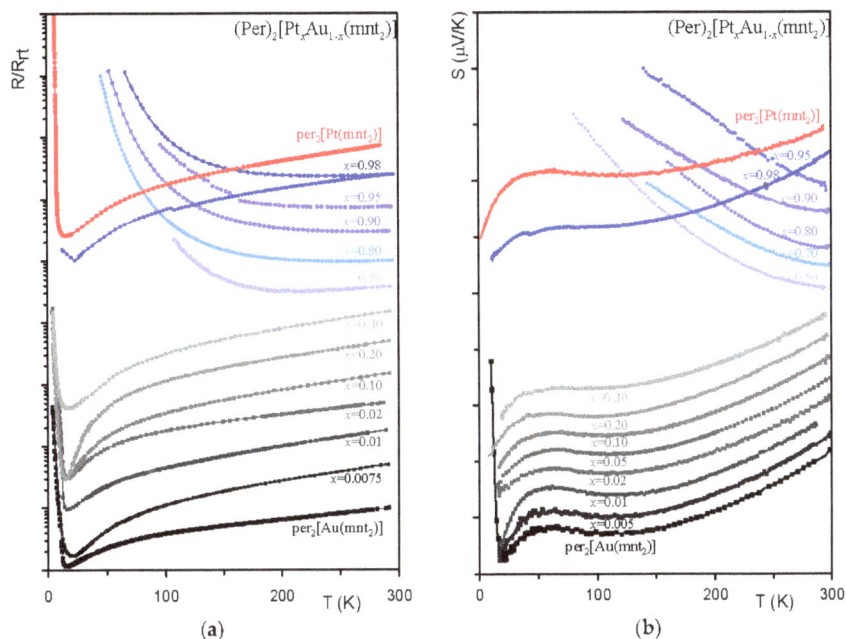

(a)

(b)

Figure 4. Electrical transport properties of $(Per)_2[Pt_x-Au_{(1-x)}(mnt)_2]$ single crystals with composition values indicated as a function of temperature, T: (**a**) Electrical resistivity, R, plotted as ratio to the room temperature (Rrt) value; (**b**) Absolute thermoelectric power (values for increasing x are off-set by 5 μV/K at room temperature for each composition).

Table 2. Metal-to-insulator transition temperatures, T_{MI}, of $(Per)_2[Pt_xAu_{1-x}(mnt)_2]$ crystals as defined by maxima in the derivative of the resistivity, dln ρ/d(1/T).

x	T_{MI} (K)
0.0	12.4
0.0075	12.3
0.01	12.2
0.02	11.8
0.05	10.8
0.1	9.7
0.2	9.6
0.4	9.1
0.5	-
0.8	-
0.9	-
0.95	-
0.98 (β)	-
0.98 (α)	8.7
1.0	8.1

The thermoelectric power results (Figure 4b) also reveal the existence of two different behaviors of the samples depending on the composition, confirming the two phases denoted by resistivity measurements. For platinum concentrations up to $x = 0.40$, the thermopower exhibited a behavior similar to that of the pure Au compound, with room temperature values of the order of ~40 µV/K decreasing upon cooling in a fashion typical of a metal, until the metal-to-insulator transition was reached. For Pt concentrations x between 0.50 and 0.95, the temperature dependence of the thermopower was clearly distinct, with room temperature values of ~35 µV/K, which increased upon cooling approximately as $1/T$ in a semiconducting behavior similar to that of the β-phases of these compounds with M = Ni, Cu and Pd [7,13,15]. For high Pt concentrations with $x = 0.98$, the temperature dependence of the thermoelectric power showed again a behavior similar to that of the α-phase of the pure Pt compound, which is also almost identical to that of the Au compound. The thermoelectric power of the α-phase crystals in the high temperature range, where a linear temperature dependence was followed, is consistent with the behavior expected for a ¾-filled band 1D system. Within the tight-binding approximation, and neglecting energy dependence of the scattering time, the thermoelectric power S of an uncorrelated 1D system is expected to follow the equation [18]:

$$S = -\frac{\pi^2 k_B^2 T}{6 |e||t|} \left[\frac{\cos(\pi\rho/2)}{\sin^2(\pi\rho/2)} \right] \tag{1}$$

where t is the transfer integral between next neighboring molecules, and ρ is the number of electrons per molecule (3/2). These approximations using the high temperature S data ($T > 100$ K) predict a bandwidth $4t = 0.58$ eV for these α-phase alloys, as for the pure compounds. This is overall in very good agreement with the values of $t = 0.143$ eV estimated from plasma frequency in reflectance [19], $t = 0.149$ eV from magnetoresistance [20,21] and $t \approx 0.148$ eV from quantum chemistry calculations under the extended Huckel approximation [22].

The semiconducting properties of the β-phases can be understood in view of the observed structural disorder. In the ordered layers of the β-phase, the regular stacking of the donors with same type of overlap and intermolecular distances comparable to those of the α-phase could lead to a similar metallic behaviour associated with an identical ¾-filled band, highly anisotropic quasi-1D system. However, the disordered layers impose an external random potential that, in this extremely anisotropic quasi-1D system, can effectively localize the electronic states.

2.3. Magnetic Susceptibility

The results of the paramagnetic susceptibility measurements of $(Per)_2[Pt_x-Au_{(1-x)}(mnt)_2]$ are shown in Figure 5. The pure Au sample, where the anions are diamagnetic, had a small Pauli-like contribution due to conduction electrons of 2.2×10^{-4} emu/mol at room temperature, which had a small decrease upon cooling and vanished at the metal-to-insulator transition, as previously reported [23] and comparable to the α-phase of the Cu compound, also with diamagnetic anions [24]. The pure Pt sample, where the anions were paramagnetic, presented a much larger paramagnetic susceptibility that in addition to a small Pauli-like contribution identical to that of the Au compound, was ascribed essentially to the paramagnetic $[Pt(mnt)_2]^-$ anions, which behaved as a chain of antiferromagnetically (AFM) coupled $S = 1/2$ spins. The maximum of paramagnetic susceptibility at ~25 K was indicative of antiferromagnetic coupling of the spins along the chains, and indeed magnetic susceptibility above the metal-to-insulator transition temperature at 8.1 K can be fairly well adjusted by the 1D Heisenberg linear chain model by Bonner-Fisher [25] with an AFM exchange interaction of $J/k_B = -15.5$ K [23]. The large paramagnetic susceptibility of the pure Pt compound experienced a sharp decrease below 8.1 K as a consequence of a spin-Peierls transition (dimerization of the anionic chains), which took place at the same temperature as the metal-to-insulator transition [26].

The main feature in the magnetic susceptibility of the solid solutions $(Per)_2[Pt_x-Au_{(1-x)}(mnt)_2]$ was that upon increasing x, the presence of paramagnetic centers in the anionic chains, gave rise to an increasing Curie-type contribution to the paramagnetism, proportional to the Pt concentration, which soon dominated over the small Pauli-like contribution of the conducting perylene stacks. At the other extreme range of concentrations, for large x, the incorporation of diamagnetic centers in the $[Pt(mnt)_2]^-$ chains soon destroyed the spin-Peierls transition as theoretically predicted [27]. For $x = 0.98$, two types of samples were obtained, as previously denoted by transport measurements, one of α- and another of β-polymorph. The one corresponding to the α-phase (inset Figure 5c)) still showed a maximum of susceptibility at ~25 K and a drop of susceptibility at 8 K, although with an increased Curie tail. This effect is comparable to that experimentally observed in other spin-Peierls systems upon incorporation of non-magnetic impurities or defects, either organic systems such as p-CyDOV [28] or inorganic compounds such as $CuGeO_3$ [29]. The study of this spin-Peierls transition with larger concentrations of diamagnetic impurities was not possible in this system due to formation of β-polymorph.

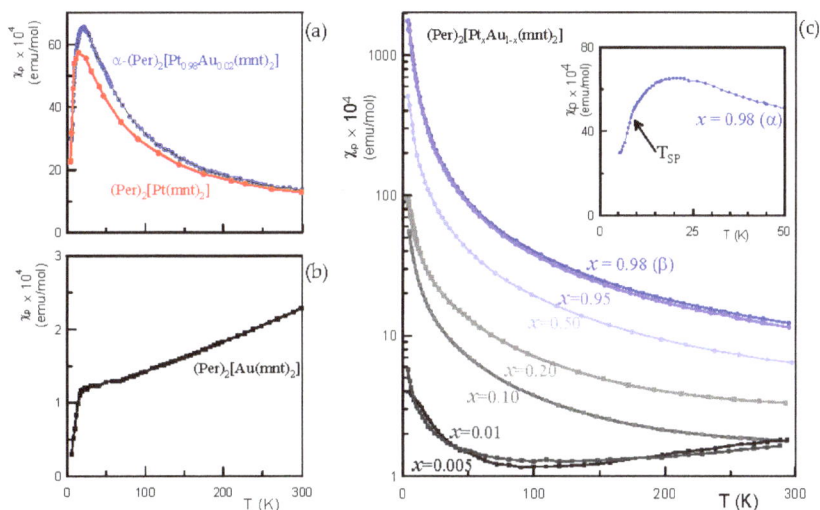

Figure 5. Paramagnetic susceptibility, χ_P, of $(Per)_2[Pt_xAu_{(1-x)}(mnt)_2]$ with composition x indicated, as a function of temperature, T. (**a**) Pt compound and $x = 0.98$, α-phase; (**b**) Au compound; (**c**) $0.005 \leq x \leq 0.95$ and 0.98, β-phase. The inset shows the behavior of a α-phase sample with $x = 0.98$.

The paramagnetic susceptibility of $(Per)_2[Pt_x-Au_{(1-x)}(mnt)_2]$ could result from donor and acceptor contributions. For the α-phase, the donor contribution, due to the delocalized conduction electrons in the perylene stacks, was expected to be identical to that of the pure Au compound. The remaining contribution of the paramagnetic $[Pt(mnt)_2]$ anions could therefore be deduced from the total paramagnetic susceptibility after subtraction of the small Pauli-like contribution of the pure Au compound. This last contribution in samples with x up to 0.95 and 0.98 of β-phase was found to follow above 50 K a Curie–Weiss law with small Weiss constants (<5 K) and Curie constant C close to that that expected for $S = 1/2$ spins with $g \approx 2.0$ (C = 0.375 emu K/mole), with no difference between α and β phases, as shown in Figure 6. A significant deviation was however observed for $x = 0.98$ of α-phase, which approached the behavior of the pure Pt compound.

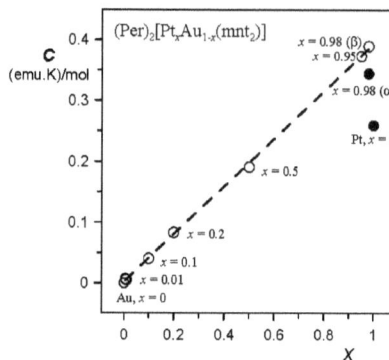

Figure 6. Curie constant, C, of the [Pt(mnt)$_2$] contribution to the paramagnetic susceptibility of (Per)$_2$[Pt$_x$-Au$_{(1-x)}$(mnt)$_2$] as a function of the composition x.

3. Materials and Methods

(Per)$_2$[Au$_{1-x}$Pt$_x$(mnt)$_2$] single crystals were obtained by electrocrystallization from dichloromethane solutions of perylene (~1.4 × 10^{-2} M) and variable relative amounts of the gold and platinum complexes as tetrabutylammonium salts (total anion concentration ~7 × 10^{-3} M) under general conditions that have been previously optimized and described in detail [11]. Special care was taken in the purification of the compounds and solvents used. Perylene (Sigma, St. Louis, MO, USA) was purified by multiple recrystallization and silica and alumina chromatography [30], followed by gradient sublimation under reduced pressure at ~110 °C. The tetrabutylammonium salts of gold and platinum were synthesized as described [31,32] and recrystallized several times from acetone under argon atmosphere. Dichloromethane was distilled, dried and passed through a column of activated alumina just before use. Galvanostatic conditions with current densities in the range of 5–20 μA/cm^2 were employed using platinum wire electrodes and two compartment cells sealed under argon. The best crystals were obtained by using a slightly increasing current over time, with an initial current of 1 μA and an increase of 1 μA every two days. The crystals were collected after 10 days and generally appeared as small black shining needles.

SEM–EDS microanalysis of the crystals were performed using a JEOL JSM 6301F microscope at CEMUP. The Au:Pt concentration ratio was estimated from the L-lines of the metals, while the K-line of sulfur was used as a control element of the analysis. Four spectra were obtained: a global spectrum in one area containing a cluster of crystals, and three others along a well-defined face of selected crystals. The typical dispersion of Pt concentration x were ±0.01. The results were found reproducible in different crystals from the same batch.

Single crystal X-ray diffraction experiments in α-(Per)$_2$[M(mnt)$_2$] crystals with M = Au and Pt and β-(Per)$_2$[Pt(mnt)$_2$] were performed with a Bruker APEX II CCD detector diffractometer, using graphite monochromated MoK$_\alpha$ radiation (λ = 0.71073 Å), in the φ and ω scans mode. A semi-empirical absorption correction was carried out using SADABS [33]. Data collection, cell refinement, and data reduction were done with the SMART and SAINT programs [34]. In the case of β-(Per)$_2$[Pt(mnt)$_2$], X-ray diffraction was also performed on a very small single crystal on a heavy-duty diffractometer at the Materials Science Beamline ID11 (λ = 0.29520 Å, ESRF, Grenoble, France) using a Frelon2K CCD detector. After conversion of the frame file format, the data were indexed using SMART, integrated with SAINT and corrected for absorption using SADABS. The structures were solved by direct methods using SIR97 [35] and refined by full matrix least-squares methods using the program SHELXL97 [36] and the winGX software package [37]. Non-hydrogen atoms were refined with anisotropic thermal parameters, whereas H atoms were placed in idealized positions and allowed to refine while riding on the parent C atom. Molecular graphics were prepared using mercury [38].

Electrical conductivity measurement in the range 4 K–300 K were performed in a helium cryostat along the long axis (b-axis of α-phase, a-axis in β-phase) using an in-line four-probe configuration, with a current of ~1 µA of frequency (77 Hz), and the voltage drop measured by a lock-in amplifier. Contacts were made to four Au pads evaporated on each crystal, connected through colloidal platinum paint to four gold wire probes (25 µm diameter). Special care was taken to ensure low unnested/nested voltages ratios, as defined by Schaffer [39].

The thermoelectric power was also measured along the needle axis of the crystals by a slow AC technique using high purity gold wires (99.99+% Goodfellow Metals) in a cell [40] similar to the one described by Chaikin [41] placed in a closed-cycle He cryostat and operated under computer control [42]. Sample voltage was measured by a Kethley 181 nanovoltmeter. The thermal gradients were measured by a previously calibrated Au (0.07% Fe)-chromel thermocouple and a Kethley 181 nanovoltmeter, and were kept below 1 K. The sample temperature was measured by an identical Au (0.07% Fe)-chromel thermocouple. Absolute thermoelectric power was calculated after correction for the absolute thermoelectric power of gold leads using the data of Huebner [43].

Magnetic susceptibility was measured in the range 5–300 K using a Faraday system (Oxford Instruments, Oxford, UK) with a 7 T superconducting magnet. Polycrystalline samples (~5–10 mg) were placed in thin-wall Teflon buckets, previously measured. The magnetic fields used were in the range of 2 to 5 T, and the force was measured with a microbalance (Sartorius S3D-V, Goettingen, Germany) applying forward and reverse gradients of 5 T/m. Under these conditions, the magnetization was found to be proportional to the applied magnetic field. Paramagnetic molar susceptibility was obtained after a correction for the diamagnetic contribution estimated from tabulated Pascal constants at 4.2×10^{-4} emu/mol.

4. Conclusions

The results concerning the evolution of the transition temperature and type of ground states as a function of the composition of the alloys $(Per)_2[Pt_x\text{-}Au_{(1-x)}(mnt)_2]$ are summarised in Figure 7. At the low Pt concentration range, the metallic systems at high temperatures underwent a Peierls transition towards a Charge Density Wave (CDW) ground state at slightly decreasing temperatures in the range 12 K to 8 K and had an increasing Curie–Weiss component of the paramagnetic susceptibility proportional to the Pt content. At the extreme range of high Pt content, the CDW ground state of the donor stacks, which coexisted with a spin-Peierls ground state in the anionic chains for the pure Pt compound, was still present at $x = 0.98$. The effect of higher concentrations of diamagnetic impurities in the magnetic chains could not be probed since in the range $x = 0.50$ to 0.95, only β-phase crystals could be obtained. The semiconducting properties of the β-phase are a consequence of localization induced by the structural disorder.

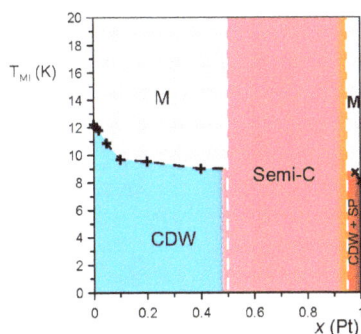

Figure 7. Phase diagram of the $(Per)_2[Pt_x\text{-}Au_{(1-x)}(mnt)_2]$ system. At left, for α-phase, it occurred a Peierls transition from a metal (M) towards a Charge Density Wave (CDW) ground state with a Curie–Weiss behaviour of the anions. At the extreme right in the α-phase, the CDW ground state of the donor stacks coexisted with a spin-Peierls (SP) ground state in the anionic chains. The spin-Peierls ground state was not suppressed by small concentrations (up to ~2%) of diamagnetic impurities. In the range x = 0.50 to 0.95, only semiconducting β-phase crystals could be obtained.

Acknowledgments: This work was partially supported by FCT (Portugal) through contracts UID/EEA/ 50008/2013, PTDC FIS/113500/2009 and UID/Multi/04349/2013.

Author Contributions: R.T.H. and M.A. conceived and designed the experiments and supervised the overall work; M.M. and M.L.A. prepared the samples; M.M., R.T.H. and G.B. performed electrical and magnetic meas44urements; I.C.S. performed the X-ray diffraction experiments and structure refinement; Data were analyzed by all authors; M.A. and M.M. wrote the paper with contributions from all other authors.

Conflicts of Interest: The authors declare no conflict of interest. The funding sponsors had no role in the design of the study; in the collection, analyses, or interpretation of data; in the writing of the manuscript, and in the decision to publish the results.

References

1. Alcácer, L.; Maki, A. Electrically Conducting Metal Dithiolate-Perylene Complexes. *J. Phys. Chem.* **1974**, *78*, 215–217. [CrossRef]

2. Alcácer, L.; Maki, A.H. Magnetic Properties of Some Electrically Conducting Perylene-Metal Dithiolate Complexes. *J. Phys. Chem.* **1976**, *80*, 1912–1916. [CrossRef]

3. Almeida, M.; Gama, V.; Henriques, R.T.; Alcácer, L. Molecular Solids with Organic Conducting Chains and Inorganic Magnetic Chains: The $(Per)_2M(mnt)_2$ Family (M = Ni, Cu, Pd, Pt, Au, Fe and Co). In *Inorganic and Organometallic Polymers with Special Properties*; Laine, R.M., Ed.; Kluwer Academic Publishers: Dordrecht, The Netherlands, 1992; pp. 163–177.

4. Almeida, M.; Henriques, R.T. Perylene Based Conductors. In *Handbook of Organic Conductive Molecules and Polymers Volume 1 "Charge Transfer Salts, Fullerenes and Photoconductors"*; Nalva, H.S., Ed.; John Wiley & Sons Ltd.: Chichester, UK, 1997; pp. 87–149.

5. Gama, V.; Henriques, R.T.; Bonfait, G.; Almeida, M.; Ravy, S.; Pouget, J.P.; Alcacer, L. The interplay between conduction electrons and chains of localized spins in the molecular metals $(Per)_2M(mnt)_2$, M = Au, Pt, Pd, Ni, Cu, Co and Fe. *Mol. Cryst. Liq. Cryst.* **1993**, *234*, 171–178. [CrossRef]

6. Pouget, J.-P.; Foury-Leylekian, P.; Almeida, M. Peierls and spin-Peierls instabilities in the $Per_2[M(mnt)_2]$ series of one-dimensional organic conductors; experimental realization of a 1D Kondo lattice for M = Pd, Ni and Pt. *Magnetocghemistry* **2017**, *3*, 13. [CrossRef]

7. Gama, V.; Almeida, M.; Henriques, R.T.; Santos, I.C.; Domingos, A.; Ravy, S.; Pouget, J.P. Low Dimensional Molecular Conductors $(Per)_2M(mnt)_2$, M = Cu and Ni: Low and High Conductivity Phases. *J. Phys. Chem.* **1991**, *95*, 4263–4267. [CrossRef]

8. Monchi, K.; Poirier, M.; Bourbonnais, C.; Matos, M.J.; Henriques, R.T. The Peierls transition in $Per_2[Au_xPt_{1-x}(mnt)_2]$: Pair-breaking field effects. *Synth. Met.* **1999**, *103*, 2228–2231. [CrossRef]

9. Matos, M.J.; Gama, V.; Bonfait, G.; Henriques, R.T. Magnetic and transport properties of the alloys $(Perylene)_2[Au_{1-x}Pt_x(mnt)_2]$. *Synth. Met.* **1993**, *56*, 1858–1863. [CrossRef]

10. Alcácer, L.; Novais, H.; Pedroso, F. Synthesis, structure and preliminary results on electrical and magnetic properties of $(Perylene)_2[Pt(mnt)_2]$. *Solid State Commun.* **1980**, *35*, 945–949. [CrossRef]

11. Afonso, M.L.; Silva, R.A.; Matos, M.; Henriques, R.T.; Almeida, M. Electrocrystallisation of $(perylene)_2$ $[M(mnt)_2]$ salts. *Phys. Status Solidi* **2012**, *9*, 1123–1126. [CrossRef]

12. Domingos, A.; Henriques, R.T.; Gama, V.; Almeida, M. Crystalline structure/transport properties relationship in the $(perylene)_2M(mnt)_2$ family (M = Au, Pd, Pt, Ni). *Synth. Met.* **1988**, *27*, 411–416. [CrossRef]

13. Afonso, M.L.; Silva, R.A.; Matos, M.; Lopes, E.B.; Coutinho, J.T.; Pereira, L.C.J.; Henriques, R.T.; Almeida, M. Growth of $(Perylene)_2[Pd(mnt)_2]$ crystals. *J. Cryst. Growth* **2012**, *340*, 56–60. [CrossRef]

14. Afonso, M.L.; Silva, R.A.; Pereira, L.C.; Coutinho, J.T.; Freitas, R.R.; Lopes, E.B.; Matos, M.; Henriques, R.T.; Viana, A.; Almeida, M. Electrocrystallisation of $(Per)_2[Pd(mnt)_2]$. *Phys. Status Solidi* **2012**, *9*, 1131–1133. [CrossRef]

15. Henriques, R.T.; Sousa, I.; Dias, J.C.; Lopes, E.B.; Almeida, M.; Matos, M. Growth of High Quality $Per_2M(mnt)_2$ Single Crystals; Evidence of β-Phase in $Per_2Pt(mnt)_2$. *J. Low Temp. Phys.* **2006**, *142*, 409–412. [CrossRef]

16. Bonfait, G.; Matos, M.J.; Henriques, R.T.; Almeida, M. The Peierls transition under high magnetic field. *Physica B* **1995**, *211*, 297–299. [CrossRef]

17. Bonfait, G.; Lopes, E.B.; Matos, M.J.; Henriques, R.T.; Almeida, M. Magnetic field dependence of the metal-insulator transition in (PER)$_2$Pt(mnt)$_2$ and (PER)$_2$Au(mnt)$_2$. *Solid State Commun.* **1991**, *80*, 391–394. [CrossRef]

18. Kwak, J.F.; Beni, G.; Chaikin, P.M. Thermoelectric power in Hubbard-model systems with different densities: *N*-methylphenazinium-tetracyanoquinodimethane (NMP-TCNQ), and quinolinium ditetracyanoquinodimethane. *Phys. Rev.* **1976**, *13*, 641–646. [CrossRef]

19. Drichko, N.; Kaiser, S.; Shewmon, R.; Eckstein, J.; Wu, D.; Dressel, M.; Matos, M.; Henriques, R.T.; Almeida, M. Infrared investigations of the one-dimensional organic conductors (perylene)$_2$M(mnt)$_2$, M = Au, Pt. *Eur. Phys. J. B* **2010**, *78*, 283–289. [CrossRef]

20. Graf, D.; Choi, E.S.; Brooks, J.S.; Matos, M.; Henriques, R.T.; Almeida, M. High Magnetic Field Induced Charge Density Wave State in a Quasi-One-Dimensional Organic Conductor. *Phys. Rev. Lett.* **2004**, *93*, 076406. [CrossRef] [PubMed]

21. Graf, D.; Brooks, J.S.; Choi, E.S.; Uji, S.; Dias, J.C.; Almeida, M.; Matos, M. Suppression of a charge-density-wave ground state in high magnetic fields: Spin and orbital mechanisms. *Phys. Rev. B* **2004**, *69*, 125113. [CrossRef]

22. Canadell, E.; Almeida, M.; Brooks, J. Electronic band structure of α-(Per)$_2$M(mnt)$_2$ compounds. *Eur. Phys. J. B* **2004**, *42*, 453–456. [CrossRef]

23. Bourbonnais, C.; Henriques, R.T.; Wzieteck, P.; Köngeter, D.; Voiron, J.; Jérome, D. Nuclear and electronic resonance approaches to magnetic and lattice fluctuations in the two-chain family of organic compounds (perylene)2[M(S2C2(CN)2)2] (M = Pt, Au). *Phys. Rev. B* **1991**, *44*, 641–651. [CrossRef]

24. Gama, V.; Henriques, R.T.; Almeida, M.; Alcácer, L. Magnetic Properties of the Low-Dimensional Systems (Per)$_2$M(mnt)$_2$ (M = Cu and Ni). *J. Phys. Chem.* **1994**, *98*, 997–1001. [CrossRef]

25. Bonner, J.C.; Fisher, M.E. Linear Magnetic Chains with Anisotropic Coupling. *Phys. Rev. A* **1964**, *135*, 640–658. [CrossRef]

26. Henriques, R.T.; Alcácer, L.; Pouget, J.P.; Jérome, D. Electrical conductivity and X-ray diffuse scattering study of the family of organic conductors (perylene)$_2$M(mnt)$_2$, (M = Pt, Pd, Au). *J. Phys. C Solid State Phys.* **1984**, *17*, 5197–5208. [CrossRef]

27. Hansen, P.; Augier, D.; Riera, J.; Poilblanc, D. Study of impurities in spin-Peierls systems including lattice relaxation. *Phys. Rev.* **1999**, *59*, 13557. [CrossRef]

28. Jamali, J.B.; Wada, N.; Shimobe, Y.; Achiwa, N.; Kuwajima, S.; Soejima, Y.; Mukai, K. The effect of non-magnetic impurities on the spin–Peierls transition of 3-(4-cyanophenyl)-1,5-dimethyl-6-oxoverdazyl radical crystal, p-CyDOV. *Chem. Phys. Lett.* **1998**, *292*, 661–666. [CrossRef]

29. Pouget, J.P.; Ravy, S.; Schoeffel, J.P.; Dhalenne, G.; Revcolevschi, A. Spin-Peierls lattice fluctuations and disorders in CuGeO$_3$ and its solid solutions. *Eur. Phys. J. B* **2004**, *38*, 581–598. [CrossRef]

30. Sangster, R.C.; Irvine, J.W., Jr. Study of Organic Scintillators. *J. Chem. Phys.* **1956**, *24*, 670–715. [CrossRef]

31. Davison, A.; Holm, R.H. Metal Complexes Derived from cis-1,2-Dicyano-ethylene-1,2-Dithiolate and Bis-Perfluoromethyl-1,2-Dithietene. *Inorg. Synth.* **1967**, *10*, 8–26.

32. Davison, A.; Edelstein, N.; Holm, R.H.; Maki, A.H. E.s.r. Studies of Four-Coordinate Complexes of Nickel, Palladium and Platinum Related by Electron Transfer Reactions. *J. Am. Chem. Soc.* **1963**, *85*, 2029–2030. [CrossRef]

33. Sheldrick, G.M. *SADABS*; Bruker AXS Inc.: Madison, WI, USA, 2004.

34. Bruker. *SMART and SAINT*; Bruker AXS Inc.: Madison, WI, USA, 2004.

35. Altomare, A.; Burla, M.C.; Camalli, M.; Cascarano, G.L.; Giacovazzo, C.; Guagliardi, A.; Moliterni, A.G.G.; Polidori, G.; Spagna, R. SIR97: A new tool for crystal structure determination and refinement. *Appl. Crystallogr.* **1999**, *32*, 115–119. [CrossRef]

36. Sheldrick, G.M. A short history of SHELX. *Acta Crystallogr. Sect. A* **2008**, *64*, 112–122. [CrossRef] [PubMed]

37. Farrugia, L.J. WinGX and ORTEP for Windows: An update. *J. Appl. Crystallogr.* **2012**, *45*, 849–854. [CrossRef]

38. Macrae, C.F.; Bruno, I.J.; Chisholm, J.A.; Edgington, P.R.; McCabe, P.; Pidcock, E.; Rodriguez-Monge, L.; Taylor, R.; van de Streek, J.; Wood, P.A. Mercury CSD 2.0—New features for the visualization and investigation of crystal structures. *J. Appl. Cryst.* **2008**, *41*, 466–470. [CrossRef]

39. Schafer, D.E.; Wudl, F.; Thomas, G.A.; Ferraris, J.P.; Cowan, D.O. Apparent giant conductivity peaks in an anisotropic medium: TTF-TCNQ. *Solid State Commun.* **1974**, *14*, 347–351. [CrossRef]

40.   Almeida, M.; Alcácer, L.; Oostra, S. Anisotropy of thermopower in *N*-methyl-*N*-ethylmorpholinium bistetracyanoquinodimethane, MEM(TCNQ)2, in the region of the high-temperature phase transitions. *Phys. Rev. B* **1984**, *30*, 2839. [CrossRef]

41.   Chaikin, P.M.; Kwak, J.F. Apparatus for thermopower measurements on organic conductors. *Rev. Sci. Instrum.* **1975**, *46*, 218. [CrossRef]

42.   Lopes, E.B. *Internal Report*; LNETI: Sacavém, Portugal, 1990.

43.   Huebner, R.P. Thermoelectric Power of Lattice Vacancies in Gold. *Phys. Rev.* **1964**, *135*, 1281–1291. [CrossRef]

magnetochemistry

MDPI

Review

Antiferromagnetic Insulating Ground State of Molecular π-d System λ-(BETS)$_2$FeCl$_4$ (BETS = Bis(ethylenedithio)tetraselenafulvalene): A Theoretical and Experimental Review

Yugo Oshima *, Heng-Bo Cui and Reizo Kato

Condensed Molecular Materials Laboratory, RIKEN, Hirosawa 2-1, Wako, Saitama 351-0198, Japan; hcui@riken.jp (H.-B.C.); reizo@riken.jp (R.K.)
* Correspondence: yugo@riken.jp; Tel.: +81-48-467-9410

Academic Editor: Manuel Almeida
Received: 23 January 2017; Accepted: 22 February 2017; Published: 24 February 2017

Abstract: The π-d molecular conductor λ-(BETS)$_2$FeCl$_4$, where BETS is bis(ethylenedithio) tetraselenafulvalene, has attracted considerable interest for the discovery of its field induced superconducting state. A mystery of this system is its antiferromagnetic insulating ground state. The point still under strong debate is whether the d spins in Fe^{3+} are ordered or not. Here, we review experimental and theoretical studies on the antiferromagnetic insulating phase in λ-(BETS)$_2$FeCl$_4$ and mention our perspective based on our ESR measurements for λ-(BETS)$_2$Fe$_x$Ga$_{1-x}$Cl$_4$. Our ESR results indicate that the π-d interaction in the system is very strong and there is no sign of paramagnetic Fe spins in the antiferromagnetic ground state.

Keywords: π-d interaction; λ-(BETS)$_2$FeCl$_4$; electron spin resonance; antiferromagnetic insulating state; metal-insulator transition

---

1. Introduction

Molecular conductors with finite π-d interactions have attracted many interests for the past few decades since interaction between the itinerant π-electron and localized d-electron yields some intriguing physical phenomena. For instance, a characteristic magnetoresistance that is related to the spin state in Fe ions was reported for (DMET)$_2$FeBr$_4$ (DMET = 4′,5′-dimethyl-4,5-(ethylenedithio)-1′,3′-diselena-1,3-dithiafulvalene) [1], and giant negative magnetoresistance was observed in Fe phthalocyanine complexes [2]. Among others, λ-(BETS)$_2$FeCl$_4$, where BETS is bis(ethylenedithio)tetraselenafulvalene, is one of the most studied and well-known π-d molecular conductors for the appearance of superconductivity in a high magnetic field [3,4]. An excellent review about the field-induced superconducting state by Uji and Brooks can be found elsewhere [5].

In contrast to the superconducting state in a high magnetic field, this system also shows cooperative conducting and magnetic properties in the low-field region thanks to the strong interaction between π- and d-electrons, namely, π-d interaction. In the high temperature phase, λ-(BETS)$_2$FeCl$_4$ is a metal where the magnetic property shows the paramagnetic behavior. At $T_N = 8.3$ K, λ-(BETS)$_2$FeCl$_4$ becomes antiferromagnetic and simultaneously shows a metal to insulator (MI) transition [6–8].

In contrast with the non-magnetic analog compound λ-(BETS)$_2$GaCl$_4$ which becomes superconducting below 6 K, the MI transition of λ-(BETS)$_2$FeCl$_4$ has been thought for a long time to be triggered by the long-range ordering of Fe spins [7,8]. On the other hand, some theoretical works claim that the Mott transition of the π-electrons is an origin of the MI transition, and then, the π-d coupling forces the Fe moments to be antiferromagnetically ordered [9,10]. However, Akiba et al. revealed a completely different role of the Fe spins based on the heat capacity measurement, where the Fe spins remain paramagnetic below T_N [11]. This 'chicken or egg' problem of the antiferromagnetic insulating (AFI) phase (which orders first, the *d*-electrons or the π-electrons), coupled with the ground state of the Fe spins is still open question under strong debate.

We here review cornerstone measurements and theoretical studies for the AFI phase of λ-(BETS)$_2$FeCl$_4$, and present our ESR measurements on the mixed compounds λ-(BETS)$_2$Fe$_x$Ga$_{1-x}$Cl$_4$ ($x = 0.2{\sim}1.0$). The main point of studying the mixed compounds by ESR is to microscopically investigate the role of Fe spins in the ground state by gradually introducing the Fe spins in the system. Our results have revealed that the Fe spins play a decisive role in the magnetic ground state, where a gradual transition from the paramagnetic to antiferromagnetic ground state is observed by increasing the Fe content. In contrast to the heat capacity measurement, the ESR measurements for the salts with $x \geq 0.6$ do not indicate any sign of paramagnetic Fe spins in the AFI phase. This suggests that more attention should be paid to a metastable state within the AFI phase to explain the previous experimental results.

This paper is organized as follows: the crystal structure and phase diagram are mentioned in the next section, and some of the previous experimental and theoretical results are described in Section 3, followed by the ESR results in Section 4. Conclusions and future prospects are presented in the final section.

2. Crystal Structure and Phase Diagram

Single crystals of λ-(BETS)$_2$FeCl$_4$ are prepared by electrochemical oxidation of BETS in organic solvent such as chlorobenzene containing 10% ethanol, with the tetraethylammonium salt of FeCl$_4^-$ as a supporting electrolyte [6]. The mixed system λ-(BETS)$_2$Fe$_x$Ga$_{1-x}$Cl$_4$ is prepared from a mixed electrolyte of [(C$_2$H$_5$)$_4$N][GaCl$_4$] and [(C$_2$H$_5$)$_4$N][FeCl$_4$] with fine tuning of the mixing ratio. Needle-shaped single crystals are obtained, and the crystallographic *c*-axis corresponds to the needle axis.

The crystal structure is presented in Figure 1a. The crystal has a triclinic unit cell with space group $P\bar{1}$. The lattice constants at 10 K were determined as $a = 15.880$, $b = 18.378$, $c = 6.529$ Å, $\alpha = 98.66°$, $\beta = 95.830°$, $\gamma = 112.13°$ [6].

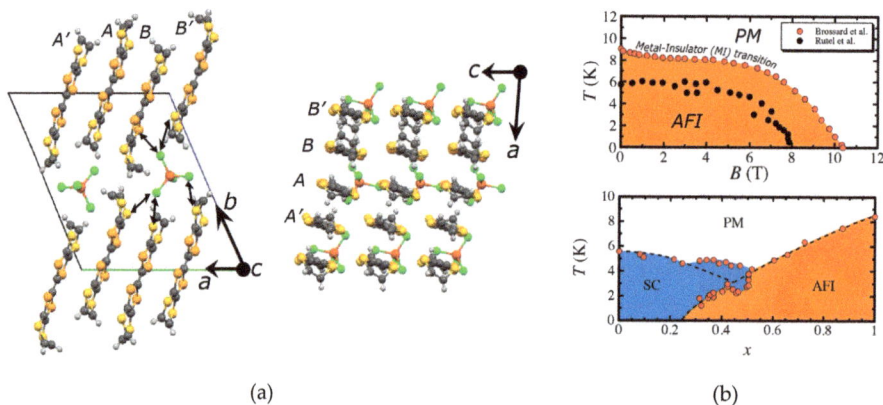

(a)   (b)

Figure 1. (a) Left: Crystal structure of λ-(BETS)$_2$FeCl$_4$. A and B denote two crystallographically independent BETS molecules, and the double headed arrows represent short Cl$\cdots$S(Se) contacts. Right: Molecular arrangement of the BETS molecules and the anions in λ-(BETS)$_2$FeCl$_4$; (b) Top: *T-B* phase diagram of λ-(BETS)$_2$FeCl$_4$ extracted from Reference [8,12]. Bottom: *T-x* phase diagram of λ-(BETS)$_2$Fe$_x$Ga$_{1-x}$Cl$_4$ extracted from Reference [13]. Antiferromagnetic insulator (AFI) and superconductor (SC) phases are determined from the metal-insulator and superconducting transitions, respectively (red circles). PM = paramagnetic metal. A metastable state in the AFI phase (see Section 3.4 for detail) was claimed from high-frequency measurements (black circles) [12].

Two BETS molecules and one $FeCl_4$ anion are crystallographically independent. Each independent BETS molecule (A and B in Figure 1a) forms a dimer, and these BETS dimers are stacked two-dimensionally in the *ac*-plane, and consequently form conducting layers. The interplanar distance between the BETS molecules A$\cdots$B is 3.683 Å, and those between A$\cdots$A', B$\cdots$B' are about 4.0 Å. The overlap integral is the strongest, 76.8×10^{-3}, between BETS molecules A$\cdots$B, and the overlap integrals are 37.1×10^{-3} and 23.1×10^{-3} for A$\cdots$A' and B$\cdots$B', respectively. The tight-binding calculation based on the extended Hückel method yields a 2D closed Fermi surface with 23% of the Brillouin zone [6]. This estimation is in good agreement with the Shubnikov-de Haas oscillation measurement [14]. On the other hand, the fairly large dimerization of the BETS molecules leads to a splitting of the HOMO (highest occupied molecular orbital) band, and the electronic state can be interpreted as effectively half-filling of the upper HOMO band. In the case of half-filling with substantial on-site Coulomb interaction U, the π-electron system could turn into a Mott insulating state. Meanwhile, the $FeCl_4$ anions, which are magnetic with a high-spin state (Fe^{3+}, $S = 5/2$), are located between the BETS layers and form 1D chains along the *c*-axis. The shortest Fe$\cdots$Fe distance is 6.593 Å at room temperature, and there are many short Cl$\cdots$S(Se) contacts of about 3.43–3.67 Å (double headed arrows in Figure 1a). Therefore, strong π-d interactions are expected in λ-$(BETS)_2FeCl_4$.

The first transport and magnetic properties of λ-$(BETS)_2FeCl_4$ were reported by Kobayashi et al. and Tokumoto et al., respectively [6,7]. A complementary study was performed by Brossard et al. [8]. This compound is metallic in the high temperature phase, and the magnetic susceptibility, where the dominant magnetic contribution originating from the Fe^{3+} ion, shows the paramagnetic behavior. With lowering temperature, λ-$(BETS)_2FeCl_4$ shows a metal to insulator (MI) transition at $T_N = 8.3$ K, while the magnetic susceptibility shows the antiferromagnetic (AF) behavior down to the lowest temperature. Furthermore, a paramagnetic metal (PM) state is recovered when all the magnetic moments are polarized in the magnetic field above $B_c = 10.5$ T. The phase diagram is presented in the top panel of Figure 1b. The field-induced superconducting state, which appears above 17 T when B is parallel to the conducting plane, is not shown in Figure 1b since it is beyond the scope of this paper. We refer to the review paper about the field-induced superconductivity in λ-$(BETS)_2FeCl_4$ for readers who might be interested [5].

As mentioned above, the non-magnetic and isostructural analog, λ-$(BETS)_2GaCl_4$, shows a superconducting transition at $T_c = 6$ K, whereas λ-$(BETS)_2FeCl_4$ becomes insulating at $T_N = 8.3$ K. Therefore, the mixed compound, λ-$(BETS)_2Fe_xGa_{1-x}Cl_4$, is an excellent system to study the effect of magnetic ion on the ground state with a minimum structural disorder effect since $FeCl_4^-$ and $GaCl_4^-$ have almost the same size. The bottom panel of Figure 1b shows the phase diagram for the mixed λ-$(BETS)_2Fe_xGa_{1-x}Cl_4$ system. The Néel temperature T_N decreases with the Fe content, and the AFI phase is not present in the region of $x < 0.25$. The superconducting transition temperature T_c in the small x region (i.e., $x < 0.25$) decreases slightly with increasing x. In the intermediate region ($0.3 < x < 0.5$), a complicated phase diagram accompanied by small changes of T_c and T_N is observed [13,15].

3. Previous Experimental and Theoretical Studies

3.1. Transport and Magnetic Properties

Temperature dependent resistivity for λ-$(BETS)_2FeCl_4$ shows a small maximum around 100 K, and decreases with lowering temperature [6]. Such a maximum of the resistivity is common in other BETS salts. As mentioned in the previous section, the MI transition is found at 8.3 K at ambient pressure. The small maximum observed around 100 K shifts to higher temperatures with increasing pressure and is completely suppressed at 2 kbar. Moreover, the MI transition temperature decreases when increasing the pressure, and the metallic state is stabilized above 3.5 kbar [6].

The first magnetoresistance measurement using a pulsed magnetic field was reported by Goze et al. [16]. The authors found that the MI transition temperature is suppressed by applying a high magnetic field, and a metallic state is recovered above 10 T. They also found that the magnetoresistance strongly depends on the direction of the magnetic field above the critical field (B_c) [16]. Brossard et al. also reported that the magnetoresistance shows hysteresis at the MI transition, which becomes larger as the magnetic field increases [8]. Our transport and magnetotransport measurements are shown in Figure 2. As mentioned by Brossard et al., the hysteresis at the MI transition becomes larger as the field increases (Figure 2a) and as temperature increases (Figure 2b), namely, near the AFI phase boundary. Additionally, kink structures in the resistance below the MI transition are observed (shown as solid circles and squares in Figure 2a).

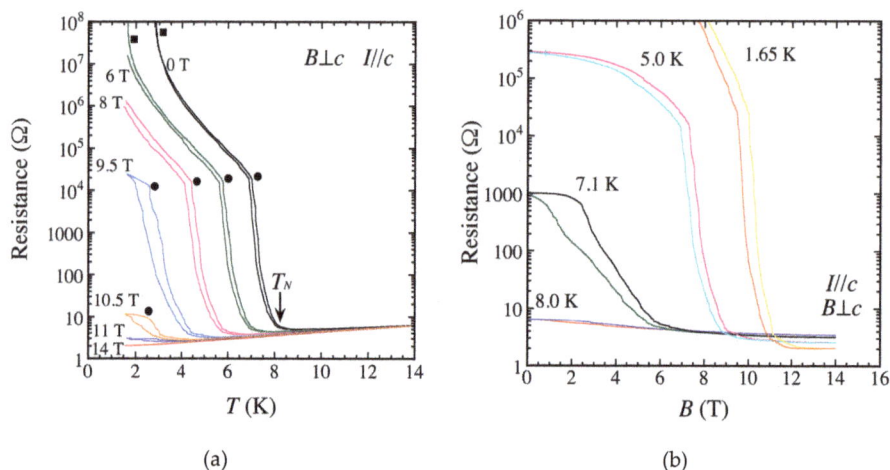

Figure 2. Resistance at zero-field and the magnetoresistance of λ-(BETS)$_2$FeCl$_4$. (**a**) resistance versus temperature; (**b**) resistance versus magnetic field. The current is applied along the *c*-axis, and the magnetic field perpendicular to the *c*-axis. Hysteresis and kink structures are observed below the MI transition temperature (solid circles and squares).

The magnetic susceptibility χ follows the Curie-Weiss law in the PM phase, where the Weiss temperature is about -15 K and the Curie constant yields an effective magnetic moment of 5.83 μ_B [7,8]. This value is very close to that of the magnetic moment of Fe^{3+} in the high spin state (i.e., 5.92 μ_B for $S = 5/2$ and $L = 0$), suggesting the dominant magnetic contribution originates from the Fe^{3+} spins. The Weiss temperature of -15 K and $S = 5/2$ leads to the exchange coupling between the spins Fe^{3+} of $J_{d\text{-}d} \sim 2.3$ K [8]. The magnetic susceptibility for the field $B//c$ shows a step-like decrease at the MI transition temperature, 8.3 K, and then, shows the behavior typical of the AF ordering for the easy-axis with lowering temperature. The step-like decrease at the MI transition is suppressed in the magnetic field $B > 2$ T, and is less pronounced for $B \perp c$. Notably, for λ-(BETS)$_2$FeBr$_4$ where a smaller π-*d* interaction is expected, the step-like structure disappears and the easy-axis behavior is observed for the field $B \perp c$ [17]. From the comparison of magnetic properties between λ-(BETS)$_2$FeCl$_4$ and λ-(BETS)$_2$FeBr$_4$, it is supposed that the peculiar magnetic behavior observed in λ-(BETS)$_2$FeCl$_4$ is due to the strong interplay between π- and *d*-electrons [8].

The field-dependence of magnetization (*M-H* curve) for the field $B//c$ shows a step-like structure around 1 T accompanied by a linear increase at higher fields. This step-like structure was interpreted as a spin-flop transition since it was not observed for the field $B \perp c$ [7,8]. The linear increase of the magnetization above the spin-flop field is due to the canted antiferromagnetic state. More precise magnetic torque measurement preformed by Sasaki et al. indicated a clear spin-flop transition around 1.2 T, where the easy-axis was found to be in the direction tilted by about 30° from the *c*-axis to the b^*-axis [18]. For low temperatures, the *M-H* curve shows another step structure accompanied with hysteresis around the critical field B_c [8,18]. A metamagnetic-like transition of the π-electrons between

AFI and PM phases was proposed since the recovery of the metallic state occurs above B_c [18]. Finally, the magnetization saturates above 10.5 T.

The mixed compound λ-(BETS)$_2$Fe$_x$Ga$_{1-x}$Cl$_4$ shows similar transport property. A small maximum of resistivity is also observed around 100 K for all mixed systems, and resistivity decreases with lowering temperature. Then, superconducting or MI transition is observed depending on the Fe content x (see bottom panel of Figure 1b) [13]. No significant difference is found also in the Fermi surface (similar cylindrical Fermi surface and similar cross-sectional area is observed by the Shubnikov–de Haas oscillation measurement) except that the Fermi surface is split by the internal field due to the π-d interaction for $x \neq 0$ [19]. The internal field seems to increase with the Fe content. The magnetic property of the mixed compound λ-(BETS)$_2$Fe$_x$Ga$_{1-x}$Cl$_4$ ($x > 0.47$) is also similar to the pure λ-(BETS)$_2$FeCl$_4$, where the magnetic susceptibility shows a step-like structure at the MI transition, and the spin-flop is observed for the field $B//c$ [20]. Hence, the overall features suggest there is no serious effect of disorder both on the transport and magnetic properties in the mixed compound.

In summary, the high temperature phase of λ-(BETS)$_2$FeCl$_4$ is a paramagnetic metal. The MI transition occurs at 8.3 K, and the system becomes antiferromagnetic simultaneously. By applying the magnetic field along the easy-axis, the spin flop transition is observed around 1.2 T, and the system goes to a canted antiferromagnetic state. The magnetization saturates (i.e., spins are fully polarized) above 10.5 T, then the metallic state recovers. The magnetic ions are essential to the AFI ground state since the non-magnetic analog λ-(BETS)$_2$GaCl$_4$ is, in contrast, a superconductor. Therefore, in the early studies on λ-(BETS)$_2$FeCl$_4$, it was thought that the insulating state (i.e., localization) of the π-electrons is triggered by the antiferromagnetic long-range ordering of the d-electrons [8].

3.2. Specific Heat Measurements and the Paramagnetic Fe Model

The first specific heat measurement of λ-(BETS)$_2$FeCl$_4$ was reported by Negishi et al. [21]. The specific heat exhibits a sharp peak due to the antiferromagnetic long-range order at T_N, and a remarkable jump of the specific heat, which they attributed to the ferroelectric transition, is observed at 70 K (T_{FM}). The ferroelectric property of λ-(BETS)$_2$FeCl$_4$ will be mentioned later in Section 3.4. Additionally, a significant excess heat capacity below T_N was reported. Negishi et al. attributed this excess specific heat to a short-range antiferromagnetic ordering since the entropy change of only 60% of $R \ln 2 + R \ln 6$ was observed at T_N.

On the other hand, Akiba et al. proposed a completely different model to explain this excess specific heat [11]. They showed from their result that the excess specific heat observed below T_N can be fit with the six-level Schottky peak of the paramagnetic Fe spins. They proposed that the sharp peak at T_N is due to the antiferromagnetic ordering of the π-electrons (Mott transition), and the $3d$ spins, which remain paramagnetic, feel the internal field from the localized π-electrons through the π-d interaction. The authors concluded that an internal field of about 4 T is created from the localized π-electrons, and this internal field induces the six energy levels of d-electron leading to the Schottky-type anomaly. Hence, this model suggests that π and $3d$ spins do not cooperatively form an antiferromagnetic order, and the Fe spins remain paramagnetic.

Akiba et al. also claimed that a model with two canted Fe spins exposed to the internal field of 4 T along the easy-axis can reproduce the results of the susceptibility measurements [22]. The specific heat measurements of the mixed compound λ-(BETS)$_2$Fe$_x$Ga$_{1-x}$Cl$_4$ ($x = 0.4\sim1.0$) can also be fit with this model, where the internal field increases with the Fe content [23]. They proposed that the difference of T_N by the Fe content is due to the fluctuation of the π-electron that is suppressed by the magnetic anisotropy introduced by the Fe spins [23]. The paramagnetic Fe model was further extended by introducing 2D Ising-like transition and double Schottky model, to explain the heat capacity behavior nearby T_N and the field dependence, respectively [24,25].

Mössbauer measurement partially supports the paramagnetic Fe model [26]. Above T_N, the Mössbauer spectra presents a single line typical of the paramagnetic Fe; and below T_N, sextet splitting that is typical of long-range ordering is observed. However, the spectrum at 8 K still shows a small fraction of about 13% of paramagnetic atoms, which can be ascribed to the slow dynamics of the transition. Moreover, two additional magnetic splittings with identical isomer shift but slightly different hyperfine fields (i.e., two sextets) are observed in the temperature range between 3.2 and 8.0 K, which suggests two different magnetic environments for the Fe sites. In addition, a significant development of the hyperfine field by decreasing temperature is also observed. Although the change in the Mössbauer spectra below T_N is indicative of a slow magnetic ordering process, these results cannot discriminate the paramagnetic model in the case of fast relaxation. Actually, the computed hyperfine fields assuming a fast relaxation model, where the Fe spins exhibit the Zeeman splitting by the internal field, are in good agreement with the specific heat measurements. The estimated internal field from the fitting is dependent on temperature, increasing from 2.45 T at 8 K to 4.2 T at 1.5 K [26].

Recent transport (including non-linear transport) and magnetic torque measurements by Sugiura et al. also support the paramagnetic model [27,28]. They observed a small dip of the magnetoresistance at 1.2 T for *B* nearly parallel to the easy axis, which can be ascribed to the spin-flop transition. The resistance dip at the spin-flop transition was explained, not with the scattering mechanism due to the antiparallel Fe, but from the electron-hole excitation model with up and down spins in the BETS layers [27]. Furthermore, Sugiura et al. investigated in detail the angular dependence of the magnetic torque that shows a sinusoidal dependence, and found the zero-crossing angle—the magnetic field direction where the magnetic torque becomes zero—significantly changes with field and temperature [28]. In comparison with the zero-crossing angle of the magnetic torque in the PM phase, and that of the non-magnetic GaCl$_4$ salt, they concluded that the origin of the change in the zero-crossing angle is due to the antiferromagnetic order of the π-electrons [28].

3.3. Theoretical Studies

The first model to explain the AFI ground state of λ-(BETS)$_2$FeCl$_4$ was proposed by Ziman in Reference [8]. A Hubbard-Kondo model—where four conduction bands associated with the BETS layers, the Coulomb repulsion U on the BETS molecule, and a Kondo coupling between the localized $S = 5/2$ spins on Fe^{3+} and the conduction π-electrons were taken into account—was introduced. For small U, the periodic potential due to the magnetic ordering of Fe^{3+} opens energy gaps at the Fermi surface, then, by applying the magnetic field, the fully-polarized magnetic moments destroy the periodic potential, restoring the Fermi surface. Although this model qualitatively explains the AFI and PM ground states of λ-(BETS)$_2$FeCl$_4$, a Kondo coupling of $J > 70$ K is needed to suppress the entire Fermi surface and the on-site Coulomb U is empirically not small [8].

Another model of the AFI phase was introduced by Hotta and Fukuyama [9]. From a mean-field calculation including the Hubbard model and the π-*d* interaction, they categorized the BETS system into a unified phase diagram, and proposed that λ-(BETS)$_2$FeCl$_4$ and λ-(BETS)$_2$GaCl$_4$ are very close to the MI boundary of the Mott insulator, and its ground state is easily influenced by the π-*d* interaction. Moreover, the authors pointed out a lack of degeneracy in the energy dispersion of the two anti-bonding HOMO bands owing to the molecular arrangement and the fairly large dimerization of the BETS molecules in the λ-type system. This splitting makes it easier to form a gap between the two anti-bonding bands where the Fermi level is situated, which means only a small on-site Coulomb U is needed to be an insulator. Therefore, the MI transition at 8.3 K is proposed to be a Mott transition of the π-electrons, and long-range order of the Fe spins is induced through the π-*d* interaction [9]. Even though this model could also drive the system into the AFI ground state, it cannot explain why a strong magnetic field restores the metallic state. To resolve this problem, Cépas et al. proposed a more effective Hubbard-Kondo model with Zeeman energies [10]. The AFI phase of λ-(BETS)$_2$FeCl$_4$ is explained by the picture where the Kondo coupling drives the system into an insulating phase in order to gain some magnetic energy. The energy of a metallic state crosses that of the insulator as the field increases, leading to a first-order transition into a PM state.

These two scenarios of the MI transition lead to different physical pictures: spin-density-wave (SDW)-like insulator and Mott insulator. Therefore, Cépas et al. proposed that the measurement of the charge gap as a function of the field would distinguish the SDW or Mott insulator scenario [10].

Furthermore, Mori and Katsuhara estimated the exchange energies in λ-(BETS)$_2$FeCl$_4$ by means of the extended Hückel method [29]. The antiferromagnetic exchange interaction between π, π-d, and d electrons are estimated to be $J_{\pi\pi}$ = 448 K, $J_{\pi d}$ = 14.6 K, and J_{dd} = 0.64 K, respectively. The internal field created by the Fe spins $H_{int} = J_{\pi d} S_d / g\mu_B$ is about 33 T in agreement with the experiment on the field-induced superconductivity [4]. Based on the mean-field calculation, the Néel temperature of T_N = 6.22 K is also deduced from these exchange couplings, which also agrees with the experimental result [29].

The above theoretical works suppose that both π- and d-electrons are long-range ordered. In this sense, the paramagnetic Fe model of Akiba et al. is quite striking. However, Ito and Shimahara examined by mean field theory a uniaxially-coupled Heisenberg antiferromagnet model with two subsystems, where the two subsystems consist of strongly interacting small spins and weakly interacting large spins [30]. In the case of $J_{\pi\pi} \gg J_{\pi d} \gg J_{dd}$, their model successfully reproduces the specific heat and the magnetic susceptibility measurements. Moreover, the magnetic anisotropy, which is essential to the easy-axis of the antiferromagnetic state, is found to originate from the π-d interaction and is described as approximately 26–27° from the c-axis which is in total agreement with the experimental result [30]. The same authors also studied the spin structure from a similar model, and discussed that a tilted canted antiferromagnetic state, where the canted spins are tilted from the magnetic field direction, can only appear in a narrow range of the magnetic field for λ-(BETS)$_2$FeCl$_4$ [31].

3.4. Other Experimental Studies

Anomalous dielectric response at T_{FM} = 70 K was first reported by Matsui et al. [32]. As mentioned above, the specific heat also shows some anomaly at this temperature [21]. By means of cavity perturbation technique at 16.3 GHz, Matsui et al. found an anomaly in the cavity response with a large dielectric constant. They ascribed it to a transition to a ferroelectric state. Although the dc resistivity indicates a highly metallic state, the microwave loss is enhanced anomalously in the range of $T_N < T < T_{FM}$. It should be noted that this work was criticized on the basis that this anomaly is due to the analysis artifact, which was soon denied by the same authors [33,34]. Moreover, the dielectric constant along the c-axis shows a broad maximum around 30 K for 44.5 GHz, which the authors attributed to a relaxor ferroelectric behavior [35]. They proposed that dielectric domains or stripes with less metallic conduction emerge inhomogeneously in the π-electron system. The X-ray diffraction data support this interpretation since the width of (007) Bragg reflection becomes broader and the peak splits around T_{FM} [36,37]. This structural anomaly was ascribed to an appearance of heterogeneous structure with dielectric relaxor domains of about 0.4 μm in size. Furthermore, unnatural values in the anisotropic atomic displacement parameters were found, although no significant evidence of a phase transition or structural changes at T_{FM} was found. The charge density map obtained by Fourier synthesis shows a distorted electron density distribution in the BETS molecule [37].

Related to those experiments in the PM phase, I-V characteristic in the AFI state was investigated by Toyota et al. [38]. A drastic non-linear transport phenomenon, known as the negative resistance effect, was observed in the I-V curve, which indicates that some carrier decondensation occurs by applying electric field. They supposed that these dielectric states are intrinsic instabilities in the charge degree of freedom of the π-electronic system, which may be strongly influenced by applying electric field. Similar to this study, Rutel and co-workers measured the high-frequency cavity response of λ-(BETS)$_2$FeCl$_4$ by sweeping the field and temperature, and observed a huge change in the cavity response inside the AFI phase [12]. The transitions from a skin-depth regime to a depolarization regime, where a huge change of the cavity response is observed, are plotted in the top panel of Figure 1b. They explained that this behavior is due to the metastable state in the π-electronic system within the AFI phase. Negishi and co-workers obtained similar results by measuring the capacitance and conductance in the AFI phase [39]. A huge dielectric change by sweeping the field was observed inside the AFI phase, called 'colossal magnetodielectricity'. The magnetic field where the divergence of the dielectric constant occurs is similar to the results on the depolarization regime by Rutel et al. (see Figure 1b) [12,39]. Such a sub-phase inside the AFI phase is confirmed from other studies [35,40].

Based on the above measurements, Negishi et al. proposed a charge ordering-induced polarization model [39]. Due to the exclusively large π-d super-exchange interaction between the π-orbitals at selective Se sites in BETS molecules labeled B (B') in Figure 1a and $3d$-orbital via Cl, the charge as well as the spin of the π-electrons are expected to be localized at these Se sites. Therefore, they expect a partial charge ordering occurs in BETS molecules A (A') and B (B'), and such charge disproportionation yields a local polarization inside the BETS molecules. The antiferromagnetic order could be associated with the energy gain of magnetic ordering of the π- and d-electrons, which overcompensates the intersite Coulomb energy V in the BETS layer. This indicates that charge ordering is the primary origin of the AFI state. At low field and low temperature, the π-electrons are locked at the Se sites to keep $J_{\pi d}$ as effectively as possible, then, these π-electrons are considered to be unlocked or melted by applying the magnetic field [39].

Endo et al. performed the first ^{1}H-NMR measurement, where a single peak splits into three asymmetric peaks at T_{FM} related to the charge disproportionation [41]. Due to local fields that are dependent on each proton site (16 independent sites) and dynamical fluctuation, the NMR line rapidly broadens and contains many different peaks below T_N, which makes the analysis difficult. ^{1}H-NMR spin-echo measurement at 9 T was performed by Wu et al. [42]. A large slow beat structure in the spin-echo decay is observed, which originates from a large inhomogeneous local field generated by the Fe^{3+} moments. They also observed a discontinuous drop in $1/T_2$ at 3.5 K (PM to AFI transition for 9 T), which is due to the change in the orientation of the Fe spins at the transition. Besides the ^{1}H-NMR, the ^{77}Se-NMR is performed by Hiraki and co-workers in the high-field PM phase and field-induced superconducting state [43]. However, it seems that neither ^{77}Se- nor ^{13}C-NMR study of the AFI state has been reported thus far. This suggests that NMR line might also be very broad even for ^{77}Se or ^{13}C due to the effects of many local fields and the disproportionation of the π-electrons nearby the phase boundary.

In summary, two types of insulating mechanisms were proposed for the AFI state of λ-(BETS)$_2$FeCl$_4$: the spin-driven insulating mechanism and the charge-driven insulating mechanism. In the former one, the Fe spins become magnetically ordered thanks to Kondo couplings, and the periodic potential of the magnetic ordering opens a gap at the Fermi surface (i.e., SDW-like insulator). This mechanism was supported in the early stage. The latter one claims that the π-electrons make the insulating state by the Mott transition or charge ordering, and then, the Fe spins become antiferromagnetic or remain paramagnetic. Recent studies seem to support the latter mechanism rather than the former one.

4. ESR Measurements

4.1. Previous ESR Studies

The first ESR measurements for λ-(BETS)$_2$FeCl$_4$ with polycrystalline and single crystal samples were reported by Kobayashi et al. and by Brossard et al., respectively [6,8]. Brossard et al. reported temperature dependence of the g-value and ESR linewidth for two magnetic field orientations ($B//c$ and $B//u$, where u is an undefined axis). The authors explained that the u-axis could not be determined due to the small sample size, but we suppose that the u-axis is in fact the b^*-axis which is perpendicular to the c-axis. At room temperature, large anisotropy was observed for both linewidth and g-factor. The ESR linewidth is about 18 mT, and the g-factor is around 2.05 for the field $B//u$. The linewidth becomes very broad (c.a. 70 mT), and the g-factor is around 2.22 for the field $B//c$ [8].

In general, such an ESR signal with large anisotropy and broad linewidth can be attributed to the paramagnetic resonance from the Fe spins, whereas the π-electrons usually give a weak ESR signal with narrow linewidth around $g \sim 2$. In fact, the ESR signal of λ-(BETS)$_2$GaCl$_4$ is $g = 1.99 \sim 2.02$ with the linewidth of about 10 mT at $T = 20$ K [44]. For λ-(BETS)$_2$FeCl$_4$, it should be noted that the ESR signal of the π-electron and the d-electron should merge to a single ESR line thanks to the strong π-d interaction known as the 'exchange narrowing'. This is due to the fast exchange between π- and d-electrons where the local field is averaged out, and the spectra merge to a single line. However, the contribution of the Fe spins should be dominant in the PM state since the magnetic moment of Fe^{3+} is larger and the π-electrons are not localized.

Besides the ESR signal around $g = 2.0 \sim 2.2$, Brossard et al. observed an additional ESR signal around $g \sim 2.6$, which clearly appears around 70 K [8]. Such a characteristic signal was also reported for the mixed compound λ-(BETS)$_2$Fe$_{0.6}$Ga$_{0.4}$Cl$_4$ by Kawamata et al., however, they did not observe the $g \sim 2.6$ signal for the pure λ-(BETS)$_2$FeCl$_4$ [44]. In that paper, they separately observed the ESR signals from the π- and d-electrons for λ-(BETS)$_2$Fe$_{0.6}$Ga$_{0.4}$Cl$_4$. Oshima et al. reported a cooling dependence of the π-d interaction in the same compound. Two ESR signals originating from the π- and d-electrons were observed for a rapid cooling (100 K/min), but these signals are merged into a single ESR line due to the exchange narrowing for a slow cooling (1 K/min) [45]. Although the detailed mechanism remains an open question, the phase separation of the normal metallic and ferroelectric domains at T_{FM} might be the origin of the additional signal around $g \sim 2.6$ [44].

Below 10 K, the broadening of ESR linewidth accompanied with a shift of g-values, typical to the development of the antiferromagnetic correlation, was observed [8,44]. Below T_N, the paramagnetic resonance near $g \sim 2$ disappears, and the antiferromagnetic resonance (AFMR) appears. Brossard et al. observed a typical bubble-like structure of the spin-flop resonance in the angular dependence of the AFMR. The spin-flop resonance was detected around 1.2 T and inclined at 25° from the c-axis, which is in good agreement with the magnetic torque measurements and theory [8,18,30]. The easy-axis and the hard-axis modes of the AFMR were also reported by Suzuki et al. and by Rutel et al. [12,46]. Although a slight deviation from the conventional AFMR mode was observed, the system seems to have an antiferromagnetic state with biaxial anisotropy [12,46]. The slight deviation might be due to the misalignment from the easy- or the hard-axis.

4.2. ESR Results of λ-(BETS)$_2$Fe$_x$Ga$_{1-x}$Cl$_4$ ($x = 0.2 \sim 1.0$)

As mentioned in the previous section, many ESR studies have already been performed for the pure λ-(BETS)$_2$FeCl$_4$ using microwave and millimeter-wave [6,8,12,44,46]. Here in this paper, we present our millimeter-wave and submillimeter-wave ESR results for the mixed compound λ-(BETS)$_2$Fe$_x$Ga$_{1-x}$Cl$_4$ ($x = 0.2$, 0.4, 0.5, 0.6, 0.8) and the pure λ-(BETS)$_2$FeCl$_4$, respectively. The millimeter-wave ESR measurements were performed by using a conventional cavity perturbation technique. The combination of an 8 T superconducting magnet and a millimeter-wave vector network analyzer (MVNA-8-350-2 of *AB millimetre*, France) was used. The MVNA includes a tunable millimeter-wave source and detector that cover the frequency range from 30 to 110 GHz. The sample was set on the end-plate of the cavity so that the oscillatory magnetic field is applied to the sample. For the mixed compound, the ESR results on the PM and AFI ground states for the field $B//a^*$ (the hard-axis in the antiferromagnetic state) are shown. We chose the a^*-axis since it usually gives the strongest ESR signal owing to the needle shape of the single crystal, and other orientations usually give poor signals when using the millimeter-wave cavity. Moreover, it is always difficult to correctly find the easy-axis of the sample, which is tilted at 30° from the c-axis to the b^*-axis. For these reasons, we discuss the results for the field $B//a^*$. For the submillimeter-wave ESR measurement, we used a simple transmission technique. The 25 T resistive magnet and the backward wave oscillator (BWO) light source were used with the transmission technique. The BWO can cover the frequency range from 200 to 700 GHz by using several vacuum tubes. The sample was placed in the Voigt configuration (i.e., the dc magnetic field is perpendicular to the propagation of the light). Therefore, the geometry of $B//c$ is convenient for the needle-shaped single crystal.

Let us start from the $x = 0.8$ and $x = 0.2$ salts of the mixed compound λ-(BETS)$_2$Fe$_x$Ga$_{1-x}$Cl$_4$. Temperature dependences of ESR spectra for $x = 0.8$ and 0.2 are presented in Figure 3a,b, respectively. As shown in the bottom panel of Figure 1b, the MI transition temperature T_N for $x = 0.8$ and the superconducting critical temperature T_c for $x = 0.2$ are about 7 K and 5 K, respectively. A single ESR line is observed for the spectrum in the high temperature region for $x = 0.8$ as shown with a black broken eye-guide line in Figure 3a. The g-value is around 2.05, which is consistent with the previous results [6,8,44]. The ESR intensity increases as temperature decreases down to 10 K, and neither g-value nor linewidth show significant change, which is a typical behavior of electron paramagnetic resonance (EPR). Although the EPR line does not shift significantly, a small shift to a higher field with decreasing temperature is observed. Such slight g-shift of the EPR line is also observed in λ-(BETS)$_2$FeCl$_4$, and is supposed to indicate the development of the spin correlation between π- and d-electrons. [8,44]. The EPR intensity starts to diminish below 10 K, and disappears around 4 K. In turn, an additional broad resonance is observed at lower magnetic field of the EPR line at $T_N = 7$ K as shown with a red broken line in Figure 3a. The intensity of this additional resonance increases and shifts to lower magnetic field with decreasing temperature. In principle, the AFMR mode for the hard-axis appears at the lower field from the EPR line. Hence, the additional peak observed below $T_N = 7$ K is attributed to the AFMR since the a^*-axis is the hard-axis. The shift of the AFMR is due to the development of the internal field. The important point is that the EPR line does not significantly shift with temperature, and the AFMR appears at the lower field than the EPR line and shifts to the lower field as temperature decreases. The ESR behavior for $x = 0.8$ is similar to that observed for the pure λ-(BETS)$_2$FeCl$_4$ (i.e., $x = 1.0$) as mentioned in the previous section. In contrast, the $x = 0.2$ salt only shows a single EPR line, intensity of which grows as the temperature decreases. Although the EPR slightly shifts to the higher field with decreasing temperature below 40 K, there is no drastic change (no g-shift and no linewidth broadening) at $T_c = 5$ K. This suggests that the magnetic flux is well penetrated in the sample even in the superconducting state. The superconducting state might be inhomogeneous with the coexistence of the paramagnetic domain since the Fe^{3+} spins are randomly introduced to the sample. As mentioned above, the ESR signals of the π- and d-electrons are averaged and merged into a single line due to the exchange narrowing. For both $x = 0.2$ and 0.8 salts, a single EPR line was observed. This suggests that the microscopic π-d interaction is strong and does not change with the content of the Fe^{3+} ions. Moreover, one can see that the S/N ratio of EPR spectra is different between the $x = 0.2$ and 0.8 salts with almost the same sample size. For instance, the $x = 0.2$ salt shows weaker EPR signal at 40 K. Furthermore, the noise level of the transmission for the $x = 0.2$ salt is larger than that for the $x = 0.8$ salt, although each EPR intensity looks almost the same. These results suggest that the EPR signal is weaker for the $x = 0.2$ salt than for the $x = 0.8$ salt. This is just because the Fe^{3+} content is smaller for the $x = 0.2$ salt. In turn, it suggests that the EPR signal from the Fe^{3+} spins is dominant.

Next, temperature dependences of the ESR spectra for the $x = 0.4$ and 0.5 salts are shown in Figure 4a,b, respectively. The MI transition occurs at about $T_N = 2.5$ K for $x = 0.4$ and 4 K for $x = 0.5$, respectively. Both salts show the EPR down to the lowest temperature, and the AFMR, which appears at the lower field of the EPR line, starts to be observed at each T_N. This suggests that the paramagnetic and antiferromagnetic states coexist. This coexistence is due to the random distribution of the Fe^{3+} spins in the anion layers. In the domain where the aggregation of the Fe^{3+} spins exist and the spin wave can propagate, the AFMR signal is observed. However, in the domain where the Fe^{3+} spins are isolated from each other, the EPR signal is observed. It means that the smaller the Fe content gets, the smaller the antiferromagnetic domain becomes. This picture can be justified from the observation of the smaller AFMR signal for the $x = 0.4$ salt in comparison with that for the $x = 0.5$ salt.

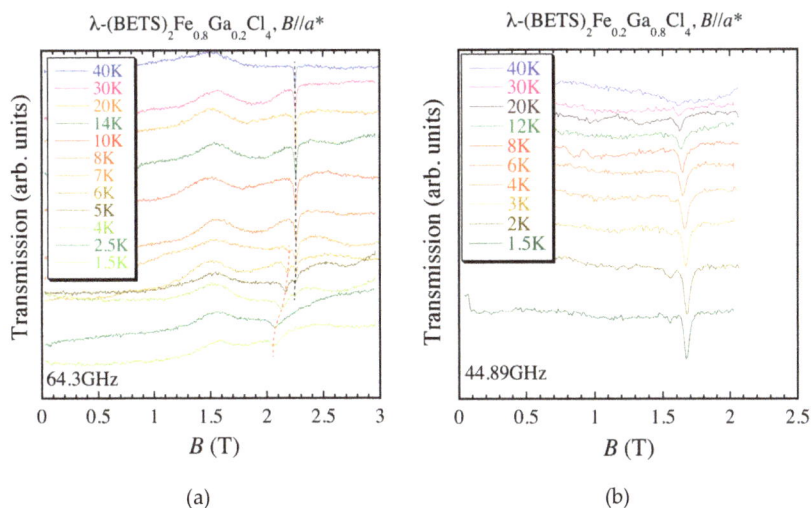

Figure 3. Temperature dependences of the ESR spectra of λ-(BETS)$_2$Fe$_x$Ga$_{1-x}$Cl$_4$ (**a**) for $x = 0.8$ and (**b**) for $x = 0.2$. The magnetic field is applied parallel to the a^*-axis. The black and red broken lines are the eye-guide of the EPR and AFMR lines, respectively.

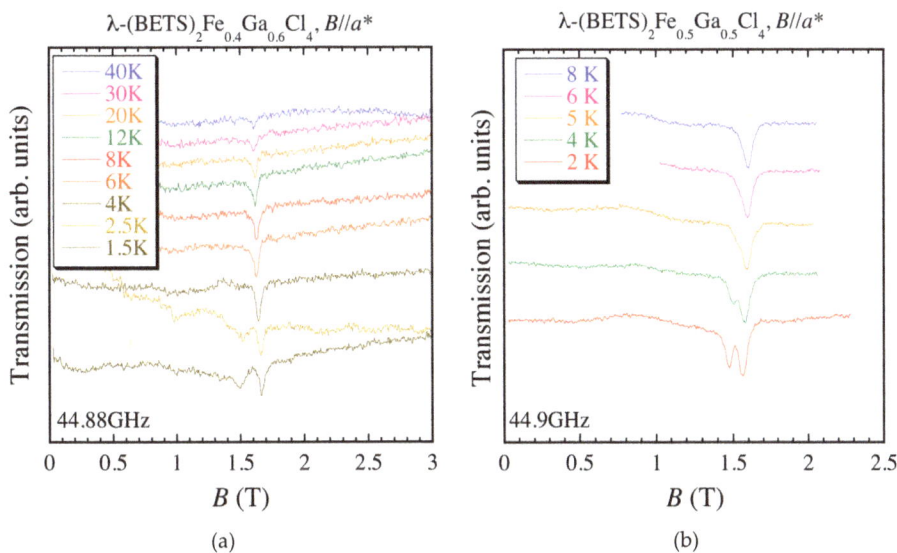

Figure 4. Temperature dependences of the ESR spectra of λ-(BETS)$_2$Fe$_x$Ga$_{1-x}$Cl$_4$ (**a**) for $x = 0.4$ and (**b**) for $x = 0.5$. The magnetic field is applied parallel to the a^*-axis.

For the $x = 0.6$ salt, where T_N is around 5 K, the EPR is observed in the high temperature region, and the AFMR starts to be observed at T_N (Figure 5). In contrast to the $x = 0.8$ salt, the difference of the resonance field between the EPR and AFMR is smaller for the $x = 0.6$ salt. This is associated with the gradual transition from EPR to AFMR in the $x = 0.6$ salt. Such a small difference of the resonance field might be due to the difference in the exchange field by the Fe content. It is clear that there is no EPR signal down to the lowest temperature (1.5 K) for the $x = 0.6$ and 0.8 salts, although there is a slow transition from the PM state to the AFI state (i.e., coexistence of the AFMR and EPR signals is observed around T_N). No EPR signal was also observed below T_N for the pure $x = 1.0$ salt [8,12,46]. The lack of EPR at the lowest temperature suggests that there is no sign of paramagnetic Fe as a ground state of the mixed compound with $x \geq 0.6$. These observations contradict the 'paramagnetic Fe model' where the Fe^{3+} spins remain paramagnetic in the AFI state. Since the intensity of the EPR signal, which mainly comes from the Fe^{3+} spins, is almost the same with that of the AFMR signal, the dominant contribution of the AFMR originates from the Fe^{3+} spins. Namely, if there is a 'paramagnetic Fe', it should be noticed from the EPR. Therefore, the lack of EPR signal below T_N suggests the Fe^{3+} spins in the λ-$(BETS)_2Fe_xGa_{1-x}Cl_4$ ($x \geq 0.6$) are antiferromagnetically ordered.

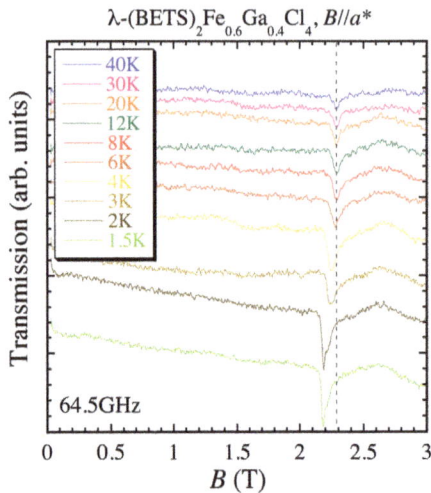

Figure 5. Temperature dependences of the ESR spectra of λ-$(BETS)_2Fe_xGa_{1-x}Cl_4$ for the $x = 0.6$ salt. The magnetic field is applied parallel to the a^*-axis. The broken line is the eye-guide of the EPR line's resonance field at 40 K.

Furthermore, a hump is observed around 1.5~1.6 T in the ESR spectra for the $x = 0.8$ salt (Figure 3a). This magnetic field position corresponds to $g\sim2.9$, which is similar to the peculiar ESR observed for the $x = 0.6$ and 1.0 salts (see the previous section) [8,44]. Kawamata et al. mentioned that such an ESR signal is due to the phase separation of the normal metallic and ferroelectric state. Although it depends on the frequency and the Fe content, similar hump structures are observed around 0.8 and 2.6 T for the $x = 0.5$ and 0.6 salts, respectively (Figures 4b and 5). These humps are observed especially for the salts with higher Fe content ($x \geq 0.5$). In contrast to the genuine ESR signal, whose transmission amplitude and phase of the millimeter-wave response change, the phase does not significantly change at the hump. Therefore, the hump is not actually an ESR signal, and is due to some high-frequency response of the sample, which might be related to the relaxor ferroelectric domain as mentioned in Section 3.4.

Let us finish this section by introducing our submillimeter-wave ESR measurements of the pure λ-$(BETS)_2FeCl_4$ in the high-field PM phase. As shown in Figure 6a, the submillimeter-wave transmission shows a huge change from 4 to 8 T, which is well below the B_c of the AFI phase. It is clear that this high-frequency response is not related to the MI transition. Although this change of the transmission is dependent on frequency and has some structures, it is similar to the observation of the depolarization regime reported by Rutel et al. [12]. With the higher magnetic field in the PM phase, where the transmission baseline becomes relatively flat, a single ESR signal is observed (arrows in Figure 6a and inset). This ESR corresponds to the EPR of the high-field PM phase. In the paramagnetic Fe model, it is expected that an internal field of about 4 T induces the splitting of the six energy levels of $S = 5/2$ [11]. The Zeeman splitting of 4 T corresponds to the energy gap of about 112 GHz (for $g = 2$). Therefore, ESR transitions with a gap of about 112 GHz at zero-field should be observed (shown as a broken line in Figure 6b). The observed ESR transition does not have a gap of 112 GHz.

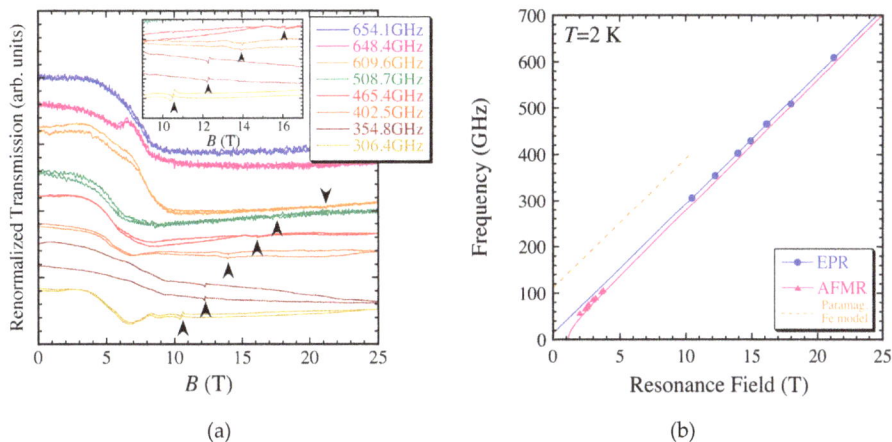

Figure 6. (**a**) Frequency dependences of the ESR spectra of λ-(BETS)$_2$FeCl$_4$ using submillimeter-wave at 2 K. The inset is the zoomed spectra around the EPR line; (**b**) The frequency-resonance field plots at 2 K. The magnetic field is applied parallel to the *c*-axis. The circles and triangles represent the EPR signal at the high-field PM phase and the AFMR signal in the AFI phase, respectively. The purple curve is the typical AFMR mode for the easy-axis, and the orange broken line is the expected ESR line for the paramagnetic Fe model (with a gap of 112 GHz~4 T at zero-field).

5. Conclusions and Future Prospects

We have revisited the previous studies on the AFI phase of λ-(BETS)$_2$FeCl$_4$, which is still under strong debate, and have presented our high-frequency ESR measurements for λ-(BETS)$_2$Fe$_x$Ga$_{1-x}$Cl$_4$ ($x = 0.2$~1.0). The $x = 0.2$ salt shows a EPR signal down to the lowest temperature since the Fe spins are isolated each other and the spin wave cannot propagate due to the low content of magnetic ions. This suggests that the ground state of the $x = 0.2$ salt is a 'paramagnetic' superconductor. Both EPR and AFMR were observed for the $x = 0.4$ and 0.5 salts, and only AFMR was observed for the $x = 0.6$ and 0.8 salts at the lowest temperature. Note that the $x = 1.0$ salt also shows only AFMR below T_N [8,12,46]. These results suggest that the 3D magnetic network starts to be formed in the system by the increase of the Fe content. The paramagnetic and antiferromagnetic domains coexist around $x = 0.4$~0.5, and finally, the system becomes totally antiferromagnetic in the $x \geq 0.6$ region. It is clear that the Fe spins play an important role in the magnetic ground state.

To explain the excess specific heat below T_N, Akiba et al. proposed that the π-electrons create an internal field of 4 T, which induces the degenarated six energy levels of Fe^{3+} ($S = 5/2$). Although such a gap should be detected by ESR, we could not observe the corresponding resonance in our submillimeter-wave measurements (estimated observation line shown in Figure 6b). Moreover, no EPR signal was found at the lowest temperature (1.5 K) in the $x \geq 0.6$ region. Therefore, the paramagnetic Fe model needs reconsideration [11].

We point out that the huge change of the high-frequency response within the AFI phase, the colossal magnetodielectricity, and nonlinear transport should be considered more seriously [12,38,39]. As shown in Figure 2a, we observed kink structures in the resistance below T_N (MI transition). Below the MI transition, the resistance gradually increases with decreasing temperature. The same kind of behavior was also observed by Toyota et al. and Sugiura et al. [27,38]. These results suggest that the

Magnetochemistry **2017**, *3*, 10

π-electrons are not fully localized below the MI transition. We think such metastable nature of the π-electrons within the AFI phase is an origin of the high frequency response. In turn, such gradual localization of the π-electrons will affect the magnetic state of the Fe^{3+} spins through the π-*d* interaction, which explains the gradual transition from PM to AFI observed in ESR, and the gradual increase of the internal field in the Mössbauer measurement. The excess specific heat also could be due to such a metastable state of the π-electrons, which affects the magnetic state of the Fe^{3+} spins. The origin of this gradual localization of π-electrons needs further investigation and theoretical support.

The 'chicken or egg problem' of the AFI phase, whether the insulating mechanism is spin-driven or charge-driven, remains an open question. It is probably worth noting that the λ-$(BETS)_2GaCl_4$, which is an isostructural analog without the 3*d* spins, does not have the AFI ground state. Furthermore, the intermediate region $(0.3 < x < 0.5)$ of the mixed compound shows a superconducting state, then, becomes insulating at the lowest temperature in association with the long-range order (see Figure 1b at the bottom). This suggests that the 3*d* spins play an important role for the MI transition, and the superconducting state is destroyed by the internal field of the antiferromagnetic long-range order. Our ESR results revealed that the π-*d* magnetic network is essential for the long-range order. If the Fe content is too small to form a magnetic network, the magnetic ground state is just paramagnetic, and the superconducting state remains at the lowest temperature. Therefore, the long-range order of the Fe spins seems to be essential for the MI transition, which favors the SDW-like transition scenario. As for the Mott transition scenario, theoretical studies suggest that the π-*d* interaction facilitates the Mott transition, which explains the different ground state between λ-$(BETS)_2GaCl_4$ and λ-$(BETS)_2FeCl_4$ [9,10]. Akiba et al. also proposed that the change of T_N by the Fe content is due to the magnetic anisotropy of Fe which suppresses the fluctuation of the π system through the π-*d* interaction [23]. Therefore, both scenarios are not decisive, and this issue remains to be solved. The charge ordering-induced polarization model also remains a strong candidate [39]. As proposed by Cépas et al., the measurement of the charge gap as a function of the field would distinguish between the SDW and Mott insulator scenarios [10]. Microscopic measurements such as NMR, ESR, and μSR should be suitable for solving this problem. At present, it is still difficult to conclude from just the temperature dependence of the ESR spectra in the low-field region. Although it is still preliminary, we have recently found that the single AFMR mode for the hard-axis splits at the high magnetic field. We suppose it is due to the delocalization of the π-electrons by the magnetic field. Hence, multi-frequency ESR (i.e., magnetic field dependence) measurements, which cover the whole AFI phase, are highly desirable to resolve this problem.

Acknowledgments: This study was performed in collaboration with the late James S. Brooks. Y.O. would like to deeply acknowledge his support and guidance for this study. Y.O. would like to also express sincere gratitude to H. Kobayashi and A. Kobayashi for their long-standing supports providing the samples. Y.O. also acknowledges T. Tokumoto, E. Jobiliong, S. A. Zvyagin, J. Krzystek, S. Uji, K. Hiraki, H. Akiba, Y. Nishio, K. Shimada, S. Sugiura for valuable discussions. This work was partly supported by the Grant-in-Aid for Scientific Research on Innovative Areas (No. 20110004) and for Young Scientist (No. 17740207).

Author Contributions: Y.O. conceived, designed, and performed the experiments and analyzed the data; H.B.C. and R.K. prepared the sample; Y.O. and R.K. wrote the paper, and all authors critically reviewed the paper.

Conflicts of Interest: The authors declare no conflict of interest.

References

1. Enomoto, K.; Yamaura, J.-I.; Miyazaki, A.; Enoki, T. Electronic and magnetic properties of organic conductors $(DMET)_2MBr_4$ (M = Fe, Ga). *Bull. Chem. Soc. Jpn.* **2003**, *76*, 945–959. [CrossRef]
2. Hanasaki, N.; Tajima, H.; Matsuda, M.; Naito, T.; Inabe, T. Giant negative magnetoresistance in quasi-one-dimensional conductor $TPP[Fe(Pc)(CN)_2]_2$: Interplay between local moments and one-dimensional conduction electrons. *Phys. Rev. B* **2000**, *62*, 5839. [CrossRef]
3. Uji, S.; Shinagawa, H.; Terashima, T.; Yakabe, T.; Terai, Y.; Tokumoto, M.; Kobayashi, A.; Tanaka, H.; Kobayashi, H. Magnetic0field-induced superconductivity in a two-dimensional organic conductor. *Nature* **2001**, *410*, 908–910. [CrossRef] [PubMed]

4. Balicas, L.; Brooks, J.S.; Storr, K.; Uji, S.; Tokumoto, M.; Tanaka, H.; Kobayashi, H.; Kobayashi, A.; Barzykin, V.; Gor'kov, L.P. Superconductivity in an organic insulator at very high magnetic field. *Phys. Rev. Lett.* **2001**, *87*, 067002. [CrossRef] [PubMed]

5. Uji, S.; Brooks, J.S. Magnetic-field-induced superconductivity in organic conductors. *J. Phys. Soc. Jpn.* **2006**, *75*, 051014. [CrossRef]

6. Kobayashi, H.; Tomita, H.; Naito, T.; Kobayashi, A.; Sakai, F.; Watanabe, T.; Cassoux, P. New BETS Conductors with Magnetic Anions (BETS = bis (ethylenedithio)tetraselenafulvalene). *J. Am. Chem. Soc.* **1996**, *118*, 368–377. [CrossRef]

7. Tokumoto, M.; Naito, T.; Kobayashi, H.; Kobayashi, A.; Laukhin, V.N.; Brossard, L.; Cassoux, P. Magnetic anisotropy of organic conductor λ-(BETS)$_2$FeCl$_4$. *Synth. Met.* **1997**, *86*, 2161–2162. [CrossRef]

8. Brossard, L.; Clerac, R.; Coulon, C.; Tokumoto, M.; Ziman, T.; Petrov, D.K.; Laukhin, V.N.; Naughton, M.J.; Audouard, A.; Goze, F.; et al. Interplay between chains of localised spins and two-dimensional sheets of organic donors in the synthetically built magnetic multilayer. *Eur. Phys. J. B* **1998**, *1*, 439–452. [CrossRef]

9. Hotta, C.; Fukuyama, H. Effects of localized spins in quasi-two dimensional organic conductors. *J. Phys. Soc. Jpn.* **2000**, *69*, 2577–2596. [CrossRef]

10. Cépas, O.; McKenzie, R.H.; Merino, J. Magnetic-field-induced superconductivity in layered organic molecular crystals with localized magnetic moments. *Phys. Rev. B* **2002**, *65*, 100502R. [CrossRef]

11. Akiba, H.; Nakano, S.; Nishio, Y.; Kajita, K.; Zhou, B.; Kobayashi, A.; Kobayashi, H. Mysterious Paramagnetic States of Fe 3d Spin in Antiferromagnetic Insulator of λ-BETS$_2$FeCl$_4$ System. *J. Phys. Soc. Jpn.* **2009**, *78*, 033601. [CrossRef]

12. Rutel, I.; Okubo, S.; Brooks, J.S.; Jobiliong, E.; Kobayashi, H.; Kobayashi, A.; Tanaka, H. Millimeter-wave investigation of the antiferromagnetic phase in λ-(BETS)$_2$FeCl$_4$ in high magnetic fields. *Phys. Rev. B* **2003**, *68*, 144435. [CrossRef]

13. Cui, H.-B.; Kobayashi, H.; Kobayashi, A. Phase diagram and anomalous constant resistivity state of a magnetic organic superconducting alloy, λ-(BETS)$_2$Fe$_x$Ga$_{1-x}$Cl$_4$. *J. Mat. Chem.* **2007**, *17*, 45–48. [CrossRef]

14. Uji, S.; Shinagawa, H.; Terakura, C.; Terashima, T.; Yakabe, T.; Terai, Y.; Tokumoto, M.; Kobayashi, A.; Tanaka, H.; Kobayashi, H. Fermi surface studies in the magnetic-field-induced superconductor λ-(BETS)$_2$FeCl$_4$. *Phys. Rev. B* **2001**, *64*, 024531. [CrossRef]

15. Sato, A.; Ojima, E.; Akutsu, H.; Kobayashi, H.; Kobayashi, A.; Cassoux, P. Temperature-Composition Phase Diagram of the Organic Alloys, λ-BETS$_2$(Fe$_x$Ga$_{1-x}$)Cl$_4$, with Mixed Magnetic and Non-Magnetic Anions. *Chem. Lett.* **1998**, *27*, 673. [CrossRef]

16. Goze, F.; Laukhin, V.N.; Brossard, L.; Audouard, A.; Ulmet, J.P.; Askenazy, S.; Naito, T.; Kobayashi, H.; Kobayashi, A.; Tokumoto, M.; et al. Magnetotransport Measurements on the λ-Phase of the Organic Conductors (BETS)$_2$MCl$_4$ (M = Ga, Fe). Magnetic-Field-Restored Highly Conducting State in λ-(BETS)$_2$FeCl$_4$. *Eur. Phys. Lett.* **1994**, *28*, 427–432, *ibid. Physica B* **1995**, *211*, 290–292. [CrossRef]

17. Akutsu, H.; Kato, K.; Ojima, E.; Kobayashi, H.; Tanaka, H.; Kobayashi, A.; Cassoux, P. Coupling of metal-insulator and antiferromagnetic transitions in the highly correlated organic conductor incorporating magnetic anions, λ-BETS$_2$FeBr$_x$Cl$_{4-x}$ [BETS = Bis (ethylenedithio) tetraselenafulvalene]. *Phys. Rev. B* **1998**, *58*, 9294–9302. [CrossRef]

18. Sasaki, T.; Uozaki, H.; Endo, S.; Toyota, N. Magnetic torque of λ-(BETS)$_2$FeCl$_4$. *Synth. Met.* **2001**, *120*, 759–760. [CrossRef]

19. Uji, S.; Terakura, C.; Terashima, T.; Yakabe, T.; Terai, Y.; Tokumoto, M.; Kobayashi, A.; Sakai, F.; Tanaka, H.; Kobayashi, H. Fermi surface and internal magnetic field of the organic conductor λ-(BETS)$_2$Fe$_x$Ga$_{1-x}$Cl$_4$. *Phys. Rev. B* **2002**, *65*, 113101. [CrossRef]

20. Sato, A.; Ojima, E.; Akutsu, H.; Nakazawa, Y.; Kobayashi, H.; Tanaka, H.; Kobayashi, A.; Cassoux, P. Magnetic properties of λ-BETS$_2$(Fe$_x$Ga$_{1-x}$)Cl$_4$ exhibiting a superconductor-to-insulator transition ($0.35 < x < 0.5$). *Phys. Rev. B* **2000**, *61*, 111–114.

21. Negishi, E.; Uozaki, H.; Ishizaki, Y.; Tsuchiya, H.; Endo, S.; Abe, Y.; Matsui, H.; Toyota, N. Specific heat studies for λ-(BEDT-TSF)$_2$FeCl$_4$. *Synth. Met.* **2003**, *133–134*, 555–556. [CrossRef]

22. Akiba, H.; Nobori, K.; Shimada, K.; Nishio, Y.; Kajita, K.; Zhou, B.; Kobayahshi, A.; Kobayashi, H. Magnetic and Thermal Properties of λ-(BETS)$_2$FeCl$_4$ System—Fe 3d Spin in Antiferromagnetic Insulating Phase—. *J. Phys. Soc. Jpn.* **2011**, *80*, 063601. [CrossRef]

23. Akiba, H.; Sugawara, H.; Nobori, K.; Shimada, K.; Tajima, N.; Nishio, Y.; Kajita, K.; Zhou, B.; Kobayashi, A.; Kobayashi, H. Paramagnetic Metal—Antiferromagnetic Insulator Transition of λ-BETS$_2$Fe$_x$Ga$_{1-x}$Cl$_4$ System. *J. Phys. Soc. Jpn.* **2012**, *81*, 053601. [CrossRef]

24. Shimada, K.; Akiba, H.; Tajima, N.; Kajita, K.; Nishio, Y.; Kato, R.; Kobayashi, A.; Kobayashi, H. Temperature Dependence of Internal Field by Analysis of Specific Heat on an Organic Conductor λ-BETS$_2$FeCl$_4$. *JPS Conf. Proc.* **2014**, *1*, 012110.

25. Shimada, K.; Tajima, N.; Kajita, K.; Nishio, Y. Effective Field Study of Canted Antiferromagnetic Insulating Phase in Magnetic Organic Conductor λ-(BETS)$_2$FeCl$_4$ through Specific Heat Measurement. *J. Phys. Soc. Jpn.* **2016**, *85*, 023601. [CrossRef]

26. Waerenborgh, J.C.; Rabaça, S.; Almeida, M.; Lopes, E.B.; Kobayashi, A.; Zhou, B.; Brooks, J.S. Mössbauer spectroscopy and magnetic transition of λ-(BETS)$_2$FeCl$_4$. *Phys. Rev. B* **2010**, *81*, 060413R. [CrossRef]

27. Sugiura, S.; Shimada, K.; Tajima, N.; Nishio, Y.; Terashima, T.; Isono, T.; Kobayashi, A.; Zhou, B.; Kato, R.; Uji, S. Charge Transport in Antiferromagnetic Insulating Phase of Two-Dimensional Organic Conductor λ-(BETS)$_2$FeCl$_4$. *J. Phys. Soc. Jpn.* **2016**, *85*, 064703. [CrossRef]

28. Sugiura, S.; Shimada, K.; Tajima, N.; Nishio, Y.; Terashima, T.; Isono, T.; Kato, R.; Uji, S. Magnetic Torque Studies in Two-Dimensional Organic Conductor λ-(BETS)$_2$FeCl$_4$. *J. Phys. Soc. Jpn.* **2017**, *86*, 014702. [CrossRef]

29. Mori, T.; Katsuhara, M. Estimation of πd-Interactions in Organic Conductors Including Magnetic Anions. *J. Phys. Soc. Jpn.* **2002**, *71*, 826–844. [CrossRef]

30. Ito, K.; Shimahara, H. Mean Field Theory of a Coupled Heisenberg Model and Its Application to an Organic Antiferromagnet with Magnetic Anions. *J. Phys. Soc. Jpn.* **2016**, *85*, 024704. [CrossRef]

31. Shimahara, H.; Ito, K. Spin-Flop Transition and a Tilted Canted Spin Structure in a Coupled Antiferromagnet. *J. Phys. Soc. Jpn.* **2016**, *85*, 043708. [CrossRef]

32. Matsui, H.; Tsuchiya, H.; Negishi, E.; Uozaki, H.; Ishizaki, Y.; Abe, Y.; Endo, S.; Toyota, N. Anomalous dielectric response in the π-d correlated metallic state of λ-(BEDT-TSF)$_2$FeCl$_4$. *J. Phys. Soc. Jpn.* **2001**, *70*, 2501–2504. [CrossRef]

33. Kitano, H.; Maeda, A. Comment on "Anomalous dielectric response in the π-d correlated metallic state of λ-(BEDT-TSF)$_2$FeCl$_4$". *J. Phys. Soc. Jpn.* **2002**, *71*, 666–667. [CrossRef]

34. Matsui, H.; Tsuchiya, H.; Toyota, N. Reply to Comment by H. Kitano and A. Maeda. *J. Phys. Soc. Jpn.* **2002**, *71*, 668–669. [CrossRef]

35. Matsui, H.; Tsuchiya, H.; Suzuki, T.; Negishi, E.; Toyota, N. Relaxor ferroelectric behavior and collective modes in the π-d correlated anomalous metal λ-(BEDT-TSF)$_2$FeCl$_4$. *Phys. Rev. B* **2003**, *68*, 155105. [CrossRef]

36. Watanabe, M.; Komiyama, S.; Kiyanagi, R.; Noda, Y.; Negishi, E.; Toyota, N. Evidence of the First-Order Nature of the Metal–Insulator Phase Transition in λ-(BEDT-TSF)$_2$FeCl$_4$. *J. Phys. Soc. Jpn.* **2003**, *72*, 452–453. [CrossRef]

37. Komiyama, S.; Watanabe, M.; Noda, Y.; Negishi, E.; Toyota, N. Relaxor-like Behavior in λ-(BEDT-TSF)$_2$FeCl$_4$ Studied by SR X-ray Diffraction. *J. Phys. Soc. Jpn.* **2004**, *73*, 2385–2388. [CrossRef]

38. Toyota, N.; Abe, Y.; Matsui, H.; Negishi, E.; Ishizaki, Y.; Tsuchiya, H.; Uozaki, H. Nonlinear electrical transport in λ-(BEDT-TSF)$_2$FeCl$_4$. *Phys. Rev. B* **2002**, *66*, 033201. [CrossRef]

39. Negishi, E.; Kuwabara, T.; Komiyama, S.; Watanabe, M.; Noda, Y.; Mori, T.; Matsui, H.; Toyota, N. Dielectric ordering and colossal magnetodielectricity in the antiferromagnetic insulating state of λ-(BEDT-TSF)$_2$FeCl$_4$. *Phys. Rev. B* **2005**, *71*, 012416. [CrossRef]

40. Oh, J.I.; Naughton, M.J.; Courcet, T.; Malfant, I.; Cassoux, P.; Tokumoto, M.; Akutsu, H.; Kobayashi, H.; Kobayashi, A. Torque anisotropy in λ-(BEDT-TSF)$_2$FeCl$_4$. *Synth. Met.* **1999**, *103*, 1861–1864. [CrossRef]

41. Endo, S.; Goto, T.; Fukase, T.; Matsui, H.; Uozaki, H.; Tsuchiya, H.; Negishi, E.; Ishizaki, Y.; Abe, Y.; Toyota, N. Anomalous Splitting of ^{1}H-NMR Spectra in λ-(BEDT-TSF)$_2$FeCl$_4$. *J. Phys. Soc. Jpn.* **2002**, *71*, 732–734. [CrossRef]

42. Wu, G.; Ranin, P.; Gaidos, G.; Clark, W.G.; Brown, S.E.; Balicas, L.; Montgomery, L.K. ^{1}H-NMR spin-echo measurements of the spin dynamic properties in λ-(BETS)$_2$FeCl$_4$. *Phys. Rev. B* **2007**, *75*, 174416. [CrossRef]

43. Hiraki, K.; Mayaffre, H.; Horvatic, M.; Berthier, C.; Uji, S.; Yamaguchi, T.; Tanaka, H.; Kobayashi, A.; Kobayashi, H.; Takahashi, T. ^{77}Se NMR Evidence for the Jaccarino–Peter Mechanism in the Field Induced Superconductor, λ-(BETS)$_2$FeCl$_4$. *J. Phys. Soc. Jpn.* **2007**, *76*, 124708. [CrossRef]

44. Kawamata, S.; Kizawa, T.; Suzuki, T.; Negishi, E.; Matsui, H.; Toyota, N.; Ishida, T. π-d Correlation in λ-(BEDT-TSF)$_2$Fe$_{1-x}$Ga$_x$Cl$_4$ by ESR Measurements. *J. Phys. Soc. Jpn.* **2006**, *75*, 104715. [CrossRef]

45. Oshima, Y.; Jobiliong, E.; Brooks, J.S.; Zvyagin, S.A.; Krzystek, J.; Tanaka, H.; Kobayashi, A.; Cui, H.; Kobayashi, H. EMR Measurements of Field-Induced Superconductor λ-(BEDT-TSF)$_2$Fe$_{1-x}$Ga$_x$Cl$_4$. *Synth. Met.* **2005**, *153*, 365–368. [CrossRef]

46. Suzuki, T.; Matsui, H.; Tsuchiya, H.; Negishi, E.; Koyama, K.; Toyota, N. Antiferromagnetic resonance in λ-(BETS)$_2$FeCl$_4$. *Phys. Rev. B* **2003**, *67*, 020408R. [CrossRef]

magnetochemistry

MDPI

Review

Recent Advances on Anilato-Based Molecular Materials with Magnetic and/or Conducting Properties

Maria Laura Mercuri [1,*], **Francesco Congiu** [2], **Giorgio Concas** [2] **and Suchithra Ashoka Sahadevan** [1]

[1]   Dipartimento di Scienze Chimiche e Geologiche, Università degli Studi di Cagliari,
      S.S. 554—Bivio per'Sestu—I09042 Monserrato (CA), Italy; suchithra.sahadevan@unica.it
[2]   Dipartimento di Fisica, Universitàdegli Studi di Cagliari, S.P. Monserrato'Sestu km
      0,700—I09042 Monserrato (CA), Italy; franco.congiu@dsf.unica.it (F.C.); giorgio.concas@dsf.unica.it (G.C.)
*   Correspondence: mercuri@unica.it; Tel./Fax: +39-70-675-4486

Academic Editor: Manuel Almeida
Received: 2 February 2017; Accepted: 7 April 2017; Published: 19 April 2017

Abstract: The aim of the present work is to highlight the unique role of anilato-ligands, derivatives of the 2,5-dioxy-1,4-benzoquinone framework containing various substituents at the 3 and 6 positions (X = H, Cl, Br, I, CN, etc.), in engineering a great variety of new materials showing peculiar magnetic and/or conducting properties. Homoleptic anilato-based molecular building blocks and related materials will be discussed. Selected examples of such materials, spanning from graphene-related layered magnetic materials to intercalated supramolecular arrays, ferromagnetic 3D monometallic lanthanoid assemblies, multifunctional materials with coexistence of magnetic/conducting properties and/or chirality and multifunctional metal-organic frameworks (MOFs) will be discussed herein. The influence of (i) the electronic nature of the X substituents and (ii) intermolecular interactions i.e., H-Bonding, Halogen-Bonding, π-π stacking and dipolar interactions, on the physical properties of the resulting material will be also highlighted. A combined structural/physical properties analysis will be reported to provide an effective tool for designing novel anilate-based supramolecular architectures showing improved and/or novel physical properties. The role of the molecular approach in this context is pointed out as well, since it enables the chemical design of the molecular building blocks being suitable for self-assembly to form supramolecular structures with the desired interactions and physical properties.

Keywords: benzoquinone derivatives; molecular magnetism; multifunctional molecular materials; spin-crossover materials; metal-organic frameworks

---

1. General Introduction

The aim of the present work is to highlight the key role of anilates in engineering new materials with new or improved magnetic and/or conducting properties and new technological applications. Only homoleptic anilato-based molecular building blocks and related materials will be discussed. Selected examples of para-/ferri-/ferro-magnetic, spin-crossover and conducting/magnetic multifunctional materials and MOFs based on transition metal complexes of anilato-derivatives, on varying the substituents at the 3,6 positions of the anilato moiety, will be discussed herein, whose structural features or physical properties are peculiar and/or unusual with respect to analogous compounds reported in the literature up to now. Their most appealing technological applications will be also reported.

Derivatives of the 2,5-dioxy-1,4-benzoquinone framework, containing various substituents at the 3 and 6 positions, constitute a well-known motif observed in many natural products showing important biological activities such as anticoagulant [1], antidiabetic [2], antioxidative [3], anticancer [4], etc. Structural modifications of the natural products afforded related compounds of relevant interest in medicinal chemistry [5,6]. Furthermore, the 2,5-dihyroxy-1,4-benzoquinone (DHBQ) represents the parent member of a family of organic compounds traditionally called anilic acids that, in their deprotonated dianionic form, act as valuable ditopic ligands towards transition metal ions [7].

Anilic acids are obtained when the hydrogens at the 3 and 6 positions of the DHBQ are replaced by halogen atoms or functional groups (see below). They can be formulated as $H_2X_2C_6O_4$ (H_2X_2An) where X indicates the substituent and C_6O_4 the anilate moiety (An). A summary of the anilic acids reported in the literature to the best of our knowledge is reported in Table 1.

Table 1. Names, molecular formulas and acronyms of the anilic acids reported in the literature to date.

X_2An^{2-}

Substituent, X	Formula	Anilic Acid Name	Acronyms	Anilate Dianion Name	Acronyms	Ref.
H	$H_4C_6O_4$	Hydranilic acid	H_2H_2An	Hydranilate	H_2An^{2-}	[8–10]
F	$H_2F_2C_6O_4$	Fluoranilic acid	H_2F_2An	Fluoranilate	F_2An^{2-}	[11]
Cl	$H_2Cl_2C_6O_4$	Chloranilic acid	H_2Cl_2An	Chloranilate	Cl_2An^{2-}	[12,13]
Br	$H_2Br_2C_6O_4$	Bromanilic acid	H_2Br_2An	Bromanilate	Br_2An^{2-}	[14]
I	$H_2I_2C_6O_4$	Iodanilic acid	H_2F_2An	Iodanilate	I_2An^{2-}	[14]
NO_2	$H_2N_2C_6O_8$	Nitranilic acid	$H_2(NO_2)_2An$	Nitranilate	$(NO_2)_2An^{2-}$	[15]
OH	$H_4C_6O_6$	Hydroxyanilic acid	$H_2(OH)_2An$	Hydroxyanilate	$(OH)_2An^{2-}$	[16–20]
CN	$H_2N_2C_8O_4$	Cyananilic acid	$H_2(CN)_2An$	Cyananilate	$(CN)_2An^{2-}$	[21,22]
Cl/CN	$H_2ClNC_7O_4$	Chlorocyananilic acid	$H_2ClCNAn$	Chlorocyananilate	$ClCNAn^{2-}$	[23]
NH_2	$H_6N_2C_6O_4$	Aminanilic acid	$H_2(NH_2)_2An$	Aminanilate	$(NH_2)_2An^{2-}$	[24]
CH_3	$H_8C_8O_4$	Methylanilic acid	H_2Me_2An	Methylanilate	Me_2An^{2-}	[25]
CH_2CH_3	$H_{12}C_{10}O_4$	Ethylanilic acid	H_2Et_2An	Ethylanilate	Et_2An^{2-}	[25]
iso-C_3H_7	$H_{16}C_{12}O_4$	Isopropylanilic acid	H_2*iso*-Pr_2An	Isopropylanilate	*iso*-Pr_2An^{2-}	[26]
C_6H_5	$H_{12}C_{18}O_4$	Phenylanilic acid	H_2Ph_2An	Phenylanilate	Ph_2An^{2-}	[10,27]
C_4H_3S	$H_8C_{14}O_4S_2$	Thiophenylanilic acid	H_2Th_2An	Thiophenylanilate	Th_2An^{2-}	[28]
$C_6H_5O_2S$	$H_{12}C_{18}O_8S_2$	3,4-ethylene dioxythiophenyl anilic acid	H_2EDOT_2An	3,4-ethylene dioxythiophenyl anilate	$EDOTAn^{2-}$	[28]
C_4H_9	$H_{20}C_{14}O_4$	2,3,5,6-tetrahydroxy-1,4-benzo quinone	H_2THBQ	2,5-di-tert-butyl-3,6-dihydroxy-1,4-benzoquinonate	$THBQ^{2-}$	[29]

The synthetic methods to obtain the anilic acids described in Table 1 are reported in Schemes 1–3 respectively.

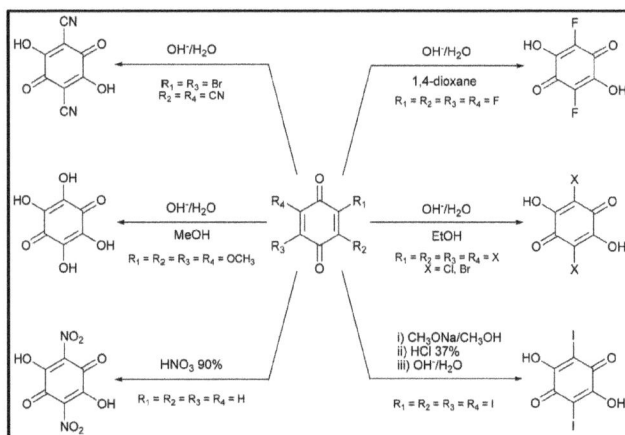

Scheme 1. Overview of the synthetic procedures for the preparation of the H_2X_2An (X = F, Cl, Br, I, NO_2, OH, CN) anilic acids. The corresponding anilate dianions, generated in solution, afford the protonated forms by simple acidification.

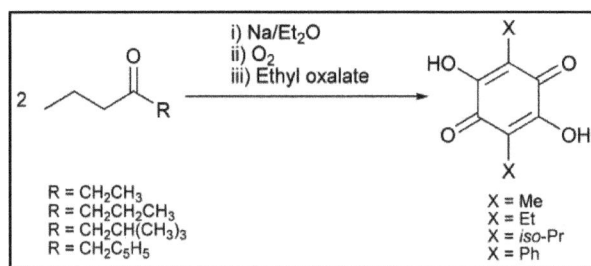

Scheme 2. Overview of the synthetic procedures for the preparation of the H_2X_2An (X = Me, Et, *iso*-Pr, Ph) anilic acids.

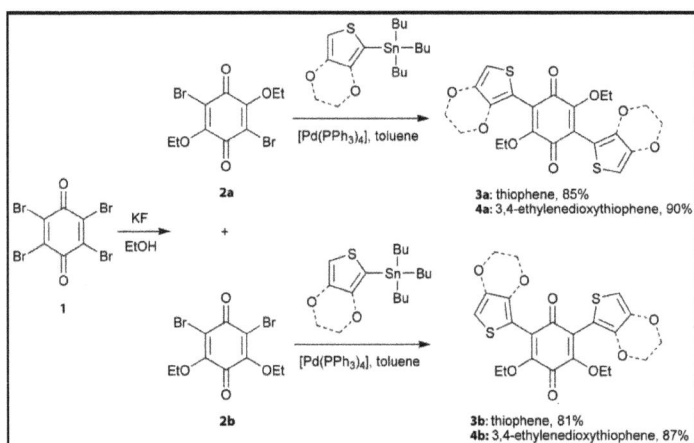

Scheme 3. Synthesis of thiophenyl (**3a,b**) and 3,4-ethylenedioxythiophenyl (**4a,b**) derivatives of 1,4-benzoquinone.

Anilic acids (**I**) undergo a mono and double deprotonation process of the two hydroxyl groups giving rise to the monoanionic (**II**) and dianionic (**III**) forms (Scheme 4: **III** prevails in aqueous media due to the strong resonance stabilization of the negative charge).

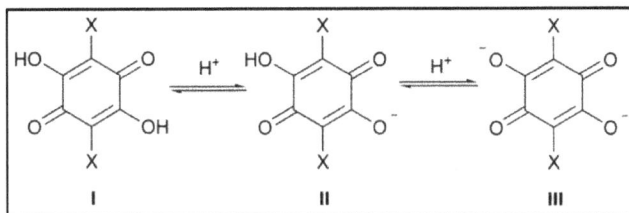

Scheme 4. Protonation equilibria for a generic anilic acid.

The molecular and crystal structures of the protonated anilic acids [24,30–38] are characterized by similar features: (i) a centrosymmetric quasi-quinonoid structure with C=O and C=C distances in the 1.215–1.235 Å and 1.332–1.355 Å ranges, respectively; (ii) a planar structure of the benzoquinone ring; and (iii) moderate-strong H-Bonding and π-stacking interactions in the crystal structure [30,32,36–38]. It should be noted that the crystal structure of $H_2(NO_2)_2An$ hexahydrate and $H_2(CN)_2An$ hexahydrate reveal the presence of hydronium nitranilates and hydronium cyananilates, respectively [35,36], as a result of their strong acidity (pK_a values for $H_2(NO_2)_2An$: -3.0 and -0.5) [11]. The structure of the nitranilic acid hexahydrate is characterized by the presence of the Zundel cation, $(H_5O_2)^{2+}$, whose proton dynamic has been recently studied by using a multi-technique approach [39]. Interestingly, the structure of $H_2(NH_2)_2An$ reveals the presence of an highly polarized zwitterionic structure with the protons located on the amino groups [24]. The molecular and crystal structures of alkali metal salts of some anilic acids have also been reported [40–47]. The X-ray analysis reveals that the carbon ring system for the anilates in their dianionic form takes the planar conformation but is not in a quinoidal form, having four C–C bonds of equal length (1.404–1.435 Å range) and two considerably longer C–C bonds (1.535–1.551 Å range) whose bond distances vary as a function of the substituents. Moreover, the four C–O bonds are of equal length (1.220–1.248 Å range). This description can be represented with four resonance structures that, in turn, can be combined in one form with delocalized π-electrons along the O–C–C(–X)–C–O bonds (Scheme 5) [34,35,38,41].

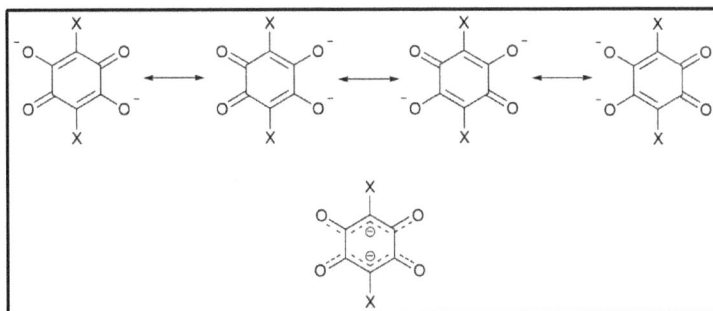

Scheme 5. Resonance structures for a generic anilate dianion. The π-electron delocalization over the O–C–C(–X) –C–O bonds is highlighted.

The crystal structures of the anilate anions are dominated by π-stacking interactions between quinoid rings. Since the dianions are characterized by (i) π-electron delocalization on the O–C–C(–X) –C–O bonds and (ii) strong repulsion due to double negative charges, their crystal structures are dominated by parallel offset π-π stacking arrangement, similarly to what found in aromatic systems. Monoanionic alkali salts are, instead, able to stack in a perfect face-to-face parallel arrangement with no offset, where single bonds are sandwiched between double bonds and vice versa, with short distances of the ring centroids (3.25–3.30 Å), as thoroughly described by Molčanov et al. [43,45,46]. This arrangement minimizes of π-electrons repulsions while maximizing σ-π and dipolar attractions [45].

An overview of the coordination modes shown by the anilate dianions is reported in Scheme 6:

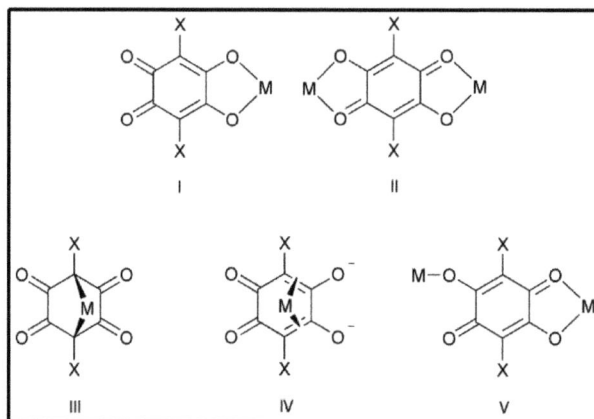

Scheme 6. Coordination modes exhibited by the anilate dianions: 1,2-bidentate (**I**), bis-1,2-bidentate (**II**), 1,4-bidentate (**III**), π-bonding (**IV**), 1,2-bidentate/monodentate (**V**).

It is noteworthy that among the described coordination modes, **I** and **II** are the most common, whereas **III**, **IV** and **V** have been only rarely observed.

Kitagawa and Kawata reported on the coordination chemistry of the anilic acids in their dianionic form (anilate ligands) with particular attention to the DHBQ^{2-} (H$_2$An^{2-}) and its chloro derivative, chloranilate (Cl$_2$An^{2-}) [7]. Since the first observation of a strong magnetic interaction between paramagnetic metal ions and the H$_2$An^{x-} (x = 3, 1) radical species, reported by Gatteschi et al. [48] several types of metal complexes ranging from finite discrete homoleptic and heteroleptic mononuclear systems to extended homoleptic and heteroleptic polymeric systems showing a large variety of peculiar crystal structures and physical properties, have been obtained so far [7,49–60]. Valence Tautomerism is an essential phenomenon in anilato-based systems [29,61–64] and it has been observed for the first time by Sato et al. [64] in the heteroleptic dinuclear complex, [(CoTPA)$_2$(H$_2$An)](PF$_6$)$_3$, (TPA = tris(2-pyridylmethyl)amine) that exhibits a valence tautomeric transition with a distinct hysteresis effect (13 K) around room temperature and photoinduced valence tautomerism under low temperature.

Slow magnetic relaxation phenomena are also one of hot topic in magnetochemistry and very recently Ishikawa, Yamashita et al. [65] reported on the first example of slow magnetic relaxation observed in a chloranilato-based system, the new field induced single-ion magnet, [Co(bpy)$_2$(Cl$_2$An)]·EtOH, (bpy = bipiridyl) a heteroleptic six-coordinate mononuclear high-spin cobalt(II) complex, formed by 1D π-π stacked chain-like structures through the bpy ligands. This compound undergoes spin-phonon relaxation of Kramers ions through two-phonon Raman and direct spin-phonon bottleneck processes and the observed slow relaxation of the magnetization is purely molecular in its origin.

The interest in the anilate chemistry has been recently renovated since uncoordinated anilic acids have been recently used as molecular building blocks for obtaining different types of functional materials such as organic ferroelectrics or as a component of charge transfer salts showing peculiar physical properties [66–69]. Anilates in fact are very challenging building blocks because of: (i) their interesting redox properties [70]; (ii) their ability to mediate magnetic superexchange interactions when the ligand coordinates two metals ions in the 1,2-bis-bidentate coordination mode; (iii) the possibility of modulating the strength of this magnetic superexchange interaction by varying the substituents (X) on the 3,6 position of the anilato-ring [71]; moreover the presence of different substituents in the anilato moiety give rise to intermolecular interactions such as H-Bonding, Halogen-Bonding, π-π stacking and dipolar interactions which may influence the physical properties of the resulting material. Therefore, these features provide an effective tool for engineering a great variety of new materials with unique physical properties.

The aim of the present work is to highlight the key role of anilates in engineering new materials with new or improved magnetic and/or conducting properties and new technological applications. Only homoleptic anilato-based molecular building blocks and related materials will be discussed. Selected examples of para-/ferri-magnetic, spin-crossover and conducting/magnetic multifunctional materials based on transition metal complexes of anilato-derivatives, on varying the substituents at the 3,6 positions of the anilato moiety, will be discussed herein, whose structural features or physical properties are peculiar and/or unusual with respect to analogous compounds reported in the literature up to now. Their most appealing technological applications will be also reported.

2. Anilato-Based Molecular Magnets

2.1. Introduction

In the design of molecule-based magnets the choice of the interacting metal ions and the bridging ligand plays a key role in tuning the nature and magnitude of the magnetic interaction between the metal ions, especially when the bridge contains electronegative groups that may act as "adjusting screws. A breakthrough in this area is represented by the preparation in 1992 by Okawa et al. [72] of the family of layered bimetallic magnets based on the oxalate ($C_2O_4^{2-}$) ligand, formulated as $[(n\text{-Bu})_4N]M^{II}Cr(C_2O_4)_3]$ (M^{II} = Mn, Fe, Co, Ni and Cu) showing the well-known 2D hexagonal honeycomb structure [73,74]. These systems show ferromagnetic order (M^{III} = Cr) with ordering temperatures ranging from 6 to 14 K, or ferrimagnetic order (M^{III} = Fe) with T_c ranging from 19 to 48 K [75–79]. In these compounds the A^+ cations play a crucial role in tailoring the assembly of the molecular building-blocks and therefore controlling the dimensionality of the resulting bimetallic framework. In addition, the substitution of these electronically innocent cations with electroactive ones can increase the complexity of these systems, adding novel properties to the final material. In the last 20 years many efforts have been addressed to add in these materials a further physical property by playing with the functionality of the A^+ cations located between the bimetallic layers. This strategy produced a large series of multifunctional molecular materials where the magnetic ordering of the bimetallic layers coexists or even interacts with other properties arising from the cationic layers, such as paramagnetism [2,76–80], non-linear optical properties [2,81,82], metal-like conductivity [83,84], photochromism [2,81,85,86], photoisomerism [87], spin crossover [88–93], chirality [94–97], or proton conductivity [2,98,99]. Moreover, it is well-established that the ordering temperatures of these layered magnets are not sensitive to the separation determined by the cations incorporated between the layers, which slightly affects the magnetic properties of the resulting hybrid material, by emphasizing its 2D magnetic character [2,75–80,95,100,101]. The most effective way to tune the magnetic properties of such systems is to act directly on the exchange pathways within the bimetallic layers. This can be achieved either by varying M^{II} and M^{III} or by modifying the bridging ligand. So far, only the first possibility has been explored, except for a few attempts at replacing the bridging oxalate ligand by with the dithioxalate one, leading to a small variations of the ordering temperatures [102–105].

In this context, anilates, larger bis-bidentate bridging ligands than oxalates, are very challenging as their coordination modes are similar to the oxalato ones and it is well-known that they are able to provide an effective pathway for magnetic exchange interactions [7].

One of the most interesting anilato-based structures obtained so far are the H_2An^{2-}- and Cl_2An^{2-}-based honeycomb layers [47,106–110]. In these 2D compounds the structure is similar to that of the oxalate honeycomb layers, but all reported systems to date are homometallic (i.e., they contain two M^{II} or two M^{III} ions of the same nature type). The layers formed with two M^{II} ions contain a 2- charge per formula, $[M^{II}_2(X_2An)_3]^{2-}$ (X = Cl, H), and, accordingly, two monocations are needed to balance the charge. The only known examples of this $[M^{II}_2L_3]^{2-}$ series are the $[M_2(H_2An)_3]^{2-}$ (M = Mn and Cd) [106] and $[M_2(Cl_2An)_3]^{2-}$ (M = Cu, Co, Cd and Zn) systems [108]. The layers formed with two M^{III} ions are neutral and the reported examples include the $[M_2(H_2An)_3]\cdot24H_2O$ (M^{III} = Y, La, Ce, Gd, Yb and Lu) [109,110], $[M_2(Cl_2An)_3]\cdot12H_2O$ (M^{III} = Sc, Y, La, Pr, Nd, Gd, Tb, Yb, Lu) [2,47,110] and $[Y_2(Br_2An)_3]\cdot12H_2O$ systems [47]. Further interest for the anilate ligands is related to their ability to form 3D structures with the (10,3)-a topology, similar to the one observed with the oxalate [111]; these structures are afforded when all the ML$_3$ units show the same chirality, in contrast with the 2D honeycomb layer, which requires alternating Λ-ML$_3$ and Δ-ML$_3$ units. This 3D structure with a (10,3)-a topology has been recently reported for the $[(n\text{-}Bu)_4N]_2[M^{II}_2(H_2An)_3]$ (M^{II} = Mn, Fe, Ni, Co, Zn and Cd) and $[(n\text{-}Bu)_4N]_2[Mn_2(Cl_2An)_3]$ systems [112], showing a double interpenetrating (10,3)-a lattice with opposite stereochemical configuration that afford an overall achiral structure. The versatility of the anilate-based derivatives is finally demonstrated by the formation of a 3D adamantane-like network in the compounds $[Ag_2(Cl_2An)]$ [113], $[H_3O][Y(Cl_2An)_3]\cdot8CH_3OH$ and $[Th(Cl_2An)_2]\cdot6H_2O$ [110]. Because these ligands are able to mediate antiferromagnetic exchange interactions, it should be expected that 2D heterometallic lattices of the $[M^{II}M^{III}(X_2An)_3]^-$ type, would afford ferrimagnetic coupling and ordering. Furthermore, if the magnetic coupling depends on the X substituents on the ligand, as expected, a change of X is expected to modify the magnetic coupling and the T_c. This is probably the most interesting and appealing advantage of the anilate ligands since they can act as the oxalate ligands, but additionally they show the possibility of being functionalized with different X groups. This should lead to a modulation of the electronic density in the benzoquinone ring, which, in turn, should result in an easy tuning of the magnetic exchange coupling and, therefore, of the magnetic properties (ordering temperatures and coercive fields) in the resulting 2D or 3D magnets. It should be highlighted that among the ligands used to produce the majority of known molecule-based magnets such as oxalato, azido, or cyano ligands, only the anilates show the this ability, to our knowledge.

A further peculiar advantage of these 2D materials is that the bigger size of anilate ligands compared with oxalate ones may enable the insertion within the anion layer of the charge-compensating counter-cation, leading to neutral layers that may be exfoliated using either mechanical or solvent-mediated exfoliation methods [114]. To date, examples of exfoliation of magnetic layered coordination polymers are rare and some of the few examples of magnetic 2D coordination polymers exfoliated so far are the Co^{2+} or Mn^{2+} 2,2-dimethylsuccinate frameworks showing antiferromagnetic ordering in the bulk [115].

2.2. Molecular Paramagnets

Two new isostructural mononuclear complexes of formula $[(Ph)_4P]_3[M(H_2An)_3]\cdot6H_2O$ (M = Fe(III) (1) or Cr(III) (2) have been obtained by reacting the hydranilate anion with the Fe(III) and Cr(III) paramagnetic metal ions [116]. The crystal structure of 1 consists of homoleptic tris-chelated octahedral complex anions $[Fe(H_2An)_3]^{3-}$ surrounded by crystallization water molecules and $(Ph)_4P^+$ cations. The metal complexes are involved in an extensive network of moderately strong hydrogen bonds (HBs) between the peripheral oxygen atoms of the ligand and crystallization water molecules; HBs are responsible, as clearly shown by the analysis of the Hirshfeld surface, for the formation of supramolecular layers that run parallel to the a crystallographic axis, showing an unprecedented H-bonded 2D architecture in the family of the anilato-based H-bonded networks [116]. DFT theoretical calculations pointed out the key role of the H substituent on the hydranilato ligand in modulating the electron density of the whole complex and favoring the electron delocalization toward the peripheral oxygen atoms of the ligands, compared with the other components of the family of halogenated tris-chelated anilato-based complexes (Figure 1); these peripheral oxygen atoms act, in turn, as suitable HB-acceptors in the observed supramolecular architecture. The magnetic properties of **1** show a typical paramagnetic behavior of quasi-isolated spin centers, while those of **2** are quite intriguing and might find their origin in some kind of charge transfer between the Cr metal ions and the hydranilate ligands, even though the observed magnetic behavior do not rule out the possibility to have extremely small magnetic coupling also mediated by HB interactions (Figure 2).

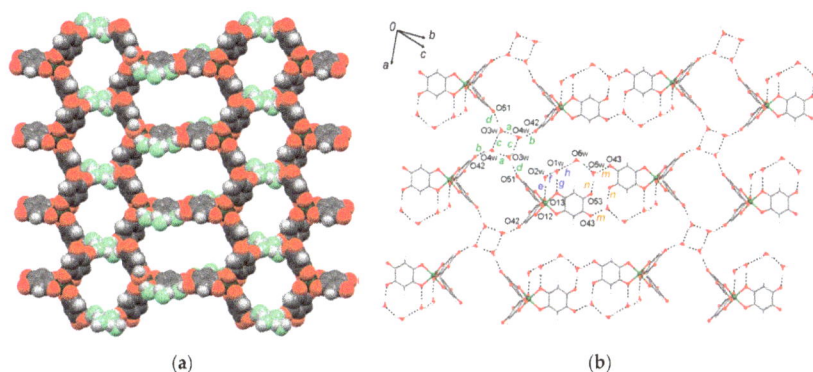

(a)   (b)

Figure 1. View of the crystal packing of **1**: (**a**) with metal complexes and water molecules in spacefill model highlighting the supramolecular topology; (**b**) highlighting the hydrogen bond (HB) interactions occurring between the water molecules and the metal complexes. The 11 HBs are indicated with colored letters. HB donors and acceptor are also indicated. Symmetry codes are omitted. Reprinted with permission from Reference [116]. Copyright 2014 American Chemical Society.

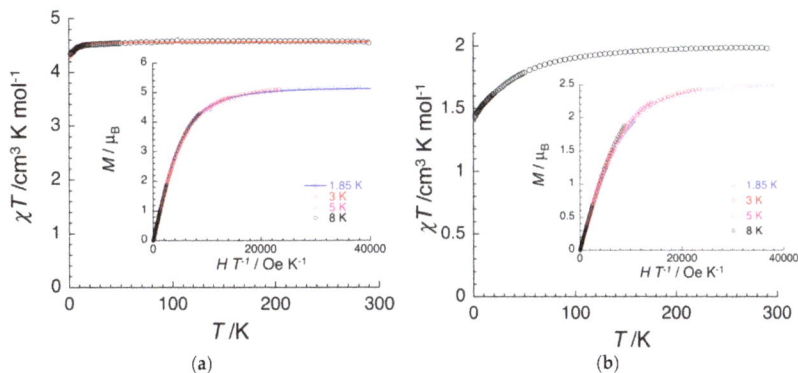

(a)   (b)

Figure 2. (**a**) Temperature dependence of χT product at 1000 Oe (where χ is the molar magnetic susceptibility equal to M/H per mole of Fe(III) complex) between 1.85 and 300 K for a polycrystalline sample of **1**. The solid line is the best fit obtained using a Curie-Weiss law. Inset shows field dependence of the magnetization for 1 between 1.85 and 8 K at magnetic fields between 0 and 7 T. The solid line is the best fit obtained using $S = 5/2$ Brillouin function; (**b**) Temperature dependence of χT product at 1000 Oe (where χ is the molar magnetic susceptibility equal to M/H per mole of Fe(III) complex) between 1.85 and 300 K for a polycrystalline sample of **2**. Inset shows field dependence of the magnetization for 2 between 1.85 and 8 K at magnetic fields between 0 and 7 T. Reprinted with permission from Reference [116]. Copyright 2014 American Chemical Society.

These compounds behave as versatile metallotectons, which are metal complexes able to be involved in well identified intermolecular interactions such as HBs and can therefore serve as building blocks for the rational construction of crystals, especially with HB-donating cations or size-tunable cationic metallotectons to afford porous coordination polymers or porous magnetic networks with guest-tunable magnetism (See Section 2.3).

By using the chloranilate ligand, Cl_2An^{2-}, the $[(TPA)(OH)Fe^{III}OFe^{III}(OH)(TPA)][Fe(Cl_2An)_3]_{0.5}$ $(BF_4)_{0.5} \cdot 1.5MeOH \cdot H_2O$ (3) [TPA = tris(2-pyridylmethyl)amine] compound has been obtained by Miller et al. [117] in an atom economical synthesis. This is the first example of the formation of homoleptic trischelated $[Fe(Cl_2An)_3]^{3-}$ mononuclear anions. The core structure of **3** consists of two (dihydroxo)oxodiiron(III) dimer dications, the tris(chloranilato)ferrate(III) anion as well as a $[BF_4]^-$ (see Figure 3).

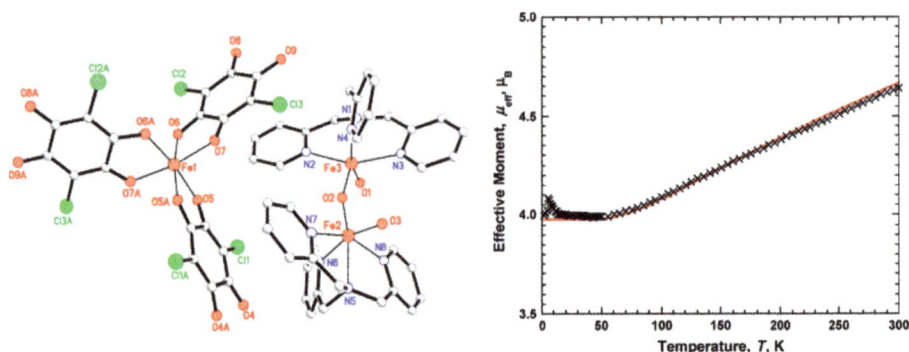

Figure 3. (left) ORTEP view of **3**. The atoms are represented by 50% probable thermal ellipsoids. Hydrogen atoms, solvent, and $[BF_4]^-$ are omitted for clarity; **(right)** $\mu_{eff}(T)$ for **3** taken at 300 Oe. The solid line is the best fit curve to the model. Reprinted with permission from [117]. Copyright 2006. American Chemical Society.

Variable-temperature magnetic measurements on a solid sample of **3** have been performed in the 2–300 K. At room temperature, the effective moment, $\mu_{eff}(T)$ is 2.93 μ_B/Fe, and $\mu_{eff}(T)$ decreases with decreasing temperature until it reaches a plateau at ca. 55 K, indicating a strong antiferromagnetic interaction within the $Fe^{III}OFe^{III}$ unit. Below 55 K, $\chi(T)$ is constant at 4.00 μ_B, which is attributed solely to $[Fe(Cl_2An)_3]^{3-}$. The $\chi(T)$ data were fit to a model for a coupled $S = 5/2$ dimer and an $S = 5/2$ Curie-Weiss term for the uncoupled $[Fe(Cl_2An)_3]^{3-}$. The best fit had J/kB of -165 K (115 cm^{-1}), $g = 2.07$, $\theta = -1$ K, and the spin impurity $\varrho = 0.05$. This experimentally determined J value for **3** is in the range observed for other oxo-bridged Fe(III) complexes with TPA as capping ligand, that is, $J = -107 \pm 10$ cm^{-1} [118].

By replacing X = H at the 3,6 positions of the benzoquinone moiety with X = Cl, Br, I, the new tris(haloanilato)metallate(III) complexes with general formula $[A]_3[M(X_2An)_3]$ (A = (n-Bu)$_4$N$^+$, (Ph)$_4$P$^+$; M = Cr(III), Fe(III); X_2An = chloranilate (Cl_2An^{2-}, see Chart 1), bromanilate (Br_2An^{2-}) and iodanilate (I_2An^{2-})), have been obtained [119]. To the best of our knowledge, except for the tris(chloranilato)ferrate(III) complex obtained by Miller et al. [117] no reports on the synthesis and characterization of trischelated homoleptic mononuclear complexes with the previously mentioned ligands are available in the literature so far.

Cation	Cl_2An^{2-}		Br_2An^{2-}		I_2An^{2-}	
	Cr(III)	Fe(III)	Cr(III)	Fe(III)	Cr(III)	Fe(III)
$(n\text{-}Bu)_4N^+$	4a	5a	6a	7a	8a	9a
$(Ph)_4P^+$	4b	5b	6b	7b	8b	9b
$(Et)_3NH^+$	4c	5c	-	-	-	-

Chart 1. $[A]_3[M(X_2An)_3]$ tris(haloanilato)metallate(III) complexes (A = $(n\text{-}Bu)_4N^+$, $(Ph)_4P^+$; M = Cr(III), Fe(III); Cl_2An^{2-} = chloranilate, Br_2An^{2-} = bromanilate and I_2An^{2-} = iodanilate).

The crystal structures of these Fe(III) and Cr(III) haloanilate complexes consist of anions formed by homoleptic complex anions formulated as $[M(X_2An)_3]^{3-}$ and $(Et)_3NH^+$, $(n\text{-}Bu)_4N$, or $(Ph_4)P^+$ cations. All complexes exhibit octahedral coordination geometry with metal ions surrounded by six oxygen atoms from three chelate ligands. These complexes are chiral according to the metal coordination of three bidentate ligands, and both Λ and Δ enantiomers are present in their crystal lattice. Interestingly the packing of $[(n\text{-}Bu)_4N]_3[Cr(I_2An)_3]$ (**8a**) shows that the complexes form supramolecular dimers that are held together by two symmetry related I···O interactions (3.092(8) Å), considerably shorter than the sum of iodine and oxygen van der Waals radii (3.50 Å). The I···O interaction can be regarded as a halogen bond (XB), where the iodine behaves as the XB donor and the oxygen atom as the XB acceptor (Figure 4a). This is in agreement with the properties of the electrostatic potential for $[Cr(I_2An)_3]^{3-}$ that predicts a negative charge accumulation on the peripheral oxygen atoms and a positive charge accumulation on the iodine. Also in $[(Ph)_4P]_3[Fe(I_2An)_3]$ (**9b**) each $[Fe(I_2An)_3]^{3-}$ molecule exchanges three I···I XBs with the surrounding complex anions. These iodine–iodine interactions form molecular chains parallel to the b axis that are arranged in a molecular layer by means of an additional I···I interaction with symmetry related I(33) atoms (3.886(2) Å), which may behave at the same time as an XB donor and acceptor. Additional XB interactions can be observed in the crystal packing of **9b** (Figure 4b).

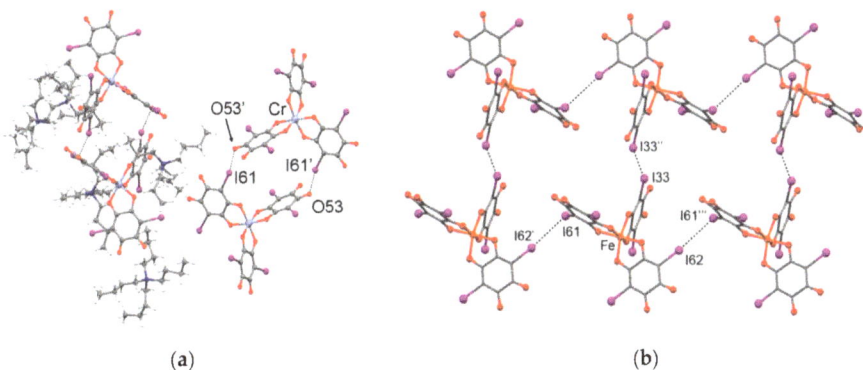

(a)　　　　　　　　(b)

Figure 4. (a) Portion of the molecular packing of 8a where four complex anions are displayed (Symmetry code ′ = 1 − x; 1 − y; 1 − z); (b) halogen bonds between the complex anions (Symmetry codes ′ = x; y + 1; z, ″ = 3/2 − x; 3/2 − y; 1 − z, ‴ = x; y − 1; z). Adapted with permission from Reference [119]. Copyright 2014 American Chemical Society.

The magnetic behaviour of all complexes, except **8a**, may be explained by considering a set of paramagnetic non-interacting Fe(III) or Cr(III) ions, taking into account the zero-field splitting effect similar to the Fe(III) hydranilate complex reported in Figure 2a. The presence of strong XB interactions in **8a** are able, instead, to promote antiferromagnetic interactions among paramagnetic centers at low temperature, as shown by the fit with the Curie-Weiss law, in agreement with the formation of halogen-bonded supramolecular dimers (Figure 5).

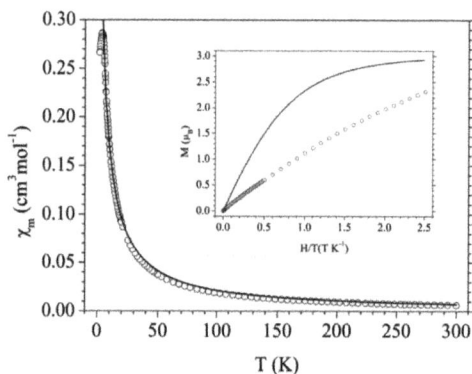

Figure 5. Thermal variation of χ_m for **8a**. Solid line is the best fit to the Curie-Weiss law. Inset shows the isothermal magnetization at 2 K. Solid line represents the Brillouin function for an isolated S = 3/2 ion with g = 2. Anions. Reprinted with permission from Reference [119]. Copyright 2014 American Chemical Society.

A simple change of one chloro substituent on the chloranilate ligand with a cyano group affects the electronic properties of the anilate moiety inducing unprecedented luminescence properties in the class of anilate-based ligands and their metal complexes. The synthesis and full characterization, including photoluminescence studies, of the chlorocyananilate ligand (ClCNAn^{2-}), a unique example of a heterosubstituted anilate ligand has been recently reported [120], along with the tris-chelated metal complexes with Cr(III), (**10**) Fe(III), (**11**) and Al(III) (**12**) metal ions, formulated as [A]$_3$[M^{III}(ClCNAn)$_3$] (A = (n-Bu)$_4$N$^+$ or Ph4P$^+$) shown in Chart 2.

Cation	ClCNAn^{2-}		
	Cr(III)	Fe(III)	Al(III)
(n-Bu)$_4$N$^+$	10a	11a	12a
(Ph)$_4$P$^+$	10b	11b	12b

Chart 2. [A]$_3$[M(X$_2$An)$_3$] tris(haloanilato)metallate(III) complexes (A = (n-Bu)$_4$N$^+$, (Ph)$_4$P$^+$; M = Cr(III), Fe(III), Al(III); ClCNAn^{2-} = chlorocyananilate).

The crystal structures of the M(III) chlorocyananilate complexes consist of homoleptic tris-chelated complex anions of formula [M(ClCNAn)$_3$]$^{3-}$ (M = Cr(III), Fe(III), Al(III)), exhibiting octahedral geometry and [(n-Bu)$_4$N]$^+$ or [Ph$_4$P]$^+$ cations. The **10a–12a** complexes are isostructural and their crystal packing is characterized by the presence of C–N··· Cl interactions between complex anions having an opposite stereochemical configuration (Λ, Δ), responsible for the formation of infinite 1D supramolecular chains parallel to the *a* crystallographic axis (Figure 6). The Cl··· N interaction can be regarded as a halogen-bond where the chlorine behaves as the halogen-bonding donor and the nitrogen atom as the halogen-bonding acceptor, in agreement with the electrostatic potential that predicts a negative charge accumulation on the nitrogen atom of the cyano group and a positive charge accumulation on the chlorine atom.

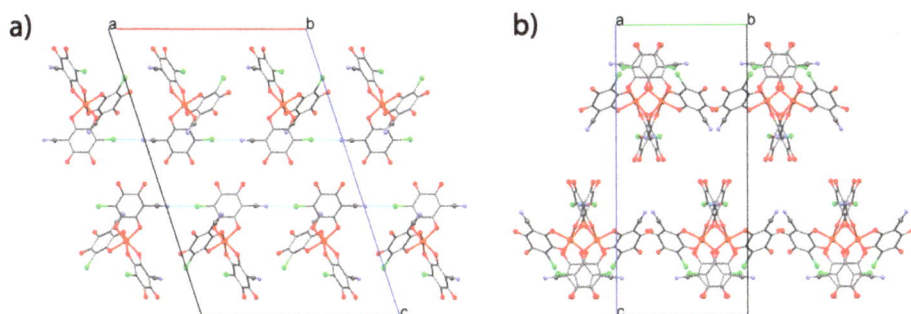

Figure 6. (**a**) Portion of the molecular packing of **11a** showing the Cl···N interactions occurring between the complex anions; (**b**) View of the supramolecular chains along the a axis. [(n-Bu)$_4$N]$^+$ cations are omitted for clarity. Reprinted with permission from Reference [120]. Copyright 2015 from The Royal Society of Chemistry.

10b–12b complexes are isostructural to the already reported analogous systems having chloranilate as ligand (vide supra). **10**, **12** (**a**, **b**) exhibit the typical paramagnetic behavior of this family of mononuclear complexes (vide supra). Interestingly TD-DFT calculations have shown that the asymmetric structure of the chlorocyananilate ligand affects the shape and energy distribution of the molecular orbitals involved in the electronic excitations. In particular, the HOMO → LUMO transition in the Vis region (computed at 463 nm) becomes partly allowed compare to the symmetric homosubstituted chloranilate and leads to an excited state associated with emission in the green region, at ca. λ_{max} = 550 nm, when exciting is in the lowest absorption band. Coordination to Al(III) (**12a**, **b**), does not significantly affect the luminescence properties of the free ligand, inducing a slight red-shift in the emission wavelength while maintaining the same emission efficiency with comparable quantum yields; thus the Al(III) complex **12a** still retains the ligand-centered emission and behaves as an appealing red luminophore under convenient visible light irradiation. **10a** and **11a** instead are essentially non-emissive, likely due to the ligand-to-metal CT character of the electronic transition in the Vis region leading to non-radiative excited states [120].

By combining [A]$_3$[M^{III}(X$_2$An)$_3$] (A = Bu$_3$MeP$^+$, (Ph)$_3$EtP$^+$; M(III) = Cr, Fe; X = Cl, Br) with alkaline metal ions (M^I = Na, K) the first examples of 2D and 3D heterometallic lattices (**13–16**) based on anilato ligands combining M(I) and a M(III) ions have been obtained by Gomez et al. [121]. (PBu$_3$Me)$_2$[NaCr(Br$_2$An)$_3$] (**13**) and (PPh$_3$Et)$_2$[KFe(Cl$_2$An)$_3$](dmf)$_2$ (**14**) show very similar 2D lattices formed by hexagonal [M^{I}M^{III}(X$_2$An)$_3$]$^{2-}$ anionic honeycomb layers with (PBu$_3$Me)$^+$ (**1**) or (PPh$_3$Et)$^+$ and dmf (**14**) charge-compensating cations inserted between the layers. While **13** and **14** show similar structures to the oxalato-based ones, a novel 3D structure, not found in the oxalato family is observed in (NEt$_3$Me)[Na(dmf)]-[NaFe(Cl$_2$An)$_3$] (**14**) formed by hexagonal layers analogous to **1** and **14** interconnected through Na$^+$ cations. (NBu$_3$Me)$_2$[NaCr(Br$_2$An)$_3$] (**16**), is the first heterometallic 3D lattice based on anilato ligands. This compound shows a very interesting topology containing two interpenetrated (10,3) chiral lattices with opposite chiralities, resulting in achiral crystals. This topology is unprecedented in the oxalato-based 3D lattices due to the smaller size of oxalateo compared to the anilato. Attempts to prepare **16** in larger quantities result in **16′**, the 2D polymorph of **16**, and as far as we know, this 2D/3D polymorphism has never been observed in the oxalato families showing the larger versatility of the anilato-ligands compared to the oxalato one. In Figure 7 the structures of **13–16** compounds are reported.

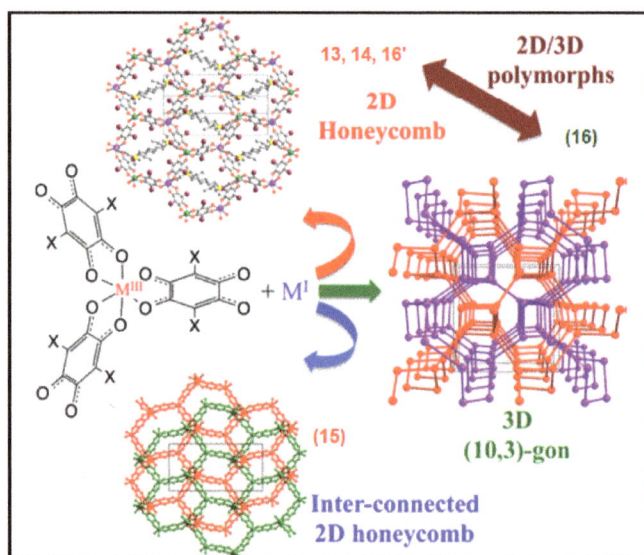

Figure 7. View of the hexagonal honeycomb layer in **13** with the (PBu$_3$Me)$^+$ cations in the hexagons. Color code: Cr = green, Na = violet, O = red, C = gray, Br = brown, and P = yellow. H atoms have been omitted for clarity. (Down) View of two adjacent layers in **15** showing the positions of the Fe(III) centers and the Na$_2$ dimers. H and Cl atoms have been omitted for clarity. (Right) Perspective view of the positions of the metal atoms in both interpenetrated sublattices (red and violet) in **16**. Adapted with permission from Reference [121]. Copyright 2015 American Chemical Society.

The magnetic measurements have been performed only on **13**, **15**, obtained and **16′** since only a few single crystals of **14** and **16** have been obtained. **13**, **15**, and **16′** show, as expected, paramagnetic behaviors that can be satisfactorily reproduced with simple monomer models including a zero field splitting (ZFS) of the corresponding S = 3/2 for Cr(III) in **13** and **16′** or S = 5/2 for Fe(III) in **15** (Figure 8a,b).

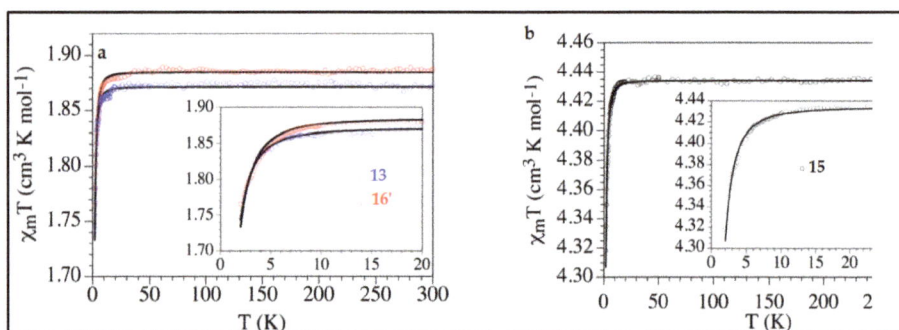

Figure 8. Thermal variation of $\chi_m T$ for (**a**) the Cr(III) compounds **13** and **16′** and (**b**) the Fe(III) compound **15**. Solid lines are the best fit to the isolated monomer models with zero field splitting (see text). Inset: low temperature regions. Reprinted with permission from Reference [121]. Copyright 2015 American Chemical Society.

In conclusion, this family of anionic complexes are versatile precursors (i) for constructing 2D molecule-based Ferrimagnets with tunable ordering temperature as a function of the halogen electronegativity (Section 2.2); (ii) as magnetic components for building up multifunctional molecular materials based on BEDT-TTF organic donors-based conductivity carriers (Section 3.2), in analogy with the relevant class of $[M(ox)_3]^{3-}$ tris-chelated complexes which have produced the first family of molecular paramagnetic superconductors [122–124]. Moreover, the ability of chlorocyananilate to work as the antenna ligand towards lanthanides, showing intense, sharp and long-lived emissions, represents a challenge due to the pletora of optical uses spanning from display devices and luminescent sensors to magnetic/luminescent multifunctional molecular materials.

2.3. Molecular Ferrimagnets

The novel family of molecule-based magnets formulated as $[Mn^{II}M^{III}(X_2An)_3]$ (A = $[H_3O(phz)_3]^+$, $(n\text{-Bu})_4N^+$, phz = phenazine; M^{III} = Cr, Fe; X = Cl, Br, I, H), namely $[(H_3O)(phz)_3][Mn^{II}M^{III}(X_2An)_3]\cdot H_2O$, with M^{III}/X = Cr/Cl (**17**), Cr/Br (**18**) and Fe/Br (**19**) and $[(n\text{-Bu}_4)N][Mn^{II}Cr^{III}(X_2An)_3]$, with X = Cl (**20**), Br (**21**), I (**22**) and H (**23**) (Chart 3) have been synthesized and fully characterized [125]. These compounds were obtained by following the so-called "complex-as-ligand approach". In this synthetic strategy, a molecular building block, the homoleptic $[M^{III}(X_2An)_3]^{3-}$ tris(anilato)metallate octahedral complex (M^{III} = Cr, Fe; X = Cl, Br, I, H), is used as ligand towards the divalent paramagnetic metal ion Mn(II). 2D anionic complexes were formed leading to crystals suitable for an X-ray characterization in the presence of the $(n\text{-Bu})_4N^+$ bulky organic cation or the $[H_3O(phz)_3]^+$ chiral adduct (Scheme 7).

Cationic Layer	Anionic Layer
$[H_3O(phz)_3]^+$	$Mn^{II}Cr^{III}$ (X-Cl) **17**
$[H_3O(phz)_3]^+$	$Mn^{II}Cr^{III}$ (X-Br) **18**
$[H_3O(phz)_3]^+$	$Mn^{II}Fe^{III}$ (X-Br) **19**
$[(n\text{-Bu}_4)N]^+$	$Mn^{II}Cr^{III}$ (X-Cl) **20**
$[(n\text{-Bu}_4)N]^+$	$Mn^{II}Cr^{III}$ (X-Br) **21**
$[(n\text{-Bu}_4)N]^+$	$Mn^{II}Cr^{III}$ (X-I) **22**
$[(n\text{-Bu}_4)N]^+$	$Mn^{II}Cr^{III}$ (X-H) **23**

Chart 3. $[Mn^{II}M^{III}(X_2An)_3]$ heterobimetallic complexes (A = $[H_3O(phz)_3]^+$, $(n\text{-Bu})_4N^+$, phz = phenazine; M^{III} = Cr, Fe; M^{II} = Mn; X = Cl, Br, I, H).

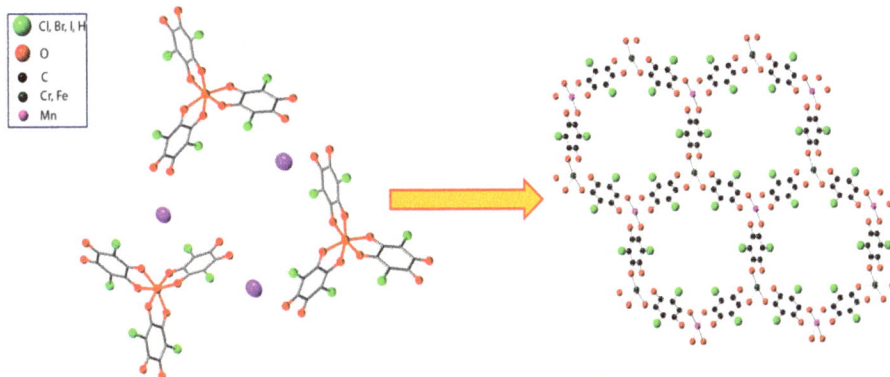

Scheme 7. Picture of the "complex-as-ligand approach" used for obtaining **17–23** compounds.

In these compounds, the monovalent cations act not only as charge-compensating counterions but also as templating agents controlling the dimensionality of the final system. In particular the chiral cation [(H₃O)(phz)₃]⁺, obtained in situ by the interaction between phenazine molecules and hydronium cations, appears to template and favor the crystallization process. In fact, most of the attempts to obtain single crystals from a mixture of the (*n*-Bu₄)N⁺ salts of the [M^{III}(X₂An)₃]³⁻ precursors and Mn(II) chloride, yielded poorly crystalline products and only the crystal structure for the [(*n*-Bu)₄N][MnCr(Cl₂An)₃] (**20**) system was obtained by slow diffusion of the two components.

Compounds [(H₃O)(phz)₃][MnCr(Cl₂An)₃(H₂O)] (**17**), [(H₃O)(phz)₃][MnCr(Br₂An)₃]·H₂O (**18**) and [(H₃O)(phz)₃][MnFe(Br₂An)₃]·H₂O (**19**) are isostructural and show a layered structure with alternating cationic and anionic layers (Figure 9). The only differences, besides the change of Cl₂An²⁻ (**17**) with Br₂An²⁻ (**18**), or Cr(III) (**18**) with Fe(III) (**19**), are (i) the presence of an inversion center in **18** and **19** (not present in **17**) resulting in a statistical distribution of the M(III) and Mn(II) ions in the anionic layers; (ii) the presence of a water molecule coordinated to the Mn(II) ions in **16** (Mn-O1w 2.38(1) Å), in contrast with compounds **18** and **19** where this water molecule is not directly coordinated.

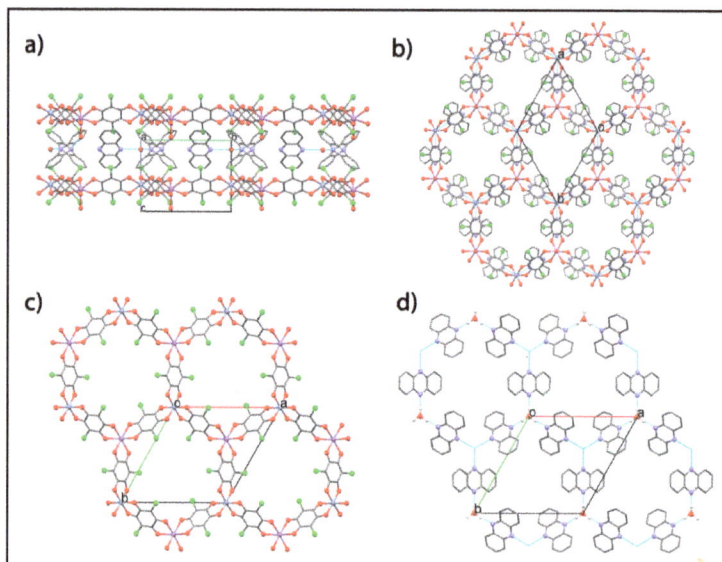

Figure 9. View of the crystal structure of **17**: (**a**) Side view of the alternating cationic and anionic layers; (**b**) Top view of the two layers; (**c**) Top view of the anionic layer; (**d**) Top view of the cationic layer showing the positions of the metal centers in the anionic layer (blue dashed hexagon). Reprinted with permission from Reference [125]. Copyright 2013 American Chemical Society.

The cationic layer is formed by chiral cations formulated as Δ-[(H₃O)(phz)₃]⁺ (Figure 9d) resulting from the association of three phenazine molecules around a central H₃O⁺ cation through three equivalent strong O–H···N hydrogen bonds. These Δ-[(H₃O)(phz)₃]⁺ cations are always located below and above the Δ-[Cr(Cl₂An)₃]³⁻ units, because they show the same chirality, allowing a parallel orientation of the phenazine and chloranilato rings (Figure 10b). This fact suggests a chiral recognition during the self-assembling process between oppositely charged [Cr(Cl₂An)₃]³⁻ and [(H₃O)(phz)₃]⁺ precursors with the same configuration (Δ or Λ).

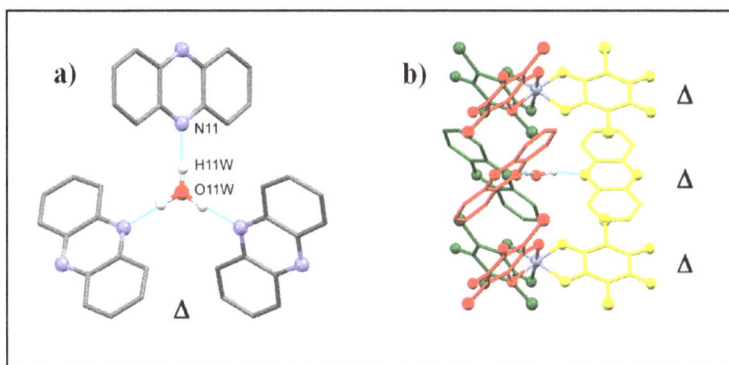

Figure 10. (a) Δ-[(H$_3$O)(phz)$_3$]$^+$ cation showing the O–H···N bonds as dotted lines; (b) side view of two anionic and one cationic layers showing Δ-[(H$_3$O)(phz)$_3$]$^+$ and the Δ-[Cr(Cl$_2$An)$_3$]$^{3-}$ entities located above and below. Parallel phenazine and anilato rings are shown with the same color. Color code: C, brown; O, pink; N, blue; H, cyan; Cl, green; Mn, yellow/orange; Cr, red. Reprinted with permission from Reference [125]. Copyright 2013 American Chemical Society.

An interesting feature of **17–19** is that they show hexagonal channels, which contain solvent molecules, resulting from the eclipsed packing of the cationic and anionic layers (Figure 11).

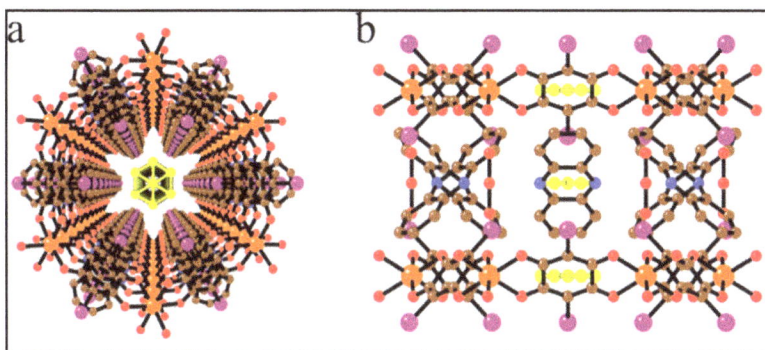

Figure 11. Structure of **18**: (a) perspective view of one hexagonal channel running along the c direction with the solvent molecules in the center (in yellow); (b) side view of the same hexagonal channel showing the location of the solvent molecules in the center of the anionic and cationic layers showing the O–H···N bonds as dotted lines. Reprinted with permission from Reference [125]. Copyright 2013 American Chemical Society.

20, the only compound with the [NBu$_4$]$^+$ cation whose structure has been solved, shows a similar layered structure as **17–19** but the main difference is the absence of hexagonal channels since the honeycomb layers are alternated and not eclipsed (Figure 12).

This eclipsed disposition of the layers generates an interesting feature of these compounds, i.e., the presence of hexagonal channels that can be filled with different guest molecules. **17–19** infact present a void volume of ca. 291 Å^3 (ca. 20% of the unit cell volume), where solvent molecules can be absorbed, opening the way to the synthesis of layered metal-organic frameworks (MOFs).

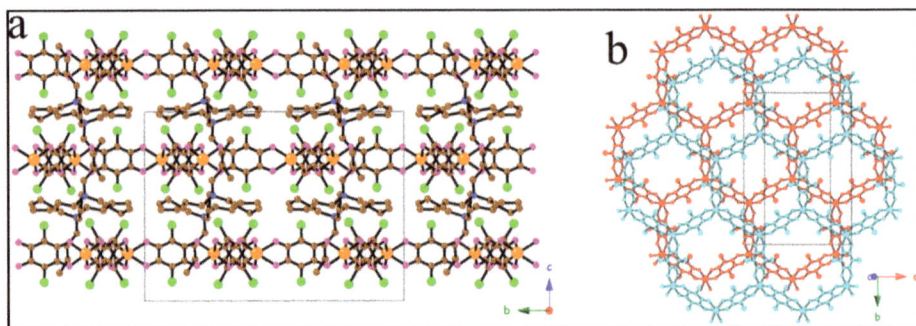

Figure 12. (a) Structure of **20**: (**a**) view of the alternating anionic and cationic layers; (**b**) projection, perpendicular to the layers, of two consecutive anionic layers showing their alternate packing. Reprinted with permission from Reference [125]. Copyright 2013 American Chemical Society.

All components of this series show ferrimagnetic long-range order as shown by susceptibility measurements, but the most interesting feature of this family is the tunability of the critical temperature depending on the nature of the X substituents: infact, as an example, an increase in Tc from ca. 5.5 to 6.3, 8.2, and 11.0 K (for X = Cl, Br, I, and H, respectively) is observed in the MnCr derivatives (Figure 13). Thus the different nature of the substituents on the bridging ligand play a key role in determining the critical temperature as shown by the linear correlation of the Tc as a function of the electronegativity of the substituents; Tc increases following the order X = Cl, Br, I, H and can be easily modulated by changing the X substituent. Both [NBu$_4$]$^+$ and Phenazinium salts, show similar magnetic behaviour showing an hysteretic behaviour with a coercive field of 5 mT.

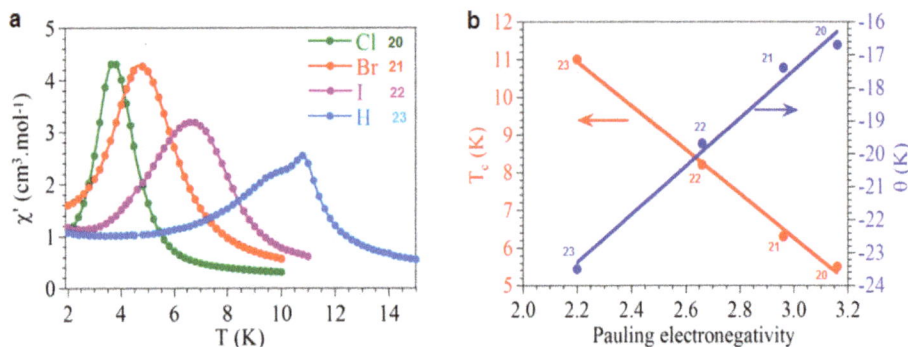

Figure 13. Magnetic properties of the [NBu$_4$][MnCr(X$_2$An)$_3$] family, X = Cl (**20**), Br (**21**), I (**22**) and H (**23**): (**a**) Thermal variation of the in phase (χm') AC susceptibility at 1 Hz.; (**b**) Linear dependence of the ordering temperature (Tc, left scale, red) and the Weiss temperature (θ, right scale, blue) with the electronegativity of the X group. Solid lines are the corresponding linear fits. Adapted with permission from Reference [125]. Copyright 2013 American Chemical Society.

17 is, therefore, the first structurally (and magnetically) characterized porous chiral layered magnet based on anilato-bridged bimetallic layers. This chirality is also expected to be of interest for studying the magnetochiral effect as well as the multiferroic properties, as has already been done in the oxalato family [94,107,121,126].

Moreover, the bigger size of the anilato compared to the oxalato ligand leads to hexagonal cavities that are twice larger than those of the oxalato-based layers. Therefore a larger library of cations can be used to prepare multifunctional molecular materials combining the magnetic ordering of the anionic layers with any additional property of the cationic one (the chirality of the phenazinium cation is only the first example). In the following section selected examples of cationic complexes spanning from chiral and/or achiral tetrathiafulvalene-based conducting networks to spin-crossover compounds to will be discussed.

3. Anilato-Based Multifunctional Molecular Materials

3.1. Introduction

π-d molecular materials, i.e., systems where delocalized π-electrons of the organic donor are combined with localized d-electrons of magnetic counterions, have attracted major interest in molecular science since they can exhibit coexistence of two distinct physical properties, furnished by the two networks, or novel and improved properties due to the interactions established between them[127–130]. The development of these π-d systems as multifunctional materials represents one of the main targets in current materials science for their potential applications in molecular electronics [78,127–130]. Important milestones in the field of magnetic molecular conductors have been achieved using as molecular building blocks the bis(ethylenedithio)tetrathiafulvalene (BEDT-TTF) organic donor[123,131–133] or its selenium derivatives, and charge-compensating anions ranging from simple mononuclear complexes $[MX_4]^{n-}$ (M = Fe^{III}, Cu^{II}; X = Cl, Br)[134–136] and $[M(ox)_3]^{3-}$ (ox = oxalate = $C_2O_4^{2-}$) with tetrahedral and octahedral geometries, to layered structures such as the bimetallic oxalate-based layers of the type $[M^{II}M^{III}(ox)_3]^-$ (M^{II} = Mn, Co, Ni, Fe, Cu; M^{III} = Fe, Cr) [123,124,131–133,137–140]. In these systems the shape of the anion and the arrangement of intermolecular contacts, especially H-bonding, between the anionic and cationic layers influence the packing motif of the BEDT-TTF radical cations, and therefore the physical properties of the obtained charge-transfer salt [141]. Typically, the structure of these materials is formed by segregated stacks of the organic donors and the inorganic counterions which add the second functionality to the conducting material. The first paramagnetic superconductor [BEDT-TTF]$_4$ [$H_3OFe^{III}(ox)_3$]·C_6H_5CN [123] and the first ferromagnetic conductor, [BEDT-TTF]$_3$[$Mn^{II}Cr^{III}(ox)_3$] [49] were successfully obtained by combining, via electrocrystallization, the mononuclear $[Fe(ox)_3]^{3-}$ and the $[Mn^{II}Cr^{III}(ox)_3]^-$ (2D honeycomb with oxalate bridges) anions with the BEDT-TTF organic donor, as magnetic and conducting carriers, respectively. Furthermore, by combining the bis(ethylenedithio)tetraselenafulvalene (BETS) molecule with the zero-dimensional $FeCl_4^-$ anion, a field-induced superconductivity with π-d interaction was observed which may be mediated through S···Cl interactions between the BETS molecule and the anion [134]. Clues for designing the molecular packing in the organic network, carrier of conductivity, were provided by the use of the paramagnetic chiral anion $[Fe(croc)_3]^{3-}$ (croc = croconate = $C_5O_5^{2-}$) as magnetic component of two systems: α-[BEDT-TTF]$_5$ [$Fe(croc)_3$]·$5H_2O$ [142], which behaves as a semiconductor with a high room-temperature conductivity (ca. 6 S cm^{-1}) and β-[BEDT-TTF]$_5$[$Fe(croc)_3$]·C_6H_5CN [143], which shows a high room-temperature conductivity (ca. 10 S cm^{-1}) and a metallic behavior down to ca. 140 K. The BEDT-TTF molecules in the α-phase are arranged in a herring-bone packing motif which is induced by the chirality of the anions. Therefore, the packing of the organic network and the corresponding conducting properties can be influenced by playing with the size, shape, symmetry and charge of the inorganic counterions. The introduction of chirality in these materials represents one of the most recent advances [144] in material science and one of the milestones is represented by the first observation of the electrical magneto-chiral anisotropy (eMChA) effect in a bulk crystalline chiral conductor [145], as a synergy between chirality and conductivity [146–148]. However, the combination of chirality with electroactivity in chiral TTF-based materials afforded several other recent important results, particularly the modulation of the structural disorder in the solid state , [130–138] and hence a difference in conductivity between the enantiopure and racemic forms [149–151] and the induction of different packing patterns and crystalline space groups in mixed valence salts of dimethylethylenedithio-TTF (DM-EDT-TTF), showing semiconducting (enantiopure forms) or metallic (racemic form) behaviour [152]. Although the first example of an enantiopure TTF derivative, namely the tetramethyl-bis(ethylenedithio)-tetrathiafulvalene (TM-BEDT-TTF), was described almost 30 years ago as the (*S*,*S*,*S*,*S*) enantiomer [153,154], the number of TM-BEDT-TTF based conducting radical cation salts is still rather limited. They range from semiconducting salts [155], as complete series of both enantiomers and racemic forms, to the [TM-BEDT-TTF]$_x$[MnCr(ox)$_3$] ferromagnetic metal [156], described only as the (*S*,*S*,*S*,*S*) enantiomer. The use of magnetic counterions, particularly interesting since they provide an additional property to the system, was largely explored in the case of the above-mentioned metal-oxalates [M(ox)$_3$]$^{3-}$ (M = Fe^{3+}, Cr^{3+}, Ga^{3+}, ox = oxalate) [124,140], present as Δ and Λ enantiomers in radical cation salts based on the BEDT-TTF donor. Other paramagnetic chiral anions, such as [Fe(croc)$_3$] [142,143] or [Cr(2,2'-bipy)(ox)$_2$]$^-$ (bipy = bipyridine) [157], have been scarcely used up to now. However, in all these magnetic conductors the tris-chelated anions were present as racemic mixtures, except for the Δ enantiomer of [Cr(ox)$_3$]$^{3-}$ [158]. As far as the π-*d* systems are concerned, the number of conducting systems based on enantiopure TTF precursors is even scarcer [156,159]. One example concerns the above-mentioned ferromagnetic metal [TM-BEDT-TTF]$_x$[MnCr(ox)$_3$] [156], while a more recent one is represented by the semiconducting paramagnetic salts [DM-BEDT-TTF]$_4$[ReCl$_6$] [159]. In this context, anilate-based metal complexes [116,119] are very interesting molecular building blocks to be used as paramagnetic counterions, also because they offer the opportunity of exchange coupling at great distance through the anilate bridge (See Section 2.3), being therefore extremely versatile in the construction of the above mentioned achiral and chiral conducting/magnetic molecule-based materials.

Furthermore multifunctional materials with two functional networks responding to an external stimulus are also very challenging in view of their potential applications as chemical switches, memory or molecular sensors [160]. For the preparation of such responsive magnetic materials two-network compounds a magnetic lattice and spin-crossover complexes as the switchable molecular component are promising candidates. These molecular complexes, which represent one of the best examples of molecular bistability, change their spin state from low-spin (LS) to high-spin (HS) configurations and thus their molecular size, under an external stimulus such as temperature, light irradiation, or pressure [161,162]. Two-dimensional (2D) and three-dimensional (3D) bimetallic oxalate-based magnets with Fe(II) and Fe(III) spin-crossover cationic complexes have been obtained, where changes in size of the inserted cations influence the magnetic properties of the resulting materials [89,90,92,93]. By combining [Fe^{III}(sal$_2$-trien)]$^+$ (sal$_2$-trien = N,N'-disalicylidene triethylenetetramine) cations with the 2D Mn^{II}Cr^{III} oxalate-based network, a photoinduced spin-crossover transition of the inserted complex (LIESST effect), has been observed unexpectedly; this property infact is very unusual for Fe(III) complexes. The bigger size of anilates has the main advantage to enable the introduction of a larger library of cations, while the magnetic network, the family of layered ferrimagnets described in Section 2.3, showing higher Tc's, can be porous and/or chiral depending on the X substituent on the anilato moiety.

Interestingly, Miller et al. [29] reported on the formation and characterization of a series of heteroleptic isostructural dicobalt, diiron, and dinickel complexes with the TPyA = tris(2-pyridylmethyl)amine ligand and bridged by the 2,5-di-*tert*-butyl-3,6-dihydroxy-1,4-benzoquinonate (DBQ^{2-} or DBQ$^{.3-}$) anilato derivative, where the more electron donating *tert*-butyl group, has been targeted to explore its influence on the magnetic properties, e.g., spin coupling and spin crossover. In particular, Co-based dinuclear complex with DBQ$^{.3-}$ has shown valence tautomeric spin crossover behavior above room temperature, while Fe-based complexes exhibit spin crossover behavior. Spin crossover behavior or ferromagnetic coupling have been also observed in the heteroleptic dinuclear Fe(II) complexes {[(TPyA)Fe^{II}(DBQ^{2-})Fe^{II}(TPyA)](BF$_4$)$_2$ and {[(TPyA)Fe^{II}(Cl$_2$An) Fe^{II}(TPyA)](BF$_4$)$_2$}, respectively [163], where the former does not exhibit thermal hysteresis, although shows ≈ room temperature SCO behavior. Thus, greater interdinuclear cation interactions are needed to induce thermal hysteresis, maybe through the introduction of interdinuclear H-bonding. Therefore 2,3,5,6-tetrahydroxy-1,4-benzoquinone (H_2THBQ) has been used as bridging ligand and the [(TPyA)Fe^{II} ($THBQ^{2-}$)FeII(TPyA)](BF_4)$_2$ obtained complex shows coexistence of spin crossover with thermal hysteresis in addition to an intradimer ferromagnetic interaction [29].

3.2. Achiral Magnetic Molecular Conductors

The first family of conducting radical cation salts based on the magnetic tris(chloranilato)ferrate(III) complex have been recently obtained by reacting the BEDT-TTF donor (D) with the tris(chloranilato)ferrate(III) complex (A), via electrocrystallization technique, by slightly changing the stoichiometric donor: anion ratio and the solvents. Three different hybrid systems formulated as [BEDT-TTF]$_3$[Fe(Cl_2An)$_3$]·$3CH_2Cl_2$·H_2O (**24**), δ-[BEDT-TTF]$_5$[Fe(Cl_2An)$_3$]·$4H_2O$ (**25**) and α'''-[BEDT-TTF]$_{18}$[Fe(Cl_2An)$_3$]$_3$·$3CH_2Cl_2$·$6H_2O$ (**26**), were obtained [164] as reported in Scheme 8.

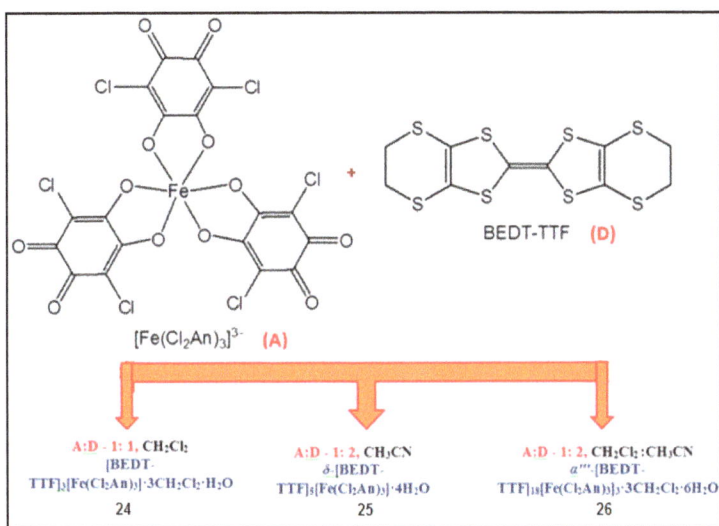

Scheme 8. Molecular structures for the complex anion [Fe(Cl_2An)$_3$]$^{3-}$ and the bis(ethylenedithio)tetrathiafulvalene (BEDT-TTF) organic donor, and experimental conditions used for obtaining **24–26** compounds.

The common structural feature for the three phases is the presence of dimerized oxidized BEDT-TTF units in the inorganic layer, very likely due to intermolecular S···Cl contacts and also electrostatic interactions. While in **24**, of 3:1 stoichiometry, the three BEDT-TTF molecules are fully oxidized in radical cations, in **25** and **26**, of 5:1 and 6:1 stoichiometry, respectively, only the donors located in the inorganic layers are fully oxidized, while those forming the organic slabs are in mixed valence state. **24** presents an unusual structure without the typical alternating organic and inorganic layers, whereas **25** and **26** show a segregated organic-inorganic crystal structure where layers formed by Λ and Δ enantiomers of the paramagnetic complex, together with dicationic BEDT-TTF dimers, alternate with layers where the donor molecules are arranged in the δ (**25**) and α''' (**26**) packing motifs.

The crystal packing of **25** and **26** plane showing the organic-inorganic layer segregation are reported in Figures 14 and 15 respectively.

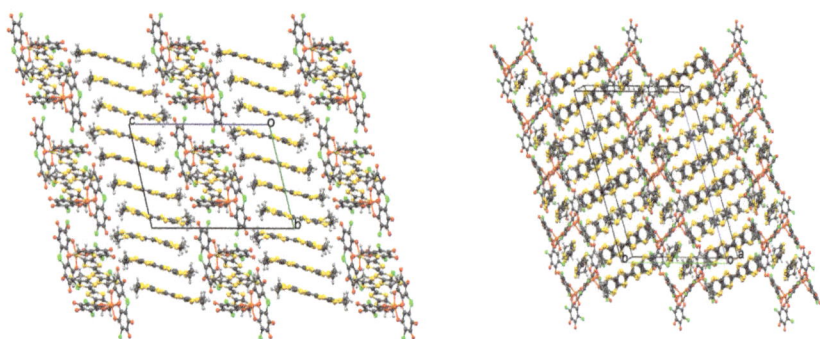

Figure 14. Crystal packing of **25** (**left**) and **26** (**right**) along the *bc* plane showing the organic-inorganic layer segregation. Crystallization water and CH$_2$Cl$_2$ molecules were omitted for clarity. Reprinted with permission from Reference [164]. Copyright 2014 American Chemical Society.

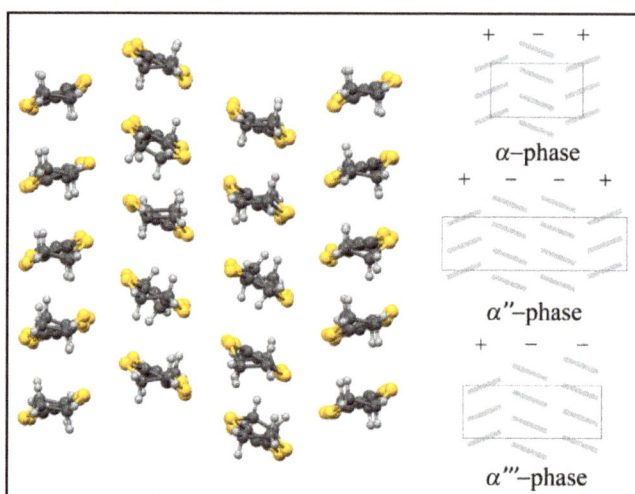

Figure 15. View of α'''-packing of **26** along the *ac* plane (**left**); schematic representation of the BEDT-TTF molecules arranged in the α, α'' and α''' packing motifs (**right**). Adapted with permission from Reference [164]. Copyright 2014 American Chemical Society.

The hybrid inorganic layers of **24**, **25** and **26** shows alternated anionic complexes of opposite chirality that surround dimers of mono-oxidized BEDT-TTF radical cations. This packing motif, shown in Figure 16 for **24**, points out the templating influence of the Cl···S interactions intermolecular interactions between the chloranilate ligand and the dimerized BEDT-TTF molecules.

The peculiar α''' structural packing motif observed in **26** is quite unusual [138,165]. In fact, the BEDT-TTF molecules stack in columns with an arrangement reminiscent of the α structural packing [165], but with a 2:1:2:1 alternation of the relative disposition of the molecules, instead of the classical 1:1:1:1 sequence (Figure 14). The α'''-phase can be regarded as 1:2 hybrid of θ- and β''- phases.

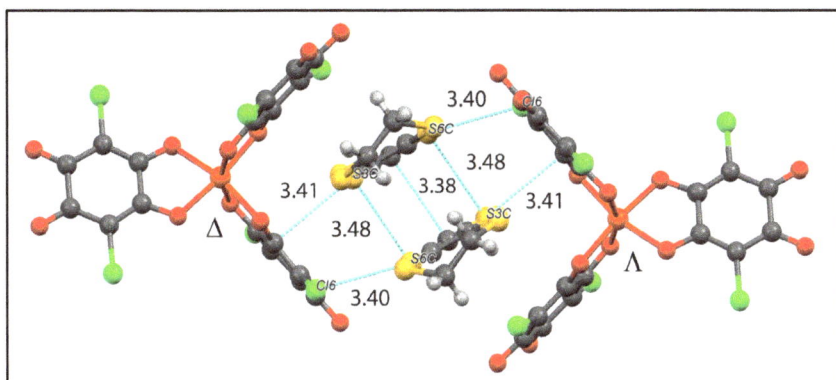

Figure 16. C-C dimer surrounded by two metal complexes of opposite chirality in **24**. Symmetry related S···S contacts and intermolecular interactions lower than the sum of the van der Waals radii between the BEDT-TTF molecules and the chloranilate ligands are highlighted. (Å): S3C···S6C 3.48, S4C···S5C 3.57, Cl6···S6C 3.40, C13C···S6C 3.41. Reprinted with permission from Reference [164]. Copyright 2014 American Chemical Society.

Single crystal conductivity measurements show semiconducting behavior for the three materials. **24** behaves as a semiconductor with a much lower conductivity due to the not-layered structure and strong dimerization between the fully oxidized donors, whereas **25** and **26** show semiconducting behaviors with high room-temperature conductivities of ca. 2 S cm^{-1} and 8 S cm^{-1}, respectively and low activation energies of 60–65 meV. Magnetic susceptibility measurements for **24** clearly indicate the presence of isolated high spin S = 5/2 Fe(III) ions, with a contribution at high temperatures from BEDT-TTF radical cations. These latter are evidenced also by EPR variable temperature measurements on single crystals of **26** (See Figure 17).

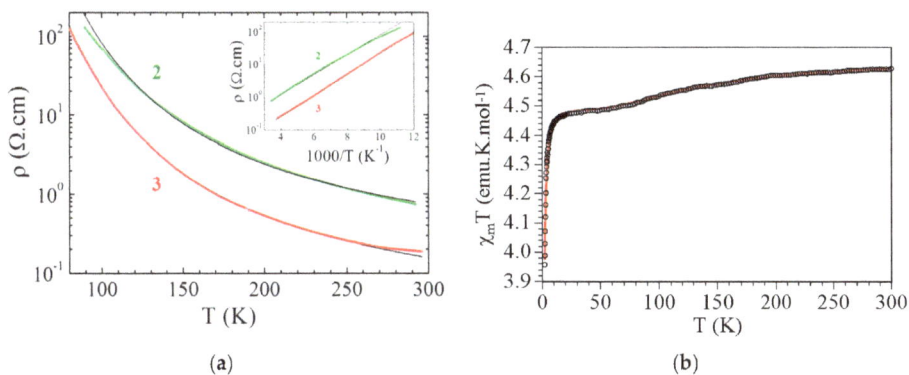

(a)

(b)

Figure 17. (**a**) Temperature dependence of the electrical resistivity ϱ for **25** and **26** single crystals. The inset shows the Arrhenius plot. The black lines are the fit to the data with the law $\varrho = \varrho_0 \exp(Ea/T)$ giving the activation energy Ea; (**b**) Thermal variation of the magnetic properties ($\chi_m T$) for **24**. Solid line is the best fit to the model (see text). Reprinted with permission from Reference [164]. Copyright 2014 American Chemical Society.

The correlation between crystal structure and conductivity behavior has been studied by means of tight-binding band structure calculations which support the observed conducting properties; and structure calculations for **25** and **26** are in agreement with an activated conductivity with low band gaps. A detailed analysis of the density of states and HOMO⋯HOMO interactions in **25** explains the origin of the gap as a consequence of a dimerization in one of the donor chains, whereas the challenging calculation of **26**, due to the presence of eighteen crystallographically independent BEDT-TTF molecules, represents a milestone in the band structure calculations of such relatively rare and complex crystal structures [164]. Recently Gomez et al. [166] has obtained a very unusual BEDT-TTF phase, called θ_{21}, by reacting the BEDT-TTF donor with the novel $(PPh_3Et)_3[Fe(C_6O_4Cl_2)_3]$ tris(chloranilato)ferrate(III) complex, via electrocrystallization technique, in the $CH_2Cl_2/MeOH$ solvent mixture. The obtained compound $[(BEDT-TTF)_6[Fe(C_6O_4Cl_2)_3]\cdot(H_2O)_{1.5}\cdot(CH_2Cl_2)_{0.5}$ (**27**) shows the same layered structure and physical properties as **26**. In Figure 18, a view of the θ_{21} BEDT-TTF packing motif is reported.

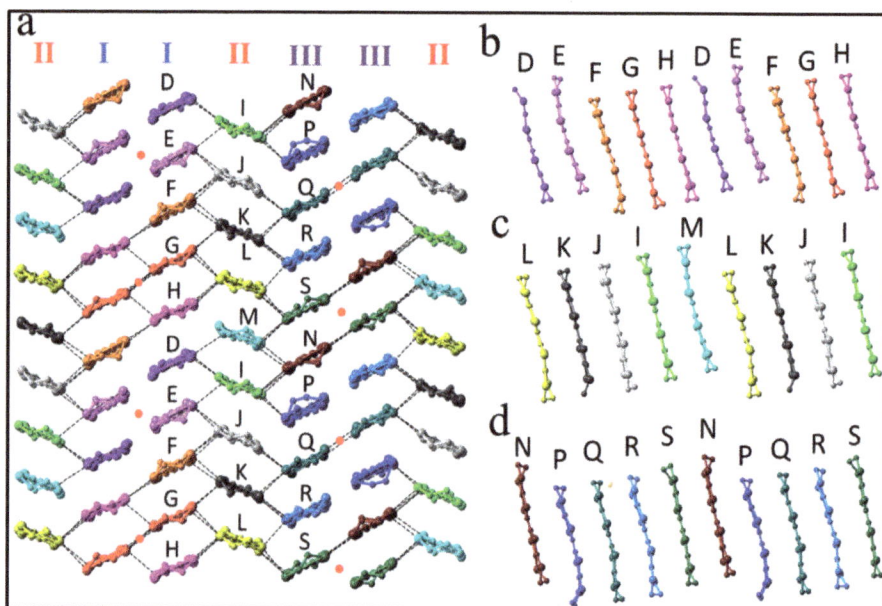

Figure 18. View of the θ_{21} BEDT-TTF packing motif in **27**. Copyright (2014) Wiley Used with permission from [166].

3.3. Chiral Magnetic Molecular Conductors

The first family of chiral magnetic molecular conductors [167] formulated as β-[(S,S,S,S)-TM-BEDT-TTF]_3PPh_4[K^IFe^{III}(Cl_2An)_3]\cdot3H_2O$ (**28**), β-[(R,R,R,R)-TM-BEDT-TTF]_3PPh_4 [K^IFe^{III}(Cl_2An)_3]\cdot3H_2O$ (**29**) and β-[(*rac*)-TM-BEDT-TTF]_3 PPh_4[K^IFe^{III}(Cl_2An)_3]\cdot3H_2O$ (**30**) have been afforded by electrocrystallization of the tetramethyl-bis(ethylenedithio)-tetrathiafulvalene (TM-BEDT-TTF) chiral donor in its forms: enantiopure (S,S,S,S)- and (R,R,R,R)- (TM-BEDT-TTF) donors, as well as the racemic mixture, in the presence of potassium cations and the tris(chloranilato)ferrate(III) $[Fe(Cl_2An)_3]^{3-}$ paramagnetic anion (Scheme 9).

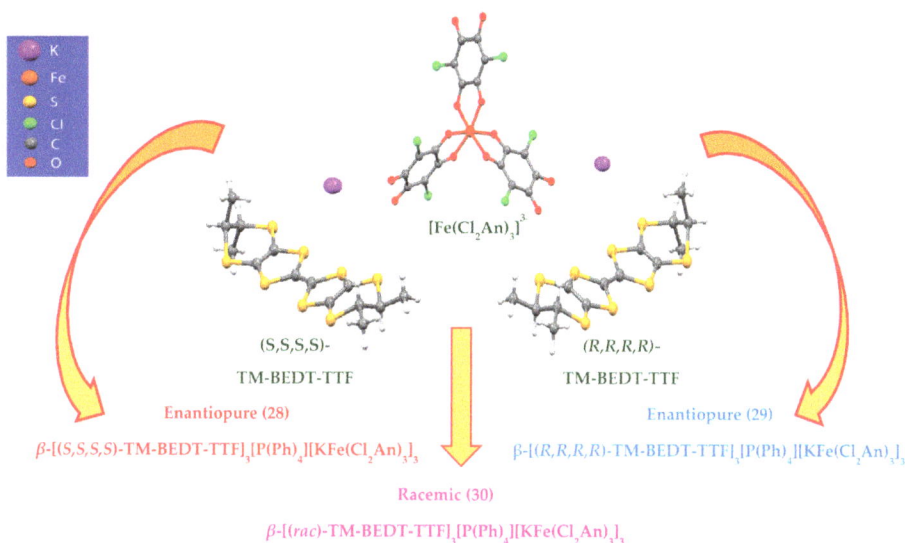

Scheme 9. Molecular structures for the $[Fe(Cl_2An)_3]^{3-}$ complex anion and the enantiopure (*S*,*S*,*S*,*S*)- and (*R*,*R*,*R*,*R*)- TM-BEDT-TTF donors, as well as the racemic mixture, in the presence of potassium cations. Adapted with permission from Reference [167]. Copyright 2015 American Chemical Society.

Compounds **28–30** are isostructural and crystallize in the triclinic space group (*P*1 for **28** and **29**, *P*-1 for **30**) showing the usual segregated organic—inorganic crystal structure, where anionic chloranilate-bridged heterobimetallic honeycomb layers obtained by self-assembling of the Λ and Δ enantiomers of the paramagnetic complex with potassium cations, alternate with organic layers where the chiral donors are arranged in the β packing motif (Figure 19).

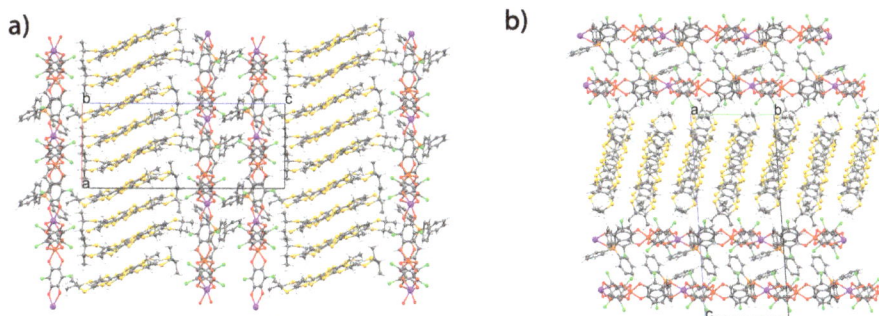

Figure 19. Crystal packing of **29** (**a**) in the *ac* plane; (**b**) in the *bc* plane, showing the organic-inorganic layer segregation. Crystallization water molecules were omitted for clarity. Reprinted with permission from Reference [167]. Copyright 2015 American Chemical Society.

The use of the "complex as ligand approach" during the electrocrystallization experiments has been successful for obtaining these systems where the self-assembling of the tris(chloranilato)ferrate(III) anion with potassium cations afforded anionic layers, that further template the structure in segregated organic and inorganic layers. The common structural features of the three systems are: (i) the presence of inorganic layers associated in double-layers, as a result of two major intermolecular interactions,

Cl···Cl and π-π stacking, between the chloranilate ligands and the $[(Ph)_4P]^+$ charge-compensating cations (Figure 19b) and (ii) the simultaneous presence of two different conformations of the TM-BEDT-TTF donor in the crystal packing, very likely due to the diverse interactions of the terminal methyl groups with the oxygen atoms of the chloranilate ligands. Therefore the molecular packing of **28**–**30** is strongly influenced by the topology of the inorganic layers. **28**–**30** behave as molecular semiconductors with room temperature conductivity values of ca. 3×10^{-4} S cm^1 and an activation energy E_a of ca. 1300–1400 K corresponding to ca. 110–120 meV, as expected from the presence of one neutral TM-BEDT-TTF donor in the crystal packing and the presence of a slight dimerization between the partially oxidized molecules. No significant difference between the enantiopure and the racemic systems is observed. Magnetic susceptibility measurements for **30** indicate the presence of quasi-isolated high spin $S = 5/2$ Fe(III) ions, since the M···M distances between paramagnetic metal centers (ca. 13.6 Å through space and ca. 16.2 Å through the bridging ligands) are too large to allow significant magnetic interactions, with a negligible contribution from the TM-BEDT-TTF radical cations (Figure 20 a,b).

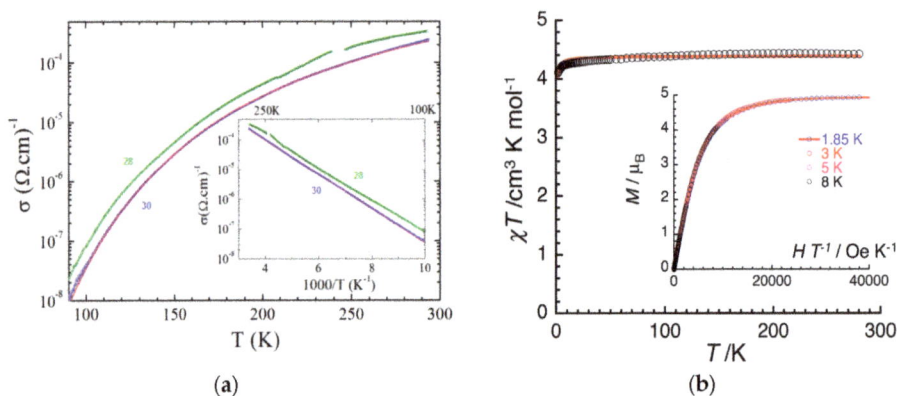

Figure 20. (a) Thermal variation of the electrical conductivity for **28** and **30**. The inset shows the Arrhenius plot. The red line is the Arrhenius fit to the data for **30**; (b) Thermal variation of magnetic properties (χT product) at 1000 Oe (where χ is the molar magnetic susceptibility equal to the ratio between the magnetization and the applied magnetic field, M/H, per mole of Fe(III) complex) between 1.85 and 280 K for a polycrystalline sample of **30**. The solid line is the Curie-Weiss best fit. Inset: M vs. H/T plot for **30** between 1.85 and 8 K at magnetic fields between 0 and 7 T. The solid line is the best fit obtained using $S = 5/2$ Brillouin function. Adapted with permission from Reference [167]. Copyright 2015 American Chemical Society.

The structural analyses and the band structure calculations are in agreement with the intrinsic semiconducting behaviour shown by the three materials (Figure 21).

This first family of isostructural chiral conducting radical cation salts based on magnetic chloranilate-bridged heterobimetallic honeycomb layers demonstrates (i) the versatility of these anions for the preparation of π-d multifunctional molecular materials where properties such as charge transport, magnetism and chirality coexist in the same crystal lattice; (ii) they are fundamental importance for a rational design of chiral conductors showing the eMChA effect as a synergy between chirality and conductivity.

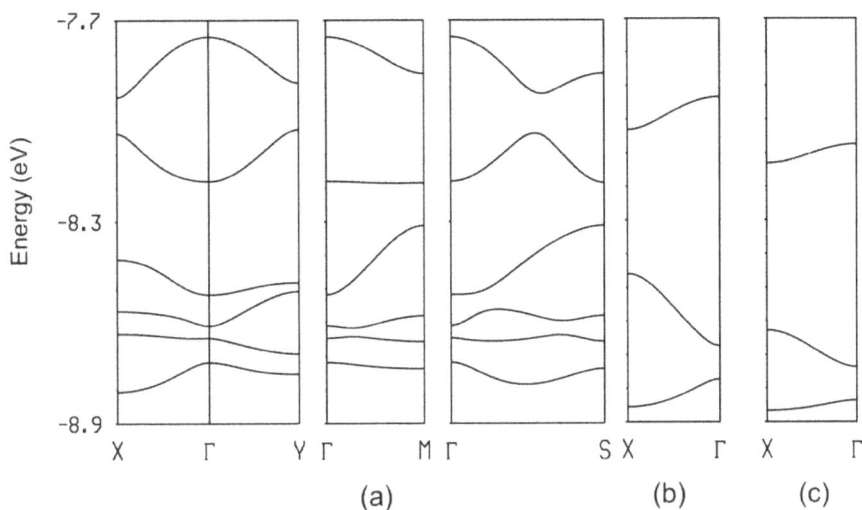

Figure 21. Electronic structure for **28**. Calculated band structure of: (**a**) the [(TM-BEDT-TTF)$_6$]$^{2+}$ donor layers; (**b**) the isolated −**B**−**E**−**C**− chains and (**c**) the isolated −**A**−**D**−**F**− chains, where Γ = (0, 0), X = ($a^*/2$, 0), Y = (0, $b^*/2$), M = ($a^*/2$, $b^*/2$) and S = ($-a^*/2$, $b^*/2$). Reprinted with permission from Reference [167]. Copyright 2015 American Chemical Society.

3.4. Spin-Crossover Complexes

The family of bimetallic Mn^{II}Cr^{III} anilate (X$_2$An; X = Cl, Br)-based ferrimagnets with inserted the following spin-crossover cationic complexes: [Fe^{III}(sal$_2$-trien)]$^+$, (X = Cl) (**31**) and its derivatives, [Fe^{III}(4-OH-sal$_2$-trien)]$^+$, (X = Cl) (**32**), [Fe^{III}(sal$_2$-epe)]$^+$, (X = Br) (**33**), [Fe^{III}(5-Cl-sal$_2$-trien)]$^+$, (X = Br) (**34**), and [Fe^{II}(tren(imid)$_3$)]$^{2+}$, (X = Cl) (**35**), (Chart 4a,b) have been prepared and fully characterized [168]. In Chart 4a, the ligands of the Fe(III) and Fe(II) spin crossover complexes are shown. The structures of **32**–**34** consist of bimetallic anionic layers with a 2D bimetallic network of formula [Mn^{II}Cr^{III}(X$_2$An)$_3$] (X = Cl, Br) with inserted Fe(III) cationic complexes and solvent molecules. The bimetallic anilate layer show the well-known honeycomb structure, which is similar to that found for other extended oxalate or anilate-based networks (Figure 22). A consequence of the replacement of oxalate by the larger anilate ligands is the presence of pores in the structures, which are filled with solvent molecules.

In contrast to the 2D compounds obtained with [Fe^{III}(sal$_2$-trien)]$^+$ and derivatives, the structure of **35** is formed by anionic 1D [Mn^{II}Cl$_2$Cr^{III}(Cl2An)$_3$]$^{3-}$ chains surrounded by [Fe^{II}(tren(imid)$_3$)]$^{2+}$, Cl$^-$ and solvent molecules. These chains are formed by [Cr^{III}(Cl$_2$An)$_3$]$^{3-}$ complexes coordinated to two Mn(II) ions through two bis-bidentate chloranilate bridges, whereas the third choranilate is a terminal one. The octahedral coordination of Mn(II) ions is completed with two chloride ions in cis. This type of structure has been found for other oxalate-based [169] and homometallic anilate-based compounds [7,170,171], but it is the first time that it is obtained for heterometallic anilate-based networks (Figure 23).

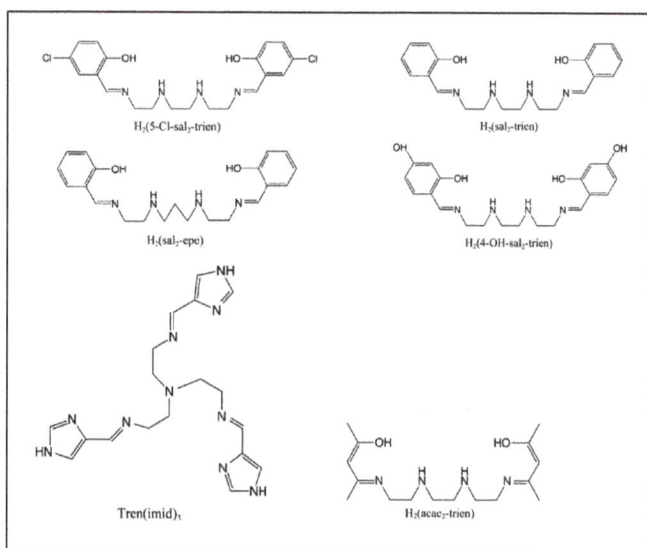

(a)

Spin CrossOver Cationic Complexes	
[Fe^{III}(sal$_2$-trien)]$^+$	MnCr(X-Cl) **31**
[Fe^{III}(4-OH-sal$_2$-trien)]$^+$	MnCr(X-Cl) **32**
[Fe^{III}(sal$_2$-epe)]$^+$	MnCr(X-Br) **33**
[Fe^{III}(5-Cl-sal$_2$-trien)]$^+$	MnCr(X-Br) **34**
[Fe^{II}(tren(imid)$_3$)]$^{2+}$	MnCr(X-Cl) **35**
[Fe^{III}(acac$_2$-trien)]$^+$	MnCr(X-Cl) **36**
	MnCr(X-Br) **37**
[Ga^{III}(acac$_2$-trien)]$^+$	MnCr(X-Br) **38**

(b)

Chart 4. Ligands of Fe(III) and Fe(II) complexes (a,b).

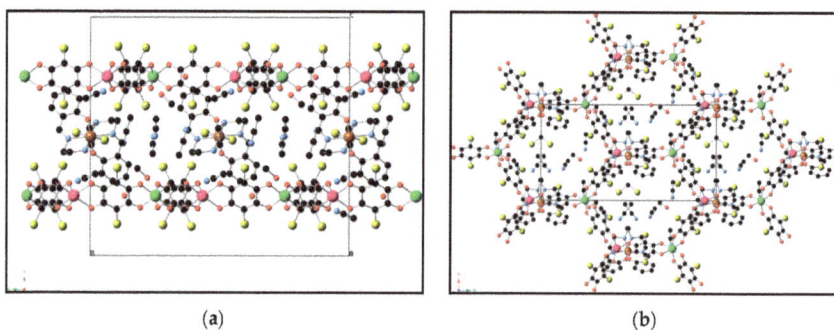

(a)   (b)

Figure 22. Projection of **31** in: (a) the *bc* plane; (b) the *ab* plane, showing one anionic layer and one cationic layer. (Fe (brown), Cr (green), Mn (pink), C (black), N (blue), O (red) Cl (yellow)). Hydrogen atoms have been omitted for clarity. Adapted with permission from Reference [168]. Copyright 2014 American Chemical Society.

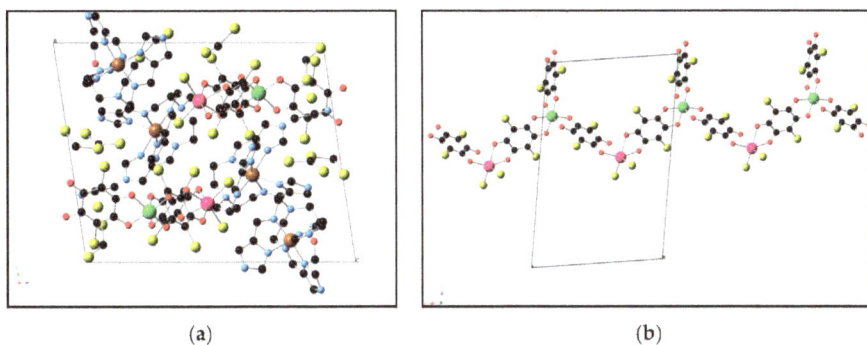

(a) (b)

Figure 23. Projection of **35** in the bc plane (**a**); $[Mn^{II}Cl_2Cr^{III}(Cl_2An)_3]^{3-}$ chains in the structure of **35** (**b**). (Fe (brown), Cr (green), Mn (pink), C (black), N (blue), O (red) Cl (yellow)). Hydrogen atoms have been omitted for clarity. Adapted with permission from Reference [168]. Copyright 2014 American Chemical Society.

Magnetic studies show that **31–34** undergo a long-range ferrimagnetic ordering at ca. 10 K (ca. 10 K for **31**, 10.4 K for **32**, 10.2 K for **33**, and 9.8 K for **34**) with most of the Fe(III) of the inserted cations in the HS state (**31–33**), or LS state (**34**). These values are much higher than those found for the $[NBu_4]^+$ and $[(H_3O)(phz)_3]^+$ salts containing similar $[Mn^{II}Cr^{III}(X_2An)_3]^-$ (X = Cl, Br) layers (5.5 and 6.3 K, respectively) (see Section 2.3), in contrast to oxalate-based 2D compounds, where Tc remains constant for a given 2D $[M^{II}M^{III}(ox)_3]^-$ lattice, independently of the inserted cation. Therefore, the magnetic coupling and, accordingly, the ordering temperatures of these heterometallic 2D anilate-based networks are much more sensitive to the changes of the inserted cations than the corresponding oxalate ones. This effect is maybe due to the presence of π-π and NH···O and NH···Cl/Br intermolecular interactions between the anilate ligands and Fe(III) complexes which may increase the Mn(II)–Cr(III) coupling constant through the anilate ligand an thus the Tc. Interestingly, this modulation of Tc with the inserted cation (or even with solvent molecules), besides the already observed modulation with the X substituents on the benzoquinone moiety, represents an additional advantage of the anilate-based networks compared with the oxalate ones.

Differently from **28–34**, **35** do not show π-π stacking interactions with the anilate ligands and therefore half of the inserted Fe(II) cations undergo a complete and gradual spin crossover from 280 to 90 K which coexists with a ferrimagnetic coupling within the chains that gives rise to a magnetic ordering below 2.6 K. The Temperature dependence of the product of the molar magnetic susceptibility times the temperature of **31–35** is reported in Figure 24.

When using the $[M^{III}(acac_2-trien)]^+$ (M^{III}—Fe or Ga complex, which has a smaller size than the $[Fe^{III}(sal_2-trien)]^+$ spin-crossover complex, three novel magnetic compounds $[Fe^{III}(acac_2-trien)][Mn^{II}Cr^{III}(Cl_2An)_3]_3(CH_3CN)_2$ (**36**), $[Fe^{III}(acac_2-trien)][Mn^{II}Cr^{III}(Br_2An)_3]_3(CH_3CN)_2$ (**37**), $[Ga^{III}(acac_2-trien)][Mn^{II}Cr^{III}(Br_2An)_3]_3(CH_3CN)_2$ (**38**), have been prepared and characterized by Coronado et al. [172]. The 2D anilate-based networks show the common honeycomb anionic packing pattern but a novel type of structure where the cations are placed into the hexagonal channels of the 2D network has been afforded due to the smaller size of the $[Fe^{III}(acac_2-trien)]^+$ or $[Ga^{III}(acac_2-trien)]^+$ complex with respect to the templating cations used in previous compounds of this type, where they are placed in between the anionic layers. An important decrease of the interlayer separation between the anilate-based layers (Figure 25a,b) is observed.

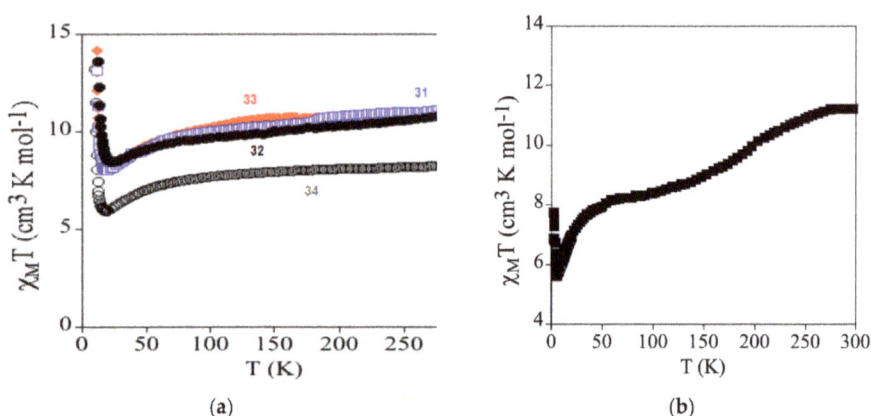

Figure 24. Thermal variation of magnetic properties ($\chi_m T$ product vs. T) at an applied field of 0.1 mT of: (**a**) **31** (empty blue, squares), **32** (full circles), **33** (full red diamonds), and **34** (empty circles); (**b**) **35**. Adapted with permission from Reference [168]. Copyright 2014 American Chemical Society.

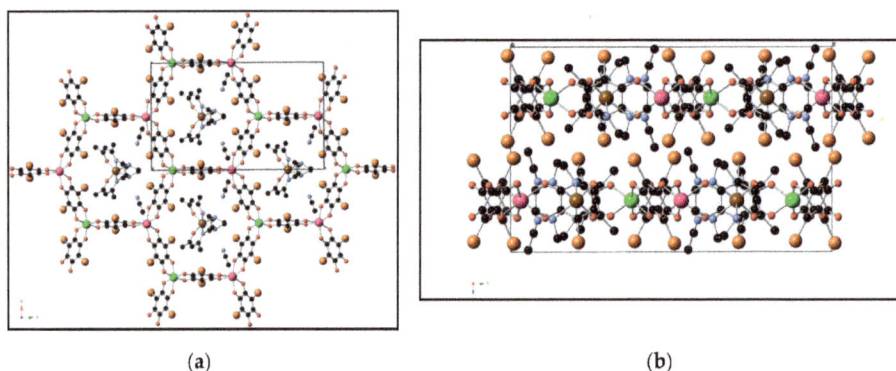

Figure 25. Projection of **36** in the *ab* plane (**a**) and in the *bc* plane (**b**). ((Fe (brown), Cr (green), Mn (pink) C (black), N (blue), O (red), Br (orange)). Hydrogen atoms have been omitted for clarity. Reprinted from Reference [172] with permission from The Royal Society of Chemistry.

The anilate-based layers with inserted [Fe^{III}(acac$_2$-trien)]$^+$ complexes may be viewed as neutral layers that interact with each other via van der Waals interactions. Thus, in **37**, the shortest contacts between neighbouring layers involve Br atoms from Br$_2$An ligands and CH$_2$ and CH$_3$ groups from [Fe^{III}(acac$_2$-trien)]$^+$ complexes of neighbouring layers. This type of structure, formed by neutral layers, has never been observed previously in oxalate or anilate-based 2D networks. The close contact of the cationic complexes with the magnetic network results in an increase of the T$_c$ (ca. 11 K) with respect to that of previous anilate-based compounds (ca. 10 K), even though to not favour the spin crossover of the inserted complexes which remain the HS state. The weak natures of the intermolecular interactions between the magnetic neutral layers play a crucial role for the exfoliation of the layers. In fact this new magnetic network is very peculiar since it can be easily exfoliated by using the so-called Scotch tape method which is a micromechanical method, capable to produce in a very efficient way, highly crystalline thin microsheets of a layered material [173–175]. To the best of our knowledge this method has never been applied to such layered materials. Flakes of **37**, with different sizes and thicknesses randomly distributed over the substrate have been obtained. AFM topography images revealed that the they show maximum lateral dimensions of ca. 5 mm, with well-defined edges and angles (Figure 26). The heights of the largest flakes of **37** are around 10–20 nm, while smaller microsheets with heights of less than 2 nm were also found.

The presence of terraces with different heights indicatse that this magnetic network is layered. Interestingly the Scotch tape method has been successful used also to exfoliate the 2D anilate-based compound $[Fe^{III}(sal_2\text{-trien})][Mn^{II}Cr^{III}(Cl_2An)_3](CH_2Cl_2)_{0.5}(CH_3OH)(H_2O)_{0.5}(CH_3CN)_5$, (**31**), described above [168], which exhibits the typical alternated cation/anion layered structure. In this case rectangular flakes of larger lateral size than those isolated in **37** (up to 20 microns) have been obtained with well-defined terraces and a minimum thickness of ca. 2 nm, which may correspond to that of a single cation/anion hybrid layer (ca. 1.2 nm).

31 and **37** have been also successfully exfoliated by solution methods. Tyndall light scattering of the colloidal suspensions of both compounds has been observed, as shown in Figure 27 for **37**, and dynamic light scattering (DLS) measurements confirm the efficiency of the liquid exfoliation.

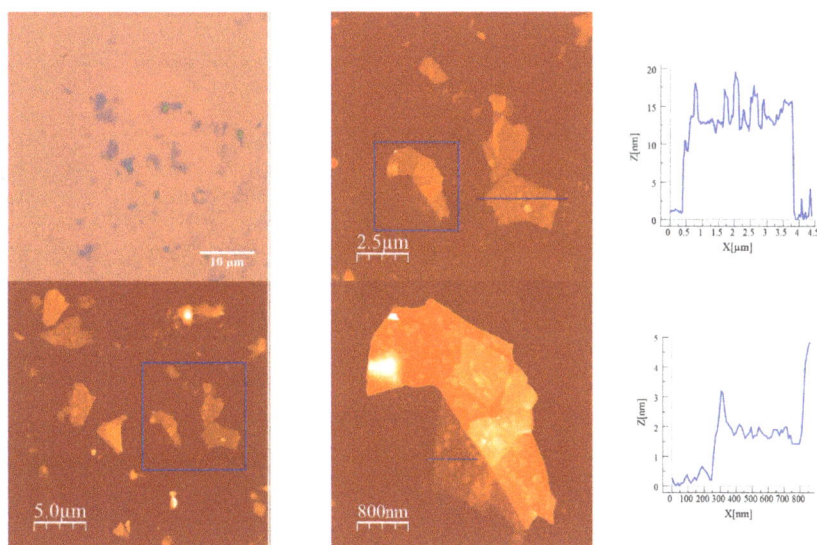

Figure 26. Images of flakes of **37**, obtained by mechanical exfoliation on a 285 nm SiO_2/Si substrate, by Optical microscopy (**left**), AFM (atomic force microscopy) (**middle**) and height profiles (**right**). Reprinted from Reference [172] with permission from The Royal Society of Chemistry.

Figure 27. Tyndall effect of crystals of **37** after suspension in acetone, ethanol or acetonitrile (1.0 mg in 1 mL) overnight and then ultrasonicating for 1 min. Reprinted from Reference [172] with permission from The Royal Society of Chemistry.

These results show that it is possible to exfoliate 2D coordination polymers formed by a 2D honeycomb anionic network and cations inserted within or between the layers. The thicknesses of the flakes obtained by micromechanical methods are clearly lower than those obtained by solution methods (ca. 5 nm), where the lateral size of the flakes is of the order of hundreds of nm (significantly smaller). The solution-based exfoliation procedure is less effective in the neutral coordination polymers which can be completely delaminated (with a thickness ca. 1–1.5 nm) [115,170,176–183]. The stronger interlayer interactions in these hybrid compounds compared with the weaker van der Waals interactions observed in neutral 2D coordination polymers could be responsible of the lower degree of exfoliation.

The hybrid nature of these layered materials, providing the opportunity to produce smart layers where the switching properties of the cationic complexes can tune the cooperative magnetism of the anionic network, represents the real challenge of these results.

3.5. Guests Intercalation of Hydrogen-Bond-Supported Layers

A successful strategy to control the molecular packing in molecule-based materials takes advantage of more flexible hydrogen bonds in combination with metal-ligand bonds (Scheme 10 to control the rigidity of a supramolecular framework.

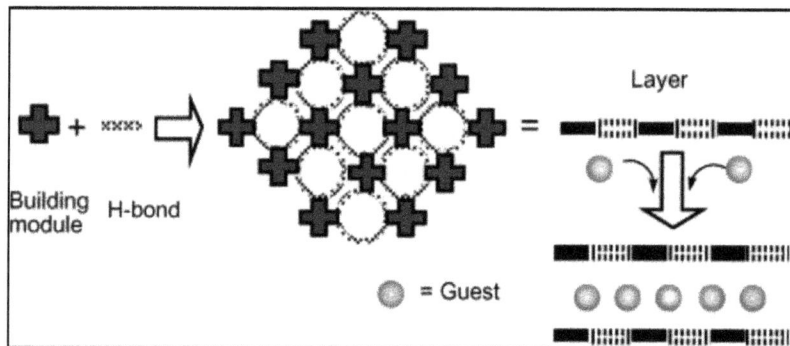

Scheme 10. Reprinted with permission from Reference [184]. Copyright 2003 American Chemical Society.

Novel intercalation compounds formed by 2D hydrogen-bond mediated Fe(III)-chloranilate layers and cationic guests, carrier of additional physical properties, $\{(H_{0.5}phz)_2[Fe(Cl_2An)_2(H_2O)2]\hat{a}2H_2O\}_n$ (**39**), $\{[Fe(Cp)_2][Fe(Cl_2An)_2(H_2O)_2]\}n$ (**40**), $\{[Fe(Cp^*)_2][Fe(Cl_2An)_2(H_2O)_2]\}_n$ (**41**), and $\{(TTF)_2[Fe(Cl_2An)_2(H_2O)_2]\}_n$ (**42**) (phz = phenazine, $[Fe(Cp)_2]$ = ferrocene, $[Fe(Cp^*)_2]$ = decamethyl ferrocene, TTF = tetrathiafulvalene) are reported by Kawata et al. [184]. The cationic guests are inserted between the $\{[Fe(Cl_2An)_2(H_2O)_2]\}^{m-}$ layers and are held together by electrostatic (**39–42**) and π-π stacking (**41, 42**) interactions. The $\{[Fe(Cl_2An)_2(H_2O)_2]\}^{m-}$ layers are very flexible and depending on the guest sizes and electronic states they can tune their charge distribution and interlayer distances. Especially **42** is a rare example of hydrogen-bonded layer of monomeric complexes, which can intercalate different charged guests, thus showing a unique electronic flexibility.

In **41** decamethylferrocene cations are stacked in tilted columns inserted in the channels created by the chlorine atoms of chloranilate dianions. In **42** TTF cations are stacked face to face with two types of S···S distances (type A; 3.579(3) Å, and type B; 3.618(3) Å) leading to 1D columns. The TTF cations in the stacked column have a head-to-tail arrangement with respect to the iron-chloranilate layer. Interestingly, slight differences are observed in the **39–42** structures built from the common anionic layer, caused by the intercalation of different types of guests that influence the crystal packing. The main difference in fact is in the interlayer distances (Fe(1)-Fe(1″)) 14.57 Å (**39**), 9.79 Å (**40**), 13.13 Å

(**41**), and 13.45 Å (**42**) as shown in Scheme 11. Interestingly chlorine atoms form channels between the layers and by changing their tilt angles and stacking distances depending on their sizes and shapes, modify layers structure (Scheme 12.)

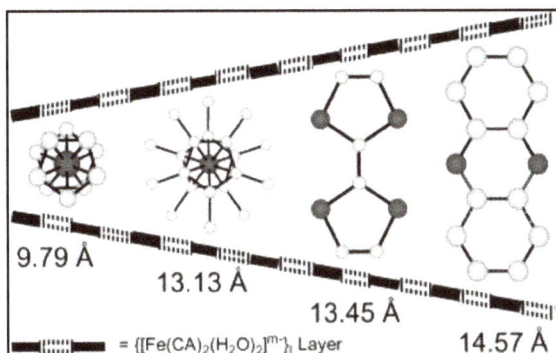

9.79 Å

13.13 Å

13.45 Å

= {[Fe(CA)$_2$(H$_2$O)$_2$]$^{m-}$}$_n$ Layer

14.57 Å

Scheme 11. Reprinted with permission from Reference [184]. Copyright 2003 American Chemical Society.

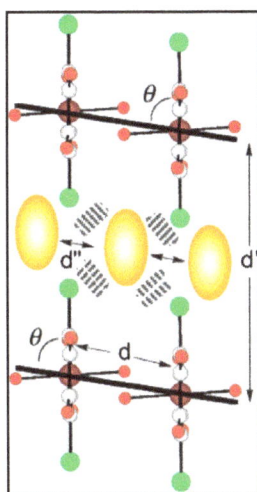

Scheme 12. Reprinted with permission from Reference [184]. Copyright 2003 American Chemical Society.

Mossbauer spectra suggests that: (i) in **41** high-spin (S = 5/2) iron(III) ions are present in {[Fe(Cl$_2$An)$_2$(H$_2$O)$_2$]}$^{m-}$ anions while low-spin (S = 1/2) iron(III) ions in [Fe(Cp*)$_2$]$^+$ cations; (ii) in **42**, the anionic layer of iron-chloranilate has a valence-trapped mixed-valence state since high-spin iron(II) and iron(III) ions are present. **39**, **40**, and **41** are EPR silent, in the 77–300 K range, whereas the EPR spectrum of **42** shows two types of signals with g = 2.008 indicating the TTF is present as radical species (Figure 28).

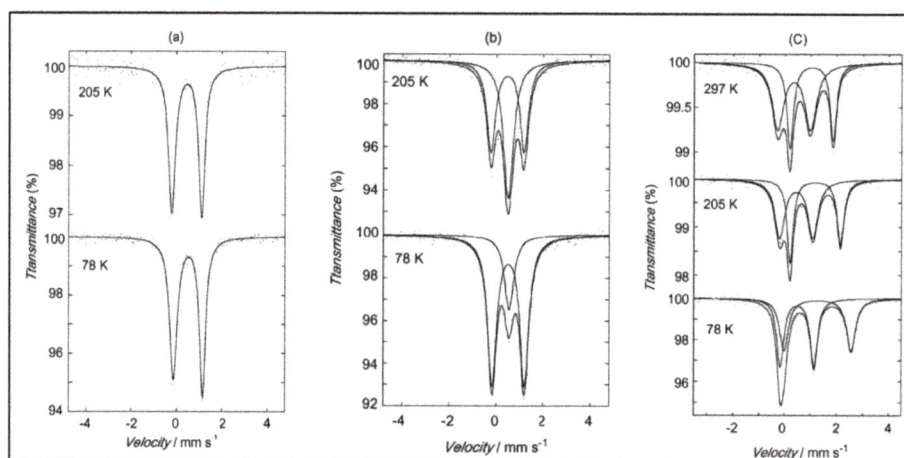

Figure 28. ^{57}Fe Mössbauer spectra of: (**a**) **39**; (**b**) **41**, showing the overlap of one singlet and one quadrupole doublet typical of low-spin iron(III) and high-spin iron(III) respectively; and (**c**) **42**. Reprinted with permission from Reference [184]. Copyright 2014 American Chemical Society.

The thermal variation of the magnetic properties ($\chi_M T$ product vs. T) for **39–42** compounds, measured in the 2–300 K temperature range, under an applied field of 0.5 T, are shown in Figure 29. The $\chi_M T$ product in **39**, **40**, and **41** shows a slight decrease with decreasing temperature and at lower temperature they show a major decrease, suggesting the existence of weak antiferromagnetic interactions.

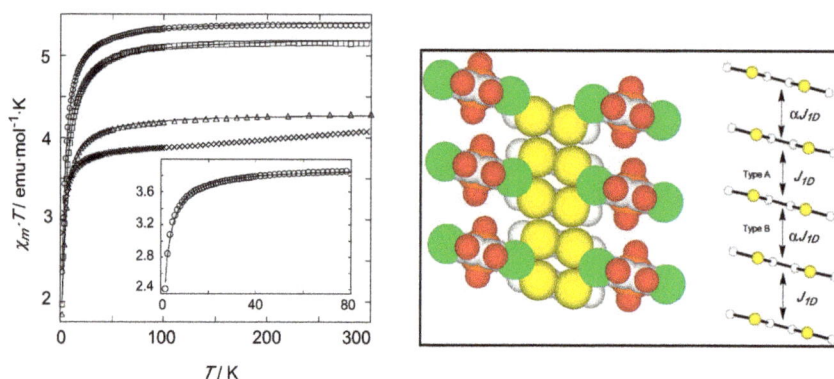

Figure 29. (**left**) Thermal variation of the magnetic properties ($\chi_M T$ vs. T) for **39** (open triangles), **40** (open squares), **41** (open circles), and **42** (cross marks). Inset, magnetic susceptibility of **42** below 80 K. Solid lines are theoretical fits of the data with the parameters listed in Reference [184]; (**right**) S···S stacking interactions of A and B types. Reprinted with permission from Reference [184]. Copyright 2003 American Chemical Society.

The observed χ_M values of **40**, **41**, and **42** are the sum of the guest and host contributions; no exchange has been observed between the two spin sublattices while are observed magnetic-field dependent susceptibility (**40, 41**) and strong antiferromagnetic coupling around 300 K (**42**). These peculiar magnetic behaviors are due to the intrachain coupling in guests which are arranged in 1D columns. In the case of **42** the antiferromagnetic interactions through the column dominates the layers structure above 80 K and this is due to the 1D columns with rare S···S contacts of TTF cations. As shown in

Figure 29 (right), the values of J_{1D} and αJ_{1D} reflect two types of S···S stacking interactions and indicate that the magnetic exchange is stronger between the stacked TTF cations located at smaller distances.

4. Anilato-Based Multifunctional Organic Frameworks (MOFs)

A rare example of Metal Organic Framework MOF composed by Fe^{III} bridged by paramagnetic linkers that additionally shows ligand mixed-valency has been reported by J. Long et al. [185]. These materials are based on 2,5-dihyroxy-1,4-benzoquinone (DHBQ) or hydranilate with R=H, the parent member of the anilates which can afford the redox processes shown in Scheme 13:

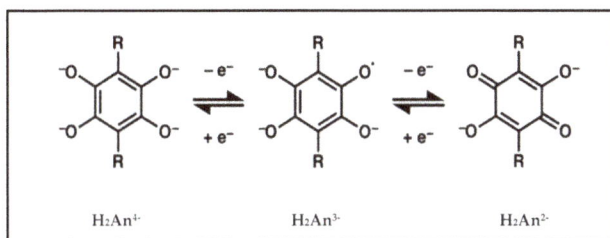

Scheme 13. Redox states of linkers deriving from 2,5-dihydroxybenzoquinone that have previously been observed in metal—organic molecules. Notably, $\mathbf{H_2An^{3-}}$ is a paramagnetic radical bridging ligand.

$(NBu_4)_2Fe^{III}{}_2(H_2An)_3$ (**43**) shows a very rare topology for H_2An^{2-}-based coordination compounds [81,112,121,125], with two interpenetrated (10,3)-a lattices of opposing chiralities where neighboring metal centers within each lattice are all of the same chirality (Figure 30b,c), generating a three-dimensional structure (Figure 30a–d). This topology differs from the classic 2D honeycomb structure type frequently observed for hydranilates and derivatives, where neighboring metal centers are of opposing chiralities [108,125].

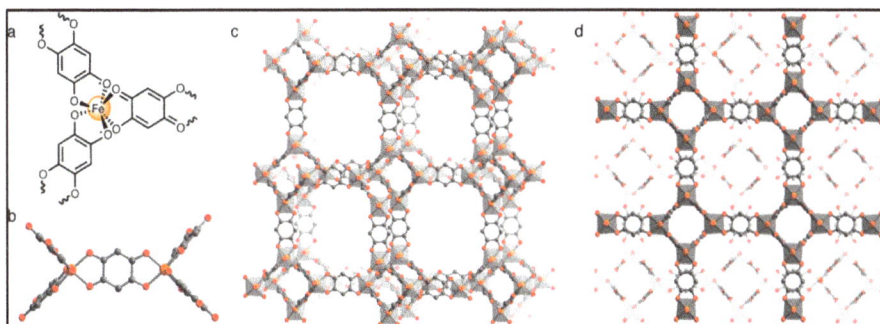

Figure 30. (**a**) Molecular structure of a single Fe^{III} center in $(NBu_4)_2Fe^{III}{}_2(H_2An)_3$, showing that two radical (H_2An^{3-}) bridging ligands and one diamagnetic (H_2An^{2-}) bridging ligand are coordinated to each metal site; (**b**) A portion of its crystal structure, showing the local environment of two H_2An^{n-}-bridged Fe^{III} centers; (**c**) A larger portion of the crystal structure, showing one of the two interpenetrated (10,3)-a nets that together generate the porous three-dimensional structure; (**d**) The two interpenetrated (10,3)-a lattices of opposing chiralities. Charge-balancing NBu_4^+ cations are not depicted for clarity. Reprinted with permission from Reference [185]. Copyright 2015 American Chemical Society.

The tetrabutylammonium countercations are crucial for templating the 3D structure, compared with other 1D or 2D hydranilate-based materials [7,108,125,186–188] since were located inside the pores and appear to fill the pores almost completely with no large voids present. Similar cation-dependent morphology changes have been observed for transition metal—oxalate coordination compounds with analogous chemical formula $[A^+]_2M^{II}_2(ox)_3$ [72,81,189–191]. The electronic absorption spectrum broad absorbance extending across the range 4500–14,000 cm^{-1}, with ν_{max} = 7000 cm^{-1}. This intense absorption features are attributed to ligand-based IVCT. Notably, a solid-state UV-vis-NIR spectrum of a molecular Fe^{III} semiquinone—catecholate compound shows a similar, though narrower, IVCT band at ν_{max} = 5200 cm^{-1} [192]. Since all the iron centers are trivalent as confirmed by Mössbauer spectroscopy, the origin of the IVCT must be the organic dhbq$^{2-/3-}$ moieties. Interestingly a very sharp absorption edge is observed at low energy (4500 cm^{-1}), one of the best-known signatures of Robin—Day Class II/III mixed-valency[193–196]. This represents the first observation of a Class II/III mixed-valency in a MOF which is also indicative of thermally activated charge transport within the lattice. **43** infact behaves as an Arrhenius semiconductor with a room-temperature conductivity of 0.16(1) S/cm and activation energy of 110 meV and it has been found to be Ohmic within ±1 V of open circuit. To the best of our knowledge, this is the highest conductivity value yet observed for a 3D connected MOF. The chemical reduction of **43** by using a stoichiometric amount of sodium naphthalenide in THF, for a stoichiometric control of the ligand redox states, affords **44**, formulated as $(Na)_{0.9}(NBu_4)_{1.8}Fe^{III}_2(H_2An)_3$, which shows a highly crystalline powder X-ray diffraction (PXRD) pattern that overlays with that simulated for **45** and is much closer to a fully H$_2$An^{3-}—bridged framework. **44** shows a lower conductivity of 0.0062(1) S/cm at 298 K and a considerably larger activation energy of 180 meV which is consistent with a further divergence of the H$_2$An^{3-}/H$_2$An^{2-} ligand ratio from the optimal mixed-valence ratio of 1:1 (Figure 31).

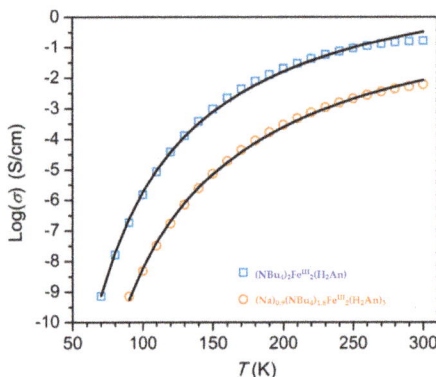

Figure 31. Variable-temperature conductivity data for **43** and **44**, shown by blue squares and orange circles, respectively. Arrhenius fits to the data are shown by black lines. Adapted with permission from Reference [185]. Copyright 2015 American Chemical Society.

Due to the presence of H$_2$An^{3-} radicals a peculiar magnetic behavior is expected on the basis of previously studied metal—organic materials with transition metals bridged by organic radicals which have shown strong magnetic coupling, leading to high temperature magnetic ordering [197]. Variable-temperature dc magnetic measurements under an applied magnetic field of 0.1 T revealed strong metal-radical magnetic interactions, leading to magnetic ordering less than 8 K (Figure 32) due to the strong magnetic coupling that has previously been observed in Fe^{III} H$_2$An^{3-} complexes [198].

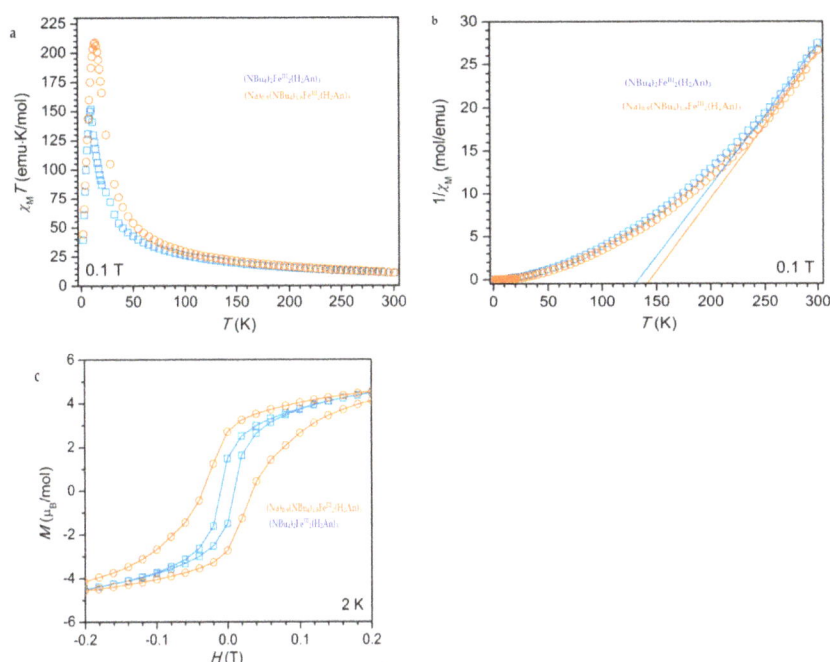

Figure 32. (**a**) DC magnetic susceptibility data for **43** (blue squares) and **44** (orange circles); (**b**) Inverse of magnetic susceptibility versus temperature for **43** and **44**. Curie-Weissfits to the data in the temperature range 250–300 K are shown by solid blue (**43**) and orange lines (**44**); (**c**) Magnetization (M) versus applied dc magnetic field (H) data for **43** and **44** in blue and orange, respectively. Hysteresis loops were recorded at a sweep rate of 2 mT/s. Solid lines are guides for the eye. Adapted with permission from Reference [185]. Copyright 2015 American Chemical Society.

Below 250 K the observed strong deviations from the Curie-Weiss behavior, observed in systems with strong π-d interactions [199], have been attributed to the competition between ferromagnetic and antiferromagnetic interactions, both maybe present in **43**, leading to magnetic glassiness [200]. A Curie-Weiss fit of the inverse magnetic susceptibility data from 250 to 300 K results in a Curie temperature of $\theta = 134$ K and a Curie constant of $C = 6.1$ emu·K/mol. The positive Curie temperature reveals that ferromagnetic interactions are dominant at high temperature and its magnitude suggests that quite high temperature magnetic coupling occurs. In contrast, the magnetic behavior at low temperature indicates that ferrimagnetic coupling predominates. Thus, the low magnetic ordering temperature is attributed to a competition of ferromagnetic and antiferromagnetic interactions that prevent true three-dimensional order, until antiferromagnetic metal—radical interactions, and thus bulk ferrimagnetic order, prevail at low temperature. **44** was expected to show an increased magnetic, ordering temperature due to the greater number of paramagnetic linkers. Indeed, a higher magnetic transition temperature of 12 K was observed and the room-temperature $\chi_M T$ product is 11.2 emu·K/mol, compared to 10.9 emu·K/mol for **43**. Finally, low temperature (2 K) magnetic hysteresis measurements shown in Figure 32c reveal that **44** is a harder magnet than **43**, with coercive fields of 350 and 100 Oe observed, respectively. These materials are rare examples of a MOF formed by metal ions bridged by paramagnetic linkers that additionally shows ligand-centered mixed valency where magnetic ordering and semiconducting behaviors stem from the same origin, the ferric semiquinoid lattice, differently from multifunctional materials based on tetrathiafulvalene derivatives with paramagnetic counterions where separate sub-lattices furnish the two distinct magnetic/conducting properties. Therefore they represent a challenge for pursuing magnetoelectric or multiferroic MOFs.

Anilates are also particularly suitable for the construction of microporous MOFs with strong magnetic coupling and they have been shown to generate extended frameworks with different dimensionality and large estimated void volumes, which could potentially give rise to materials with permanent porosity [112]. Very recently Harris et al. [201] have reported on the synthesis and full characterization of $(Me_2NH_2)_2[Fe_2L_3]\cdot 2H_2O\cdot 6DMF$ (**45**), the first structurally characterized example of a microporous magnet containing the $Cl_2An^{3-}\cdot$chloranilate radical species, where solvent-induced switching from Tc = 26 to 80 K has been observed. **45** shows the common 2D honeycomb layered packing where the layers are eclipsed along the crystallographic c axis, with a H_2O molecule located between Fe centers, leading to the formation of 1D hexagonal channels (Figure 33). These channels are occupied by disordered DMF molecules, as was confirmed by microelemental analysis and thermogravimetric analysis (TGA).

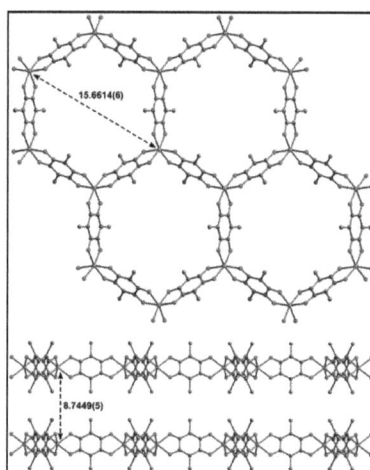

Figure 33. View of the crystal structure of $[Fe_2L_3]^{2-}$, along the crystallographic c axis (upper) and b axis (lower), with selected Fe$\cdots$Fe distances (Å). Orange = Fe, green = Cl, red = O, and gray = C. Adapted with permission from Reference [201]. Copyright 2015 American Chemical Society.

X-ray diffraction, Raman spectra and and Mo··ssbauer spectra confirm the presence of Fe^{III} metal ions and mixed-valence ligands which can be formulated as $[(Cl_2An)_3]^{8-}$, obtained through a spontaneous redox reaction from Fe^{II} to Cl_2An^{2-}. Upon removal of DMF and H_2O solvent molecules, the desolvated phase $(Me_2NH_2)_2Fe_2(Cl_2An)_3]$, **45a**, showing a slight structural distortion respect to **45**, has been obtained. **45a** gives a a Brunauer-Emmett-Teller (BET) surface area of 885(105) m^2/g, from a fit to N_2 adsorption data, at 77 K, confirming the presence of permanent microporosity. This value is the second highest reported for a porous magnet, overcome only with a value of 1050 m^2/g reported for a lactate-bridged Co^{II} material [202]. Finally, the structural distortion is fully reversible and similar "breathing" behavior has been previously observed in MOFs [203–205]. The thermal variation of the magnetization shows that **45** and **45a** have a spontaneous magnetization below 80 and 26 K with magnetic hysteresis up to 60 and 20 K (Figure 34).

To precisely determine the Tcs of **45** and **45a**, variable-temperature ac susceptibility data under zero applied field were collected at selected showing for **45a** slightly frequency dependent peak in both in-phase (χ_M') and out-of-phase (χ_M'') susceptibility and give a Tc = 80 K. The frequency dependence can be quantified by the Mydosh parameter, in this case $\phi = 0.023$, which is consistent with glassy magnetic behavior. Such glassiness can result from factors such as crystallographic disorder and spin frustration arising from magnetic topology. In contrast, the plot of χ_M' vs. T for **45a** exhibits a sharp, frequency-independent peak with a maximum at 26 K, indicating that **45a** undergoes long-range magnetic ordering at Tc = 26 K. The magnetic data demonstrate that **45** and **45a** behave as magnets that involve dominant intralayer antiferromagnetic interactions between adjacent spins. Moreover these results demonstrate that the incorporation of semiquinone radical ligands into an extended can generate a 2D magnet with Tc = 80 K, with permanent porosity and activated phase undergoing a slight structural distortion and associated decrease in magnetic ordering temperature to Tc = 26 K.

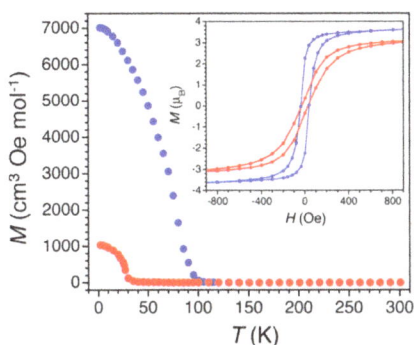

Figure 34. Thermal variation of the magnetization for **45** (blue) and **45a** (red), collected under an applied dc field of 10 Oe. Inset:Variable-field magnetization data for **45** at 60 K (blue) and **45a** at 10 K (red). Reprinted with permission from Reference [201] Copyright 2015 American Chemical Society.

The first example of a 3D monometallic lanthanoid assembly, the $Na_5[Ho(H_2An^{4-})_2]_3$ $7H_2O$ (**46**) complex, showing ferromagnetism with a Curie temperature of 11 K, has been reported by Ohkoshi et al. [206]. In this compound the Ho^{3+} ion adopts a dodecahedron (D_4d) coordination geometry and each Ho^{3+} ion is connected to eight O atoms of four bidentate H_2An^{4-} ligands directed toward the *a* and *b* axes, which resulted in a 3-D network with regular square-grid channels (Figure 35a–c). These channels (4.9 × 4.9 Å) was occupied by Na^+ ions and noncoordinated water molecules.

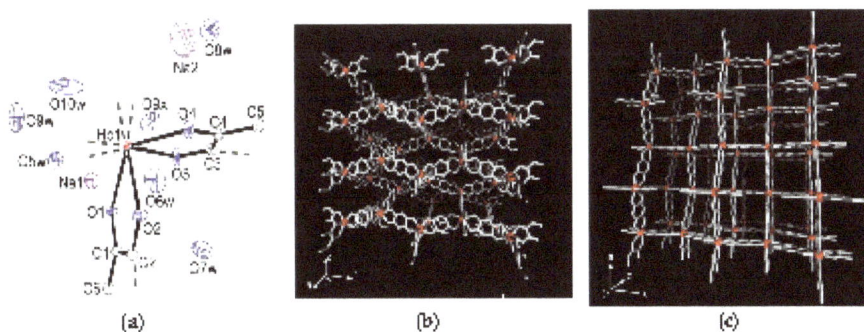

Figure 35. (a) Thermal ellipsoid plots (50% probability level) of the molecular structure of **46**. All independent atoms, including water and Na^+ ions, are labeled. H atoms are omitted for clarity. Ho, Na, O, and C atoms, are represented by red, purple, blue, and light gray colours respectively; (b) X-ray crystal structure along the *a* axis; (c) X-ray crystal structure along the *c* axis. Water molecules, Na^+ ions and H atoms are omitted for clarity. Adapted with permission from Reference [206]. Copyright 2009 American Chemical Society.

The thermal variation of the magnetic properties (χ_MT product vs. T) is reported in Figure 36. The χ_MT value at room temperature was 14.4 cm^3 K mol^{-1}, which nearly corresponds to the expected value of 13.9 cm^3 K mol^{-1} for Ho^{3+} ion (J = 8, L = 6, S = 2, and g = 5/4). The thermal variation of the field-cooled magnetization (FCM) and the remnant magnetization (RM) showed that a spontaneous magnetization has been observed at T_C = 11 K (Figure 36a) with a M-H hysteresis loop at 2 K indicating a value of 170 Oe for H_C and a value of 6.4 μ_B at 50 kOe for M (Figure 36b), close to the expected value of 6.8 μ_B [207].

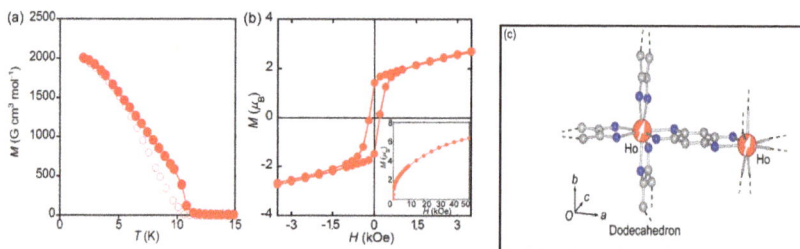

Figure 36. Thermal variation of magnetic properties of **46**. (**a**) field-cooled magnetization (FCM) obtained with decreasing temperature at an applied field of 10 Oe (red filled circles). RM obtained with increasing temperature without an applied field; (**b**) M-H hysteresis plot. Inset shows the magnetic field dependence of magnetization at 2 K; (**c**) Schematic representation of magnetic ordering. Red, blue, and light gray represent Ho, O, and C atoms, respectively. Adapted with permission from Reference [206]. Copyright 2009 American Chemical Society.

The ferromagnetic ordering is due to the effective mediation of the magnetic interactions between Ho^{3+} ions by the π orbitals of H$_2$An^{4-}. Interestingly this material opens the way to explore the chemistry and physical properties of related systems with different lanthanides which can result in very challenging luminescent/magnetic microporous materials.

The molecular packing and physical properties of all compounds discussed in this work are summarized in Table 2.

Table 2. Molecular Packing and Physical Properties of anilato-based magnetic/conducting molecular materials.

Compound	Molecular Packing	Physical Properties	Ref.
[(Ph)$_4$P]$_3$[Fe(H$_2$An)$_3$]·6H$_2$O **1**	Homoleptic tris-chelated octahedral complex. strong HBs between oxygen atoms of the ligand and crystallization water molecules	PM J/k_B = −0.020 K	[116]
[(Ph)$_4$P]$_3$[Cr(H$_2$An)$_3$]·6H$_2$O **2**	Homoleptic tris-chelated octahedral complex. π-π interactions	PM Weak magnetic coupling due to charge transfer between the Cr metal ions and the hydranilate ligands	[116]
[(TPA)(OH)Fe^{III}OFe^{III}(OH) (TPA)][Fe(Cl$_2$An)$_3$]$_{0.5}$(BF$_4$)$_{0.5}$· 1.5MeOH·H$_2$O **3**	Homoleptic trischelated complex	μ_{eff}(RT) = 2.93 μ_B Strong AFM interaction within Fe^{III}OFe^{III} with a plateau at 55 K. Below 55 K, μ(T) is constant at 4.00 μ_B J/k_B = −165 K	[117]
[(n-Bu)$_4$N]$_3$[Cr(Cl$_2$An)$_3$] **4a**	Homoleptic tris-chelated octahedral complex.	PM ZFS	[119]
[(Ph)$_4$P]$_3$ [Cr(Cl$_2$An)$_3$] **4b**	Homoleptic tris-chelated octahedral complex. π-π interactions	PM ZFS	[119]

Table 2. *Cont.*

Compound	Molecular Packing	Physical Properties	Ref.
[(Et)$_3$NH]$_3$ [Cr(Cl$_2$An)$_3$] **4c**	Homoleptic tris-chelated octahedral complex.	PM ZFS	[119]
[(n-Bu)$_4$N]$_3$[Fe(Cl$_2$An)$_3$] **5a**	Homoleptic tris-chelated octahedral complex.	Curie-Weiss PM J/k$_B$ = −2.2 K	[119]
[(Ph)$_4$P]$_3$ [Fe(Cl$_2$An)$_3$] **5b**	Homoleptic tris-chelated octahedral complex. π-π interactions	PM ZFS	[119]
[(Et)$_3$NH]$_3$ [Fe(Cl$_2$An)$_3$] **5c**	Homoleptic tris-chelated octahedral complex.	PM ZFS	[119]
[(n-Bu)$_4$N]$_3$[Cr(Br$_2$An)$_3$] **6a**	Homoleptic tris-chelated octahedral complex.	PM ZFS	[119]
[(Ph)$_4$P]$_3$ [Cr(Br$_2$An)$_3$] **6b**	Homoleptic tris-chelated octahedral complex. π-π interactions	PM ZFS	[119]
[(n-Bu)$_4$N]$_3$[Fe(Br$_2$An)$_3$] **7a**	Homoleptic tris-chelated octahedral complex.	Curie-Weiss PM (r.t.−4.1 K), AFM coupling via halogen-bonding between the complexes forming the dimers	[119]
[(Ph)$_4$P]$_3$ [Fe(Br$_2$An)$_3$] **7b**	Homoleptic tris-chelated octahedral complex. π-π interactions	Curie-Weiss PM	[119]
[(n-Bu)$_4$N]$_3$[Cr(I$_2$An)$_3$] **8a**	Supramolecular dimers that are held together by two symmetry-related I···O interactions	Curie-Weiss PM (r.t.−4.1 K), AFM coupling via halogen-bonding between the complexes forming the dimers	[119]
[(Ph)$_4$P]$_3$ [Cr(I$_2$An)$_3$] **8b**	Homoleptic tris-chelated octahedral complex. π-π interactions	Curie-Weiss PM	[119]
[(n-Bu)$_4$N]$_3$Fe(I$_2$An)$_3$] **9a**	Homoleptic tris-chelated octahedral complex.	Curie-Weiss PM J/k$_B$ = 0.011 K	[119]
[(Ph)$_4$P]$_3$[Fe(I$_2$An)$_3$] **9b**	Homoleptic tris-chelated octahedral complex. iodine–iodine interactions XB interactions π-π interactions	Curie-Weiss PM J/k$_B$ = 0.34 K	[119]
[(n-Bu)$_4$N]$_3$[Cr(ClCNAn)$_3$] **10a**	Homoleptic tris-chelated octahedral complex. C–N···Cl interactions between complex anions having an opposite stereochemical configuration (Λ, Δ)	Curie-Weiss PM J/k$_B$ = 0.0087 K	[120]
[(Ph)$_4$P]$_3$ [Cr(ClCNAn)$_3$] **10b**	Homoleptic tris-chelated octahedral complex. π-π interactions	Curie-Weiss PM J/k$_B$ = −0.24 K	[120]
[(n-Bu)$_4$N]$_3$[Fe(ClCNAn)$_3$] **11a**	Homoleptic tris-chelated octahedral complex. C–N···Cl interactions between complex anions having an opposite stereochemical configuration (Λ, Δ)	Curie-Weiss PM	[120]
[(Ph)$_4$P]$_3$ [Fe(ClCNAn)$_3$] **11b**	Homoleptic tris-chelated octahedral complex. π-π interactions	Curie-Weiss PM	[120]
[(n-Bu)$_4$N]$_3$[Al(ClCNAn)$_3$] **12a**	Homoleptic tris-chelated octahedral complex. C–N···Cl interactions between complex anions having an opposite stereochemical configuration (Λ, Δ)	Red luminophore Ligand centred emission	[120]
[(Ph)$_4$P]$_3$ [Al(ClCNAn)$_3$] **12b**	Homoleptic tris-chelated octahedral complex.	Red luminophore Ligand centred emission	[120]

Table 2. *Cont.*

Compound	Molecular Packing	Physical Properties	Ref.
$(PBu_3Me)_2[NaCr(Br_2An)_3]$ **13**	2D lattice Heterometallic anionic Honeycomb layers alternated with cationic layer in alternated manner	PM ZFS	[121]
$(PPh_3Et)_2[KFe(Cl_2An)_3]$ $(dmf)_2$ **14**	2D lattice Heterometallic anionic Honeycomb layers alternated with cationic layer in alternated manner	PM ZFS	[121]
$(NEt_3Me)[Na(dmf)]-[NaFe(Cl_2An)_3]$ **15**	Inter-connected 2D honeycomb	PM ZFS	[121]
$(NBu_3Me)_2[NaCr(Br_2An)_3]$ **16**	3D lattice	PM ZFS	[121]
$[(H_3O)(phz)_3][MnCr(Cl_2An)_3(H_2O)]$ **17**	Eclipsed Heterometallic anionic Honeycomb layers alternated with cationic layers	Ferrimagnet Tc = ca. 5.0 K	[125]
$[(H_3O)(phz)_3][MnCr(Br_2An)_3]$ $\cdot H_2O$ **18**	Eclipsed Heterometallic anionic Honeycomb layers alternated with cationic layers	Ferrimagnet Tc = ca. 5.0 K	[125]
$[(H_3O)(phz)_3][MnFe(Br_2An)_3]$ $\cdot H_2O$ **19**	Eclipsed Heterometallic anionic honeycomb layers	Weak FM due to long-range AF ordering with spin canting at ca. 3.5 K	[125]
$[(n\text{-}Bu)_4N]_3[MnCr(Cl_2An)_3(H_2O)]$ **20**	Alternated Heterometallic anionic honeycomb layers	Ferrimagnet Tc = 5.5 K $J/k_B = -8.7$ K	[125]
$[(n\text{-}Bu)_4N]_3[MnCr(Br_2An)_3(H_2O)]$ **21**	Alternated Heterometallic anionic honeycomb layers	Ferrimagnet Tc = 6.3 K $J/k_B = -8.7$ K	[125]
$[(n\text{-}Bu)_4N]_3[MnCr(I_2An)_3(H_2O)]$ **22**	Alternated heterometallic anionic honeycomb layers	Ferrimagnet Tc = 8.2 K $J/k_B = -10$ K	[125]
$Bu)_4N]_3[MnCr(H_2An)_3(H_2O)]$ **23**	Alternated Heterometallic anionic honeycomb layers	Ferrimagnet Tc = 11.0 K $J/k_B = -12$ K	[125]
$[BEDT\text{-}TTF]_3[Fe(Cl_2An)_3]\cdot 3CH_2Cl_2\cdot H_2O$ **24**	BEDT-TTF dimers not-layered structure Cl$\cdots$S interactions	PM with a contribution at high temperatures from BEDT-TTF radical cations semiconductor $\sigma_{RT} = 3 \times 10^{-4}$ S cm^{-1} Intradimer Coupling Constant $J_{CC} = -2.6 \times 10^{33}$ K	[164]
$\delta\text{-}[BEDT\text{-}TTF]_5[Fe(Cl_2An)_3]\cdot 4H_2O$ **25**	organic-inorganic layers segregation δ packing of BEDT TTF Cl$\cdots$S interactions	PM with a contribution at high temperatures from BEDT-TTF radical cations Semiconductor $\sigma_{RT} = 2$ S cm^{-1}	[164]
$\alpha'''\text{-}[BEDT\text{-}TTF]_{18}[Fe(Cl_2An)_3]_3\cdot 3CH_2Cl_2\cdot 6H_2O$ **26**	organic-inorganic layers segregation α''' packing of BEDT TTF Cl$\cdots$S interaction	PM with a contribution at high temperatures from BEDT-TTF radical cations Semiconductor $\sigma_{RT} = 8$ S cm^{-1}	[164]
$[(BEDT\text{-}TTF)_6[Fe(Cl_2An)_3]\cdot(H_2O)_{1.5}\cdot(CH_2Cl_2)_{0.5}$ **27**	organic-inorganic layers segregation θ^{21} phase of BEDT TTF Cl$\cdots$S interaction	PM with Pauli PM contribution Semiconductor σ_{RT} = ca. 10 S cm^{-1}	[166]
$\beta\text{-}[(S,S,S,S)\text{-}TM\text{-}BEDT\text{-}TTF]_3$ $PPh_4[K^IFe^{III}(Cl_2An)_3]\cdot 3H_2O$ **28** $\beta\text{-}[(R,R,R,R)\text{-}TM\text{-}BEDT\text{-}TTF]_3$ $PPh_4[K^IFe^{III}(Cl_2An)_3]\cdot 3H_2O$ **29**	heterobimetallic anionic honeycomb layers alternated with cationic chiral donors Cl$\cdots$Cl contact, π-π stacking terminal CH$_3\cdots$O contacts (segregated columns of cations and anions) β packing of TM-BEDT-TTF	Curie-Weiss PM Semiconductors $\sigma_{RT} = 3 \times 10^{-4}$ S cm^{-1}	[167]

Table 2. *Cont.*

Compound	Molecular Packing	Physical Properties	Ref.
β-[(*rac*)-TM-BEDT-TTF]$_3$ PPh$_4$[K^{I}Fe^{III}(Cl$_2$An)$_3$]·3H$_2$O **30**	heterobimetallic anionic honeycomb layers alternated with cationic chiral -(*rac*)-donors Cl···Cl contact, π-π stacking terminal CH$_3$···O contacts (segregated columns of cations and anions) β packing of TM-BEDT-TTF	Curie-Weiss PM Semiconductors σ$_{RT}$ = 3 × 10^{-4} S cm^{-1}	[167]
[Fe^{III}(sal$_2$-trien)]MnCr (Cl$_2$An)$_3$ **31**	2D Honeycomb bimetallic anionic layers with inserted Fe(III) cationic complexes and solvent molecules.	FerriM Inserted HS Fe(III) cations Tc = 10K J/k$_B$ = −10 K Exfoliation	[168]
[Fe^{III}(4-OH-sal$_2$-trien)] MnCr(Cl$_2$An)$_3$ **32**	2D Honeycomb bimetallic anionic layers with inserted Fe(III) cationic complexes and solvent molecules.	FerriM Inserted HS Fe(III) cations Tc = 10.4 K J/k$_B$ = −7.2 K	[168]
[Fe^{III}(sal$_2$-epe)] MnCr (Br$_2$An)$_3$ **33**	2D Honeycomb bimetallic anionic layers with inserted Fe(III) cationic complexes and solvent molecules.	FerriM Inserted HS Fe(III) cations Tc = 10.2 K J/k$_B$ = −6.5 K	[168]
[Fe^{III}(5-Cl-sal$_2$-trien)] MnCr(Br$_2$An)$_3$ **34**	2D honeycomb bimetallic anionic layers with inserted Fe(III) cationic complexes and solvent molecules.	FerriM Inserted LS Fe(III) cations Tc = 9.8 K J/k$_B$ = −6.7 K	[168]
[Fe^{II}(tren- (imid)$_3$)]$_2$ Mn^{II}Cl$_2$Cr^{III}(Cl$_2$An)$_3$]Cl· solvent **35**	1D anionic chain formed by Cr^{III}complexes bonded to two Mn(II) ions through two bis-bidentate chloranilate bridges, and terminal third choranilate.	FerriM coupling within the chains that gives rise to a magnetic ordering below 2.6 K	[168]
[Fe^{III}(acac$_2$-trien)] [Mn^{II}Cr^{III}(Cl$_2$An)$_3$]$_3$ (CH$_3$CN)$_2$ **36**	Neutral layers formed by 2D honeycomb bimetallic anionic layers with cationic complexes inside the hexagonal channels. van der Waals interactions between the layers.	FerriM at ca. 10.8 K, inserted HS Fe(III) cations Exfoliation	[172]
[Fe^{III}(acac$_2$-trien)] [Mn^{II}Cr^{III}(Br$_2$An)$_3$]$_3$ (CH$_3$CN)$_2$ **37**	Neutral layers formed by 2D Honeycomb bimetallic anionic layers with cationic complexes inside the hexagonal channels. Van der Waals interactions between the layers	FerriM at ca. 11.4 K, inserted HS Fe(III) cations Exfoliation	[172]
[Ga^{III}(acac$_2$-trien)] [Mn^{II}Cr^{III}(Br$_2$An)$_3$]$_3$ (CH$_3$CN)$_2$ **38**	Neutral layers formed by 2D Honeycomb bimetallic anionic layers with cationic complexes inside the hexagonal channels. Van der Waals interactions between the layers	FerriM at ca. 11.6 K	[172]
{(H$_{0.5}$phz)$_2$[Fe(Cl$_2$An)$_2$(H$_2$O)$_2$]· 2H$_2$O}$_n$ **39**	Supramolecular Framework Novel Intercalation Compounds Electrostatic interactions	Interlayer distances (Fe(1)-Fe(1″))14.57 Å In 77–300 K temperature range, EPR silent. Intralayer AFM exchange via Hydrogen-Bonds and stacking interactions among [Fe(Cl$_2$An)$_2$(H$_2$O)$_2$]$^-$ monomers J$_{2D}$/k$_B$ = −0.10 K	[184].
{[Fe(Cp)$_2$][Fe(Cl$_2$An)$_2$ (H$_2$O)$_2$]}n **40**	Supramolecular Framework Novel Intercalation Compounds Electrostatic interactions	Interlayer distances (Fe(1)-Fe(1″)) 9.79 Å In 77–300 K temperature range, EPR silent. Intralayer AFM exchange via Hydrogen-Bonds and stacking interactions among [Fe(Cl$_2$An)$_2$(H$_2$O)$_2$]$^-$ monomers and Heisenberg AFM intrachain stacking interactions in 1D arrays of [Fe(Cp)$_2$]$^+$ cations J$_{2D}$/k$_B$ = −0.13 K J$_{1D}$/k$_B$ = −2.4 K	[184]

Table 2. *Cont.*

Compound	Molecular Packing	Physical Properties	Ref.
[Fe(Cp*)$_2$][Fe(Cl$_2$An)$_2$ (H$_2$O)$_2$]]$_n$ **41**	Supramolecular Framework Novel Intercalation Compounds Electrostatic interactions π-π stacking tilted columns of stacked decamethylferrocene cations	Interlayer distances (Fe(1)-Fe(1″)) 13.13 Å. In 77–300 K temperature range, EPR silent High-spin (S = 5/2)Fe(III) ions in {[Fe(Cl$_2$An)$_2$(H$_2$O)$_2$]}$^{m-}$ anions Low-spin (S = 1/2) Fe(III) ions in [Fe(Cp*)$_2$]$^+$ cations Intralayer AFM exchange via hydrogen-bonds and stacking interactions among [Fe(Cl$_2$An)$_2$(H$_2$O)$_2$]$^-$ monomers and Heisenberg AFM intrachains stacking interaction in 1D arrays of [Fe(Cp)$_2$]$^+$ cations J/k$_B$ = −9.5 K J$_{1D}$/k$_B$ = −1.9 K	[184]
{(TTF)$_2$[Fe(Cl$_2$An)$_2$(H$_2$O)$_2$]}$_n$ **42**	Novel intercalation compounds formed by the 2D hydrogen-bond supported layers and functional guests. Electrostatic interactions π-π stacking Face to Face stacking of TTF cations in columnar structure S···S distances (type A; 3.579(3) Å, and type B; 3.618(3) Å). Head-to-Tail arrangement for TTF cations in the stacked column	Interlayer distances (Fe(1)-Fe(1″)) 13.45 Å. EPR active with g = 2.008 (2 signals) indicating TTF is present as radical species High-spin Fe(II) and Fe(III) ions (the iron-chloranilate anionic layer has a valence-trapped mixed-valence state) Isotropic intralayer AFM exchange via hydrogen-bonds and stacking interaction among iron(III)- and iron(II)-chloranilate monomers (1:1) Heisenberg alternating AFM linear chain for isotropic exchange in the 1D array of TTF cations via intrachain stacking interactions J/k$_B$ = −6.5 K J$_{1D}$/k$_B$ = −443 K	[184]
(NBu$_4$)$_2$Fe^{III}$_2$(H$_2$An)$_3$ **43**	3D structure MOF with Robin-Day Class II/III mixed-valency ligand	Curie-Weiss PM J$_{1D}$/k$_B$ = 0.89 K (250–300 K) High T- FM Low T- FerriM interactions Arrhenius semiconductor σ$_{RT}$ = 0.16(1) S cm^{-1}	[185]
(Na)$_{0.9}$(NBu$_4$)$_{1.8}$Fe^{III}$_2$ (H$_2$An)$_3$ **44**	Isostructural to **46** (PXRD) MOF with Robin-Day Class II/III mixed-valency ligand	Curie-Weiss PM J$_{1D}$/k$_B$ = 0.95 K (250-300K) Arrhenius semiconductor σ$_{RT}$ = 0.0062(1) S cm^{-1}	[185]
(Me2NH2)2[Fe2Cl$_2$An3]· 2H2O·6DMF **45**	Eclipsed 2D honeycomb layered packing with a H$_2$O between Fe centers, leading to the formation of 1D hexagonal channels	2D Microporous magnet with strong magnetic coupling. Intralayer AFM interactions Tc = 80 K, glassy Magnet, Mydosh parameter, φ = 0.023	[201]
(Me2NH2)2[Fe2Cl$_2$An3] **45a**	Eclipsed 2D honeycomb layered packing Desolvated phase of **48**	Intralayer AFM interactions Tc = 26 K. Permanent porosity with BET surface area of 885(105) m^2/g	[201]
Na$_5$[Ho(H$_2$An^{4-})$_2$]$_3$ 7H$_2$O **46**	3D monometallic lanthanoid assembly Ho^{3+} ion adopts a dodecahedron (D4d) geometry with regular square-grid channels	FM with a Curie Temperature of 11 K	[206]

PM = Paramagnet; FM = Ferromagnet; FerriM = Ferrimagnet; AFM = Antiferromagnet.

5. Conclusions

The compounds described in this work are summarized in Table 2. It can be envisaged that the real challenge of anilate-based materials is due to their peculiar features: (i) easy to modify or functionalize by the conventional synthetic methods of organic and coordination chemistry, with no influence on their coordination modes (ii) easy to tune the magnetic exchange coupling between the coordinated metals by a simple change of the X substituent (X = H, F, Cl, Br, I, NO$_2$, OH, CN, Me, Et, etc.) at the 3,6 positions of the anilato moiety; (iii) influence of the electronic nature of the X substituents on the intermolecular interactions and thus the physical properties of the resulting materials. The novel family of complexes of the anilato-derivatives containing the X=Cl, Br, I, H, and Cl/CN substituents with d-transition Fe(III) and Cr(III) metal ions (**3–10**) are a relevant example of the crucial role played by halogens n their physical properties, either at the electronic level, by varying the electron density on the anilate ring, or at the supramolecular level, affecting the molecular packing via halogen-bonding interactions. It is noteworthy that halogen-bonding interactions observed in **9b** are responsible for a unique magnetic behaviour in this family. Moreover the anilato derivatives having Cl/CN substituents and their complexes with Al(III) metal ions (**12a,b**), show unprecedented properties such as luminescence in the visible region (green and red luminophores, respectively), never observed in this family to the best of our knowledge. The paramagnetic anionic complexes has shown to be excellent building blocks for constructing via the "complex as-ligand approach": (i) new 2D and 3D heterometallic lattices with alkaline M(I) and d-transition M(III) metal ions (**13–16**); (ii) 2D layered molecular ferrimagnets (**17–23**) which exhibit tunable ordering temperature as a function of the halogen electronegativity; it is noteworthy how subtle changes in the nature of the substituents (X = Cl, Br, I, H) have been rationally employed as "adjusting screws" in tuning the magnitude of the magnetic interaction between the metals and thus the magnetic properties of the final material; the additional peculiarity of these molecular magnets is that they form void hexagonal channels and thus can behave as layered chiral magnetic MOFs with tunable size which depends, in turn, on the halogen size. The paramagnetic anionic complexes worked well as magnetic components of multifunctional molecular materials based on BEDT-TTF organic donors (**24–27**) which has furnished the pathway for combining electrical conductivity with magnetic properties, in analogy with the relevant class of [M(ox)$_3$]$^{3-}$ (ox = oxalate) tris-chelated complexes which have produced the first family of molecular paramagnetic superconductors. The introduction of chirality in the BEDT-TTF organic donor has been successful and a complete series of radical-cation salts have been obtained by combining the TM-BEDT-TTF organic donor in its (S,S,S,S) and (R,R,R,R) enantiopure forms, or their racemic mixture (rac), with 2D heterobimetallic anionic layers formed "in situ" by self-assembling of the tris(chloranilato)ferrate(III) metal complexes in the presence of potassium cations in the usual honey-comb packing pattern (**28–30**). Another advantage of anilato-ligands compared to the oxalato ones is their bigger size leading to hexagonal cavities that are twice larger than those of the oxalato-based layers, where a large library of cationic complexes can be inserted. When using spin crossover cations such as [Fe^{III}(sal$_2$-trien)]$^+$, (X = Cl) (**31**) and its derivatives and the [M^{III}(acac$_2$-trien)]$^+$, M = Fe or Ga complex (**32–38**), which has a smaller size than the [Fe^{III}(sal$_2$-trien)]$^+$ complex, 2D anilate-based materials have been obtained showing in the former the typical alternated cation/anion layered structure while in the latter neutral layers never observed previously in oxalate or anilate-based 2D networks, where the spin crossover cations are inserted in the centre of the hexagonal channels. This novel type of structure opens the way to the synthesis of a new type of multifunctional materials in which small templating cations are confined into the 1D channels formed by 2D anilate-based networks and could be a useful strategy for the introduction of other properties such as electric or proton conductivity in addition to the magnetic ordering of the anilate-based network. Interestingly this type of magnetic hybrid coordination polymers can be considered as graphene related magnetic materials. In fact being formed by a 2D anionic network and cations inserted within or between the layers, with interlayer ionic or weak van der Waals interactions, they have been successfully exfoliated using either the micromechanical Scotch tape method leading to good quality micro-sheets of these layered magnets or solvent-mediated exfoliation methods. Interestingly the hybrid nature of these magnetic layers provides the unique opportunity to generate smart layers where the switching properties of the inserted complexes can modulate the cooperative magnetism of the magnetic network. **39–42** are interesting examples of host-guest intercalation compounds formed by the common 2D hydrogen-bond supported layers and ferrocene/decamethylferrocene and TTF functional guests showing intralayer AFM exchange via hydrogen-bonds and stacking interactions among $[Fe(Cl_2An)_2(H_2O)_2]^-$ monomers and the Heisenberg AFM intrachain stacking interactions in 1D arrays of ferrocene/decamethylferrocene and TTF cations. Interestingly these compounds shows how hydrogen-bond-supported anionic layers based on iron-chloranilate mononuclear complexes can be used as inorganic hosts for the intercalation of guest cations to construct new types of multilayered inorganic-organic hybrid materials. It is noteworthy that these layers are so flexible that they can include and stabilize various kinds of guests in the channels showing the versatility of the anilate building blocks and the challenge of the molecular approach as synthetic procedure.

Finally **43** and **44** are rare examples of a MOF formed by metal ions bridged by paramagnetic linkers, the hydranilates, that additionally shows ligand-centered Robin—Day Class II/III mixed valency, observed for the first time in a MOF. Interestingly **43** exhibits a conductivity of $0.16 \pm 0.01\,S/cm$ at 298 K, one of the highest values yet observed in a MOF and the origin of this electronic conductivity is determined to be ligand mixed-valency. In these materials the magnetic ordering and semiconducting behaviors stem from the same origin, the ferric semiquinoid lattice, differently from multifunctional materials based on tetrathiafulvalene derivatives with paramagnetic counterions where separate sub-lattices furnish the two distinct magnetic/conducting properties. Therefore they represent a challenge for pursuing magnetoelectric or multiferroic MOFs. **45** represents also the first structurally characterized example of a microporous magnet containing the Cl_2An^{3-}·chloranilate radical species, showing Tc = 80 K and solvent-induced switching from Tc = 26 to 80 K. Upon removal of DMF and H_2O solvent molecules, this compound undergoes a slight structural distortion, which is fully reversible, to give the desolvated phase $(Me_2NH_2)_2Fe_2(Cl_2An)_3]$, **45a**, and a fit to N_2 adsorption data, at 77 K, of this activated compound gives a BET surface area of $885(105)\,m^2/g$ (the second highest reported for a porous magnet up to now) confirming the presence of permanent microporosity. These results highlight the ability of redox-active anilate ligands to generate 2D magnets with permanent porosity. **46** is another interesting example of a monometallic lanthanoid assembly which consist of a 3-D network framework showing regular square-grid channels and ferromagnetism with a curie Temperature of 11 K. This is the first structurally characterized example of magnetic lanthanoid assemblies opening the way to the preparation of 3D magnetic/luminescent MOFs.

Acknowledgments: This work was supported by the Fondazione di Sardegna—Convenzione triennale tra la Fondazione di Sardegna e gli Atenei Sardi, Regione Sardegna—L.R. 7/2007 annualità 2016—DGR 28/21 del 17.05.2015 "Innovative Molecular Functional Materials for Environmental and Biomedical Applicatons". Special Thanks are due to the Guest Editor, Manuel Leite Almeida C2TN, Instituto Superior Técnico, Universidade de Lisboa, Portugal, for its kind invitation to give a contribution to this Special Issue "Magnetism of Molecular Conductors".

Conflicts of Interest: The authors declare no conflict of interest.

References

1. Khanna, J.M.; Malone, M.H.; Euler, K.L.; Brady, L.R. Atromentin, anticoagulant from hydnellum diabolus. *J. Pharm. Sci.* **1965**, *54*, 1016–1020. [CrossRef] [PubMed]
2. Zhang, B.; Salituro, G.; Szalkowski, D.; Li, Z.; Zhang, Y.; Royo, I.; Vilella, D.; Diez, M.T.; Pelaez, F.; Ruby, C.; et al. Discovery of a small molecule insulin mimetic with antidiabetic activity in mice. *Science* **1999**, *284*, 974–977. [CrossRef] [PubMed]
3. Tsukamoto, S.; Macabalang, A.D.; Abe, T.; Hirota, H.; Ohta, T. Thelephorin a: A new radical scavenger from the mushroom thelephora vialis. *Tetrahedron* **2002**, *58*, 1103–1105. [CrossRef]
4. Puder, C.; Wagner, K.; Vettermann, R.; Hauptmann, R.; Potterat, O. Terphenylquinone inhibitors of the src protein tyrosine kinase from *stilbella* sp. *J. Nat. Prod.* **2005**, *68*, 323–326. [CrossRef] [PubMed]

5.   Liu, K.; Xu, L.; Szalkowski, D.; Li, Z.; Ding, V.; Kwei, G.; Huskey, S.; Moller, D.E.; Heck, J.V.; Zhang, B.B.; et al. Discovery of a potent, highly selective, and orally efficacious small-molecule activator of the insulin receptor. *J. Med. Chem.* **2000**, *43*, 3487–3494. [CrossRef] [PubMed]

6.   Wood, H.B., Jr.; Black, R.; Salituro, G.; Szalkowski, D.; Li, Z.; Zhang, Y.; Moller, D.E.; Zhang, B.; Jones, A.B. The basal sar of a novel insulin receptor activator. *Bioorgan. Med. Chem. Lett.* **2000**, *10*, 1189–1192. [CrossRef]

7.   Kitagawa, S.; Kawata, S. Coordination compounds of 1,4-dihydroxybenzoquinone and its homologues. Structures and properties. *Coord. Chem. Rev.* **2002**, *224*, 11–34. [CrossRef]

8.   Barltrop, J.A.; Burstall, M.L. 435. The synthesis of tetracyclines. Part I. Some model diene reactions. *J. Chem. Soc. (Resumed)* **1959**, 2183–2186. [CrossRef]

9.   Jones, R.G.; Shonle, H.A. The preparation of 2,5-dihydroxyquinone. *J. Am. Chem. Soc.* **1945**, *67*, 1034–1035. [CrossRef]

10.   Viault, G.; Gree, D.; Das, S.; Yadav, J.S.; Grée, R. Synthesis of a Focused Chemical Library Based on Derivatives of Embelin, a Natural Product with Proapoptotic and Anticancer Properties. *Eur. J. Org. Chem.* **2011**, *7*, 1233–1244. [CrossRef]

11.   Wallenfels, K.; Friedrich, K. Über fluorchinone, ii. Zur hydrolyse und alkoholyse des fluoranils. *Chem. Ber.* **1960**, *93*, 3070–3082. [CrossRef]

12.   Stenhouse, J. On chloranil and bromanil. *J. Chem. Soc.* **1870**, *23*, 6–14. [CrossRef]

13.   Benmansour, S.; Vallés-García, C.; Gómez-García, C.J. A H-bonded chloranilate chain with an unprecedented topology. *Struct. Chem. Crystallogr. Commun.* **2015**, *1*, 1–7.

14.   Torrey, H.A.; Hunter, W.H. The action of iodides on bromanil. Iodanil and some of its derivatives. *J. Am. Chem. Soc.* **1912**, *34*, 702–716. [CrossRef]

15.   Meyer, H.O. Eine neue Synthese der Nitranilsäure. *Berichte der Deutschen Chemischen Gesellschaft (A and B Series)* **1924**, *57*, 326–328. [CrossRef]

16.   Fatiadi, A.J.; Sager, W.F. Tetrahydroxyquinone. *Org. Synth.* **1962**, *42*, 90.

17.   Gelormini, O.; Artz, N.E. The oxidation of inosite with nitric acid. *J. Am. Chem. Soc.* **1930**, *52*, 2483–2494. [CrossRef]

18.   Hoglan, F.A.; Bartow, E. Preparation and properties of derivatives of inositol. *J. Am. Chem. Soc.* **1940**, *62*, 2397–2400. [CrossRef]

19.   Junek, H.; Unterweger, B.; Peltzmann, R.Z. Notizen: Eine einfache synthese von tetrahydroxybenzochinon-1.4/A simple synthesis of tetrahydroxy-benzoquinone-1,4. *Z. Naturforsch. B* **1978**, *33B*, 1201–1203. [CrossRef]

20.   Preisler, P.W.; Berger, L. Preparation of tetrahydroxyquinone and rhodizonic acid salts from the product of the oxidation of inositol with nitric acid. *J. Am. Chem. Soc.* **1942**, *64*, 67–69. [CrossRef]

21.   Zaman, B.M.; Morita, Y.; Toyoda, J.; Yamochi, H.; Sekizaki, S.; Nakasuji, K. Convenient preparation and properties of 2,5-dichloro- and 2,5-dibromo-3,6-dicyano-1,4-benzoquinone (cddq and cbdq): Ddq analogs with centrosymmetry. *Mol. Cryst. Liq. Cryst. Sci. Technol. Section A Mol. Cryst. Liq. Cryst.* **1996**, *287*, 249–254. [CrossRef]

22.   Wallenfels, K.; Bachmann, G.; Hofmann, D.; Kern, R. Cyansubstituierte chinone—II: 2,3-, 2,5-2,6-dicyanchinone und tetracyanbenzochinon. *Tetrahedron* **1965**, *21*, 2239–2256. [CrossRef]

23.   Rehwoldt, R.E.; Chasen, B.L.; Li, J.B. 2-chloro-5-cyano-3,6-dihydroxybenzoquinone, a new analytical reagent for the spectrophotometric determination of calcium(II). *Anal. Chem.* **1966**, *38*, 1018–1019. [CrossRef]

24.   Akutagawa, T.; Nakamura, T. Crystal and electronic structures of hydrogen-bonded 2,5-diamino-3,6-dihydroxy-p-benzoquinone. *Cryst. Growth Des.* **2006**, *6*, 70–74. [CrossRef]

25.   Kögl, F.; Lang, A. Über den mechanismus der fichterschen synthese von dialkyl-dioxy-chinonen. *Eur. J. Inorg. Chem.* **1926**, *59*, 910–913. [CrossRef]

26.   Fichter, F.; Willmann, A. Ueber synthesen dialkylirter dioxychinone durch ringschluss. *Ber. Deutsch. Chem. Ges.* **1904**, *37*, 2384–2390. [CrossRef]

27.   Fichter, F. Ueber synthetische p-dialkylirte dioxychinone. *Justus Liebigs Ann. Chem.* **1908**, *361*, 363–402. [CrossRef]

28.   Atzori, M.; Pop, F.; Cauchy, T.; Mercuri, M.L.; Avarvari, N. Thiophene-benzoquinones: Synthesis, crystal structures and preliminary coordination chemistry of derived anilate ligands. *Org. Biomol. Chem.* **2014**, *12*, 8752–8763. [CrossRef] [PubMed]

29. Min, K.S.; DiPasquale, A.G.; Rheingold, A.L.; White, H.S.; Miller, J.S. Observation of redox-induced electron transfer and spin crossover for dinuclear cobalt and iron complexes with the 2,5-di-tert-butyl-3,6-dihydroxy-1,4-benzoquinonate bridging ligand. *J. Am. Chem. Soc.* **2009**, *131*, 6229–6236. [CrossRef] [PubMed]

30. Semmingsen, D. The crystal and molecular structure of 2,5-dihydroxybenzoquinone at-1620c. *Acta Chem. Scand. B* **1977**, *31*, 11–14. [CrossRef]

31. Munakata, M.; Wu, L.P.; Kuroda-Sowa, T.; Yamamoto, M.; Maekawa, M.; Moriwaki, K. Assembly of a mixed-valence cu(i/ii) system coupled by multiple hydrogen bonding through tetrahydroxybenzoquinone. *Inorg. Chim. Acta* **1998**, *268*, 317–321. [CrossRef]

32. Klug, A. The crystal structure of tetrahydroxy-p-benzoquinone. *Acta Crystallogr.* **1965**, *19*, 983–992. [CrossRef]

33. Robl, C. Crystal structure and hydrogen bonding of 2,5-dihydroxy-3,6-dimethyl-p-benzoquinone. *Z. Krist. Cryst. Mater.* **1988**, *184*, 289–293. [CrossRef]

34. Andersen, E.K.; Andersen, I.G.K. The crystal and molecular structure of hydroxyquinones and salts of hydroxyquinones. Vii. Hydronium cyananilate (cyananilic acid hexahydrate) and hydronium nitranilate (a redetermination). *Acta Crystallogra. Sect. B* **1975**, *31*, 379–383. [CrossRef]

35. Andersen, E.K.; Andersen, I.G.K. The crystal and molecular structure of hydroxyquinones and salts of hydroxyquinones. VIII. Fluoranilic acid. *Acta Crystallogr. Sect. B* **1975**, *31*, 384–387. [CrossRef]

36. Andersen, E. The crystal and molecular structure of hydroxyquinones and salts of hydroxyquinones. I. Chloranilic acid. *Acta Crystallogr.* **1967**, *22*, 188–191. [CrossRef]

37. Andersen, E. The crystal and molecular structure of hydroxyquinones and salts of hydroxyquinones. II. Chloranilic acid dihydrate. *Acta Crystallogr.* **1967**, *22*, 191–196. [CrossRef]

38. Andersen, E. The crystal and molecular structure of hydroxyquinones and salts of hydroxyquinones. V. Hydronium nitranilate, nitranilic acid hexahydrate. *Acta Crystallogr.* **1967**, *22*, 204–208. [CrossRef]

39. Molcanov, K.; Stare, J.; Vener, M.V.; Kojic-Prodic, B.; Mali, G.; Grdadolnik, J.; Mohacek-Grosev, V. Nitranilic acid hexahydrate, a novel benchmark system of the zundel cation in an intrinsically asymmetric environment: Spectroscopic features and hydrogen bond dynamics characterised by experimental and theoretical methods. *Phys. Chem. Chem. Phys.* **2014**, *16*, 998–1007. [CrossRef] [PubMed]

40. Andersen, E. The crystal and molecular structure of hydroxyquinones and salts of hydroxyquinones. III. Ammonium chloranilate monohydrate. *Acta Crystallogr.* **1967**, *22*, 196–201. [CrossRef]

41. Andersen, E. The crystal and molecular structure of hydroxyquinones and salts of hydroxyquinones. IV. Ammonium nitranilate. *Acta Crystallogr.* **1967**, *22*, 201–203. [CrossRef]

42. Biliskov, N.; Kojic-Prodic, B.; Mali, G.; Molcanov, K.; Stare, J. A Partial Proton Transfer in Hydrogen Bond O–H⋯O in Crystals of Anhydrous Potassium and Rubidium Complex Chloranilates. *J. J. Phys. Chem. A* **2011**, *115*, 3154–3166. [CrossRef] [PubMed]

43. Molcanov, K.; Sabljic, I.; Kojic-Prodic, B. Face-to-face p-stacking in the multicomponent crystals of chloranilic acid, alkali hydrogenchloranilates, and water. *CrystEngComm* **2011**, *13*, 4211. [CrossRef]

44. Molcanov, K.; Juric, M.; Kojic-Prodic, B. Stacking of metal chelating rings with π-systems in mononuclear complexes of copper(II) with 3,6-dichloro-2,5-dihydroxy-1,4-benzoquinone (chloranilic acid) and 2,2′-bipyridine ligands. *Dalton Trans.* **2013**, *42*, 15756–15765. [CrossRef] [PubMed]

45. Molcanov, K.; Kojic-Prodic, B. Face-to-face stacking of quinoid rings of alkali salts of bromanilic acid. *Acta Crystallogr. Section B* **2012**, *68*, 57–65. [CrossRef] [PubMed]

46. Molcanov, K.; Kojic-Prodic, B.; Meden, A. π-stacking of quinoid rings in crystals of alkali diaqua hydrogen chloranilates. *CrystEngComm* **2009**, *11*, 1407–1415. [CrossRef]

47. Robl, C. Complexes with substituted 2,5-dihydroxy-p-benzoquinones: The inclusion compounds [Y(H$_2$O)$_3$]$_2$ (C$_6$Cl$_2$O$_4$)$_3$·6H$_2$O and [Y(H$_2$O)$_3$]$_2$ (C$_6$Br$_2$O$_4$)$_3$·6H$_2$O. *Mater. Res. Bull.* **1987**, *22*, 1483–1491. [CrossRef]

48. Dei, A.; Gatteschi, D.; Pardi, L.; Russo, U. Tetraoxolene radical stabilization by the interaction with transition-metal ions. *Inorg. Chem.* **1991**, *30*, 2589–2594. [CrossRef]

49. Coronado, E.; Galan-Mascaros, J.R.; Gomez-Garcia, C.J.; Laukhin, V. Coexistence of ferromagnetism and metallic conductivity in a molecule-based layered compound. *Nature* **2000**, *408*, 447–449. [CrossRef] [PubMed]

50. Anil Reddy, M.; Vinayak, B.; Suresh, T.; Niveditha, S.; Bhanuprakash, K.; Prakash Singh, S.; Islam, A.; Han, L.; Chandrasekharam, M. Highly conjugated electron rich thiophene antennas on phenothiazine and phenoxazine-based sensitizers for dye sensitized solar cells. *Synth. Metals* **2014**, *195*, 208–216. [CrossRef]

51. Schweinfurth, D.; Klein, J.; Hohloch, S.; Dechert, S.; Demeshko, S.; Meyer, F.; Sarkar, B. Influencing the coordination mode of tbta (tbta = tris[(1-benzyl-1H-1,2,3-triazol-4-yl)methyl]amine) in dicobalt complexes through changes in metal oxidation states. *Dalton Trans.* **2013**, *42*, 6944–6952. [CrossRef] [PubMed]

52. Schweinfurth, D.; Khusniyarov, M.M.; Bubrin, D.; Hohloch, S.; Su, C.-Y.; Sarkar, B. Tuning spin–spin coupling in quinonoid-bridged dicopper(II) complexes through rational bridge variation. *Inorg. Chem.* **2013**, *52*, 10332–10339. [CrossRef] [PubMed]

53. Baum, A.E.; Lindeman, S.V.; Fiedler, A.T. Preparation of a semiquinonate-bridged diiron(II) complex and elucidation of its geometric and electronic structures. *Chem. Commun.* **2013**, *49*, 6531–6533. [CrossRef] [PubMed]

54. Nie, J.; Li, G.-L.; Miao, B.-X.; Ni, Z.-H. Syntheses, Structures and Magnetic Properties of Dinuclear Cobalt(II) Complexes [Co$_2$(TPEA)$_2$(DHBQ)](ClO$_4$)$_2$ and [Co$_2$(TPEA)$_2$(DHBQ)](PF$_6$)$_2$. *J. Chem. Crystallogr.* **2013**, *43*, 331. [CrossRef]

55. Wu, D.-Y.; Huang, W.; Wang, L.; Wu, G. Synthesis, structure, and magnetic properties of a dinuclear antiferromagnetically coupled cobalt complex. *Z. Anorg. Allg. Chem.* **2012**, *638*, 401–404. [CrossRef]

56. Chatterjee, P.B.; Bhattacharya, K.; Kundu, N.; Choi, K.-Y.; Clérac, R.; Chaudhury, M. Vanadium-induced nucleophilic ipso substitutions in a coordinated tetrachlorosemiquinone ring: Formation of the chloranilate anion as a bridging ligand. *Inorg. Chem.* **2009**, *48*, 804–806. [CrossRef] [PubMed]

57. Bruijnincx Pieter, C.A.; Viciano-Chumillas, M.; Lutz, M.; Spek, A.L.; Reedijk, J.; van Koten, G.; Klein Gebbink, R.J.M. Oxidative double dehalogenation of tetrachlorocatechol by a bio-inspired cu ii complex: Formation of chloranilic acid. *Chemistry* **2008**, *14*, 5567–5576. [CrossRef] [PubMed]

58. Ghumaan, S.; Sarkar, B.; Maji, S.; Puranik, V.G.; Fiedler, J.; Urbanos, F.A.; Jimenez-Aparicio, R.; Kaim, W.; Lahiri, G.K. Valence-state analysis through spectroelectrochemistry in a series of quinonoid-bridged diruthenium complexes [(acac)$_2$Ru(μ-l)ru(acac)2]$_n$ (n = +2, +1, 0, −1, −2). *Chemistry* **2008**, *14*, 10816–10828. [CrossRef] [PubMed]

59. Guo, D.; McCusker, J.K. Spin exchange effects on the physicochemical properties of tetraoxolene-bridged bimetallic complexes. *Inorg. Chem.* **2007**, *46*, 3257–3274. [CrossRef] [PubMed]

60. Min, K.S.; Rheingold, A.L.; DiPasquale, A.; Miller, J.S. Characterization of the chloranilate(·3−) π radical as a strong spin-coupling bridging ligand. *Inorg. Chem.* **2006**, *45*, 6135–6137. [CrossRef] [PubMed]

61. Yu, F.; Xiang, M.; Wu, Q.-G.; He, H.; Cheng, S.-Q.; Cai, X.-Y.; Li, A.-H.; Zhang, Y.-M.; Li, B. Valence tautomerism and photodynamics observed in a dinuclear cobalt-tetraoxolene compound. *Inorg. Chim. Acta* **2015**, *426*, 146–149. [CrossRef]

62. Li, B.; Chen, L.-Q.; Tao, J.; Huang, R.-B.; Zheng, L.-S. Unidirectional charge transfer in di-cobalt valence tautomeric compound finely tuned by ancillary ligand. *Inorg. Chem.* **2013**, *52*, 4136–4138. [CrossRef] [PubMed]

63. Li, B.; Tao, J.; Sun, H.-L.; Sato, O.; Huang, R.-B.; Zheng, L.-S. Side-effect of ancillary ligand on electron transfer and photodynamics of a dinuclear valence tautomeric complex. *Chem. Commun.* **2008**, 2269–2271. [CrossRef] [PubMed]

64. Tao, J.; Maruyama, H.; Sato, O. Valence tautomeric transitions with thermal hysteresis around room temperature and photoinduced effects observed in a cobalt—Tetraoxolene complex. *J. Am. Chem. Soc.* **2006**, *128*, 1790–1791. [CrossRef] [PubMed]

65. Ishikawa, R.; Horii, Y.; Nakanishi, R.; Ueno, S.; Breedlove, B.K.; Yamashita, M.; Kawata, S. Field-induced single-ion magnetism based on spin-phonon relaxation in a distorted octahedral high-spin cobalt(II) complex. *Eur. J. Inorg. Chem.* **2016**, 3233–3239. [CrossRef]

66. Horiuchi, S.; Kumai, R.; Tokura, Y. High-temperature and pressure-induced ferroelectricity in Hydrogen-bonded supramolecular crystals of anilic acids and 2,3-di(2-pyridinyl)pyrazine. *J. Am. Chem. Soc.* **2013**, *135*, 4492–4500. [CrossRef] [PubMed]

67. Kagawa, F.; Horiuchi, S.; Minami, N.; Ishibashi, S.; Kobayashi, K.; Kumai, R.; Murakami, Y.; Tokura, Y. Polarization switching ability dependent on multidomain topology in a uniaxial organic ferroelectric. *Nano Lett.* **2014**, *14*, 239–243. [CrossRef] [PubMed]

68. Murata, T.; Yakiyama, Y.; Nakasuji, K.; Morita, Y. Proton-transfer salts between an EDT-TTF derivative having imidazole-ring and anilic acids: Multi-dimensional networks by acid-base hydrogen-bonds, pi-stacks and chalcogen atom interactions. *Crystengcomm* **2011**, *13*, 3689–3691. [CrossRef]

69. Horiuchi, S.; Kumai, R.; Tokura, Y. Room-temperature ferroelectricity and gigantic dielectric susceptibility on a supramolecular architecture of phenazine and deuterated chloranilic acid. *J. Am. Chem. Soc.* **2005**, *127*, 5010–5011. [CrossRef] [PubMed]

70. Ward, M.D.; McCleverty, J.A. Non-innocent behaviour in mononuclear and polynuclear complexes: Consequences for redox and electronic spectroscopic properties. *J. Chem. Soc. Dalton Trans.* **2002**, 275–288. [CrossRef]

71. Tinti, F.; Verdaguer, M.; Kahn, O.; Savariault, J.M. Interaction between copper(II) ions separated by 7.6 A. Crystal structure and magnetic properties of the μ-iodanilato bis[n, n, n′, n′ tetramethylethylenediamine copper(II)] diperchlorate. *Inorg. Chem.* **1987**, *26*, 2380–2384. [CrossRef]

72. Tamaki, H.; Zhong, Z.J.; Matsumoto, N.; Kida, S.; Koikawa, M.; Achiwa, N.; Hashimoto, Y.; Ōkawa, H. Design of metal-complex magnets. Syntheses and magnetic properties of mixed-metal assemblies {NBu₄[MCr(ox)₃]}ₓ (NBu₄⁺ = tetra(n-butyl)ammonium ion; ox²⁻ = oxalate ion; M = Mn²⁺, Fe²⁺, Co²⁺, Ni²⁺, Cu²⁺, Zn²⁺). *J. Am. Chem. Soc.* **1992**, *114*, 6974–6979. [CrossRef]

73. Decurtins, S.; Schmalle, H.W.; Oswald, H.R.; Linden, A.; Ensling, J.; Gütlich, P.; Hauser, A. A polymeric two-dimensional mixed-metal network. Crystal structure and magnetic properties of {[P(Ph)₄][MnCr(ox)₃]}. *Inorg. Chim. Acta* **1994**, *216*, 65–73. [CrossRef]

74. Atovmyan, L.O.; Shilov, G.V.; Lyubovskaya, R.N.; Zhilyaeva, E.I.; Ovanesyan, N.S.; Pirumova, S.I.; Gusakovskaya, I.G.; Morozov, Y.G. Crystal-structure of the molecular ferromagnet NBu4[MnCr(C₂O₄)₃] (Bu = N-C₄H₉). *J. Exp. Theor. Phys.* **1993**, *58*, 766–769.

75. Mathonière, C.; Nuttall, C.J.; Carling, S.G.; Day, P. Ferrimagnetic mixed-valency and mixed-metal tris(oxalato)iron(III) compounds: Synthesis, structure, and magnetism. *Inorg. Chem.* **1996**, *35*, 1201–1206. [CrossRef] [PubMed]

76. Coronado, E.; Galan-Mascaros, J.R.; Gómez-García, C.J.; Ensling, J.; Gutlich, P. Hybrid molecular magnets obtained by insertion of decamethylmetallocenium cations in layered bimetallic oxalate complexes. Syntheses, structure and magnetic properties of the series [Z^III Cp*₂][M^II M^III (ox)₃] (Z^III = Co, Fe; MIII = Cr, Fe; MII = Mn, Fe, Co, Ni, Cu; Cp* = pentamethylcyclopentadienyl). *Eur. J. Inorg. Chem.* **2000**, *6*, 552–563.

77. Coronado, E.; Galán-Mascarós, J.R.; Gómez-García, C.J.; Martínez-Agudo, J.M. Increasing the coercivity in layered molecular-based magnets A[M^II M^III (ox)₃] (M^II = Mn, Fe, Co, Ni, Cu; M^III = Cr, Fe; ox = oxalate; A = organic or organometallic cation). *Adv. Mater.* **1999**, *11*, 558–561. [CrossRef]

78. Coronado, E.; Clemente-Leon, M.; Galan-Mascaros, J.R.; Gimenez-Saiz, C.; Gomez-Garcia, C.J.; Martinez-Ferrero, E. Design of molecular materials combining magnetic, electrical and optical properties. *J. Chem. Soc. Dalton Trans.* **2000**, 3955–3961. [CrossRef]

79. Clemente-Leon, M.; Coronado, E.; Galan-Mascaros, J.R.; Gomez-Garcia, C.J. Intercalation of decamethylferrocenium cations in bimetallic oxalate-bridged two-dimensional magnets. *Chem. Commun.* **1997**, 1727–1728. [CrossRef]

80. Coronado, E.; Galán-Mascarós, J.R.; Gómez-García, C.J.; Martínez-Agudo, J.M.; Martínez-Ferrero, E.; Waerenborgh, J.C.; Almeida, M. Layered molecule-based magnets formed by decamethylmetallocenium cations and two-dimensional bimetallic complexes [M^II Ru^III (ox)₃]-(M^II =;Mn, Fe, Co, Cu and Zn; ox = oxalate). *J. Solid State Chem.* **2001**, *159*, 391–402. [CrossRef]

81. Bénard, S.; Yu, P.; Audière, J.P.; Rivière, E.; Clément, R.; Guilhem, J.; Tchertanov, L.; Nakatani, K. Structure and NLO properties of layered bimetallic oxalato-bridged ferromagnetic networks containing stilbazolium-shaped chromophores. *J. Am. Chem. Soc.* **2000**, *122*, 9444–9454. [CrossRef]

82. Bénard, S.; Rivière, E.; Yu, P.; Nakatani, K.; Delouis, J.F. A photochromic molecule-based magnet. *Chem. Mater.* **2001**, *13*, 159–162. [CrossRef]

83. Alberola, A.; Coronado, E.; Galán-Mascarós, J.R.; Giménez-Saiz, C.; Gómez-García, C.J. A molecular metal ferromagnet from the organic donor bis(ethylenedithio)tetraselenafulvalene and bimetallic oxalate complexes. *J. Am. Chem. Soc.* **2003**, *125*, 10774–10775. [CrossRef] [PubMed]

84. Aldoshin, S.M.; Nikonova, L.A.; Shilov, G.V.; Bikanina, E.A.; Artemova, N.K.; Smirnov, V.A. The influence of an n-substituent in the indoline fragment of pyrano-pyridine spiropyran salts on their crystalline structure and photochromic properties. *J. Mol. Struct.* **2006**, *794*, 103–109. [CrossRef]

85. Aldoshin, S.M.; Sanina, N.A.; Minkin, V.I.; Voloshin, N.A.; Ikorskii, V.N.; Ovcharenko, V.I.; Smirnov, V.A.; Nagaeva, N.K. Molecular photochromic ferromagnetic based on the layered polymeric tris-oxalate of Cr(III), Mn(II) and 1-[(1′,3′,3′-trimethyl-6-nitrospiro[2H-1-benzopyran-2,2′-indoline]-8-yl)methyl]pyridinium. *J. Mol. Struct.* **2007**, *826*, 69–74. [CrossRef]

86. Kida, N.; Hikita, M.; Kashima, I.; Okubo, M.; Itoi, M.; Enomoto, M.; Kato, K.; Takata, M.; Kojima, N. Control of charge transfer phase transition and ferromagnetism by photoisomerization of spiropyran for an organic—Inorganic hybrid system, (SP)[Fe^{II}Fe^{III}(dto)$_3$] (SP = spiropyran, dto = C$_2$O$_2$S$_2$). *J. Am. Chem. Soc.* **2009**, *131*, 212–220. [CrossRef] [PubMed]

87. Sieber, R.; Decurtins, S.; Stoeckli-Evans, H.; Wilson, C.; Yufit, D.; Howard, J.A.K.; Capelli, S.C.; Hauser, A. A thermal spin transition in [Co(bpy)$_3$][LiCr(ox)$_3$] (ox = C$_2$O$_4{}^{2-}$; bpy = 2,2′-bipyridine). *Chemistry* **2000**, *6*, 361–368. [CrossRef]

88. Clemente-León, M.; Coronado, E.; López-Jordà, M.; Waerenborgh, J.C. Multifunctional magnetic materials obtained by insertion of spin-crossover Fe^{III} complexes into chiral 3D bimetallic oxalate-based ferromagnets. *Inorg. Chem.* **2011**, *50*, 9122–9130. [CrossRef] [PubMed]

89. Clemente-León, M.; Coronado, E.; López-Jordà, M.; Mínguez Espallargas, G.; Soriano-Portillo, A.; Waerenborgh, J.C. Multifunctional magnetic materials obtained by insertion of a spin-crossover Fe^{III} complex into bimetallic oxalate-based ferromagnets. *Chemistry* **2010**, *16*, 2207–2219. [CrossRef] [PubMed]

90. Clemente-Leon, M.; Coronado, E.; Lopez-Jorda, M.; Desplanches, C.; Asthana, S.; Wang, H.; Letard, J.-F. A hybrid magnet with coexistence of ferromagnetism and photoinduced Fe(III) spin-crossover. *Chem. Sci.* **2011**, *2*, 1121–1127. [CrossRef]

91. Clemente-Leon, M.; Coronado, E.; Lopez-Jorda, M. 2D and 3D bimetallic oxalate-based ferromagnets prepared by insertion of different Fe^{III} spin crossover complexes. *Dalton Trans.* **2010**, *39*, 4903–4910. [CrossRef] [PubMed]

92. Clemente-León, M.; Coronado, E.; Giménez-López, M.C.; Soriano-Portillo, A.; Waerenborgh, J.C.; Delgado, F.S.; Ruiz-Pérez, C. Insertion of a spin crossover Fe^{III} complex into an oxalate-based layered material: Coexistence of spin canting and spin crossover in a hybrid magnet. *Inorg. Chem.* **2008**, *47*, 9111–9120. [CrossRef] [PubMed]

93. Train, C.; Gheorghe, R.; Krstic, V.; Chamoreau, L.-M.; Ovanesyan, N.S.; Rikken, G.L.; Gruselle, M.; Verdaguer, M. Strong magneto-chiral dichroism in enantiopure chiral ferromagnets. *Nat. Mater.* **2008**, *7*, 729–734. [CrossRef] [PubMed]

94. Gruselle, M.; Train, C.; Boubekeur, K.; Gredin, P.; Ovanesyan, N. Enantioselective self-assembly of chiral bimetallic oxalate-based networks. *Coord. Chem. Rev.* **2006**, *250*, 2491–2500. [CrossRef]

95. Clemente-León, M.; Coronado, E.; Dias, J.C.; Soriano-Portillo, A.; Willett, R.D. Synthesis, structure, and magnetic properties of [(s)-[PhCH(CH$_3$)n(CH$_3$)$_3$]][Mn(CH$_3$CN)$_{2/3}$Cr(ox)$_3$]·(CH$_3$CN)_(solvate), a 2D chiral magnet containing a quaternary ammonium chiral cation. *Inorg. Chem.* **2008**, *47*, 6458–6463. [CrossRef] [PubMed]

96. Brissard, M.; Gruselle, M.; Malézieux, B.; Thouvenot, R.; Guyard-Duhayon, C.; Convert, O. An anionic {[MnCo(ox)$_3$]$^-$}$_n$ network with appropriate cavities for the enantioselective recognition and resolution of the hexacoordinated monocation [Ru(bpy)$_2$(ppy)]$^+$ (bpy = bipyridine, ppy = phenylpyridine). *Eur. J. Inorg. Chem.* **2001**, *2001*, 1745–1751. [CrossRef]

97. Sadakiyo, M.; Ōkawa, H.; Shigematsu, A.; Ohba, M.; Yamada, T.; Kitagawa, H. Promotion of low-humidity proton conduction by controlling hydrophilicity in layered metal–organic frameworks. *J. Am. Chem. Soc.* **2012**, *134*, 5472–5475. [CrossRef] [PubMed]

98. Ōkawa, H.; Shigematsu, A.; Sadakiyo, M.; Miyagawa, T.; Yoneda, K.; Ohba, M.; Kitagawa, H. Oxalate-bridged bimetallic complexes {NH(prol)$_3$}[MCr(ox)$_3$] (M = Mn^{II}, Fe^{II}, Co^{II}; NH(prol)$^{3+}$ = tri(3-hydroxypropyl)ammonium) exhibiting coexistent ferromagnetism and proton conduction. *J. Am. Chem. Soc.* **2009**, *131*, 13516–13522. [CrossRef] [PubMed]

99. Fishman, R.S.; Clemente-León, M.; Coronado, E. Magnetic compensation and ordering in the bimetallic oxalates: Why are the 2D and 3D series so different? *Inorg. Chem.* **2009**, *48*, 3039–3046. [CrossRef] [PubMed]

100. Clément, R.; Decurtins, S.; Gruselle, M.; Train, C. Polyfunctional two- (2D) and three- (3D) dimensional oxalate bridged bimetallic magnets. *Mon. Chem. Chem. Mon.* **2003**, *134*, 117–135. [CrossRef]

101. Kojima, N.; Aoki, W.; Itoi, M.; Ono, Y.; Seto, M.; Kobayashi, Y.; Maeda, Y. Charge transfer phase transition and Ferromagnetism in a mixed-valence iron complex, (N-C$_3$H$_7$)$_4$n[Fe^{II}Fe^{III}(dto)$_3$] (dto = C$_2$O$_2$S$_2$). *Solid State Commun.* **2001**, *120*, 165–170. [CrossRef]

102. Hisashi, O.; Minoru, M.; Masaaki, O.; Masahito, K.; Naohide, M. Dithiooxalato(dto)-bridged bimetallic assemblies {NPr4[MCr(dto)3]}$_x$ (M = Fe, Co, Ni, Zn; NPr4 = tetrapropylammonium ion): New complex-based ferromagnets. *Bull. Chem. Soc. Jpn* **1994**, *67*, 2139–2144.

103. Carling, S.G.; Bradley, J.M.; Visser, D.; Day, P. Magnetic and structural characterisation of the layered materials AMnFe(C$_2$S$_2$O$_2$)$_3$. *Polyhedron* **2003**, *22*, 2317–2324. [CrossRef]

104. Bradley, J.M.; Carling, S.G.; Visser, D.; Day, P.; Hautot, D.; Long, G.J. Structural and physical properties of the ferromagnetic tris-dithiooxalato compounds, A[M^{II}Cr^{III}(C$_2$S$_2$O$_2$)$_3$], with A$^+$ = N(n-C$_n$H$_{2n+1}$)$^{4+}$ (n = 3–5) and P(C$_6$H$_5$)$_4$$^+$ and MII = Mn, Fe, Co, and Ni. *Inorg. Chem.* **2003**, *42*, 986–996. [CrossRef] [PubMed]

105. Weiss, A.; Riegler, E.; Robl, C. Polymeric 2,5-dihydroxy-1,4-benzoquinone transition metal complexes Na$_2$(H$_2$O)$_{24}$[M$_2$(C$_6$H$_2$O$_4$)$_3$] (M = Manganese(2+), Cadmium(2+). *Z. Naturforsch. Teil B Anorg. Chem.Org. Chem.* **1986**, *41*, 1501–1505.

106. Shilov, G.V.; Nikitina, Z.K.; Ovanesyan, N.S.; Aldoshin, S.M.; Makhaev, V.D. Phenazineoxonium chloranilatomanganate and chloranilatoferrate: Synthesis, structure, magnetic properties, and mössbauer spectra. *Russ. Chem. Bull.* **2011**, *60*, 1209–1219. [CrossRef]

107. Luo, T.-T.; Liu, Y.-H.; Tsai, H.-L.; Su, C.-C.; Ueng, C.-H.; Lu, K.-L. A novel hybrid supramolecular network assembled from perfect π-π stacking of an anionic inorganic layer and a cationic hydronium-ion-mediated organic layer. *Eur. J. Inorg. Chem.* **2004**, *2004*, 4253–4258. [CrossRef]

108. Abrahams, B.F.; Coleiro, J.; Hoskins, B.F.; Robson, R. Gas hydrate-like pentagonal dodecahedral M$_2$(H$_2$O)$_{18}$ cages (M = lanthanide or y) in 2,5-dihydroxybenzoquinone-derived coordination polymers. *Chem. Commun.* **1996**, 603–604. [CrossRef]

109. Abrahams, B.F.; Coleiro, J.; Ha, K.; Hoskins, B.F.; Orchard, S.D.; Robson, R. Dihydroxybenzoquinone and chloranilic acid derivatives of rare earth metals. *J. Chemical. Soc. Dalton Trans.* **2002**, 1586–1594. [CrossRef]

110. Coronado, E.; Galán-Mascarós, J.R.; Gómez-García, C.J.; Martínez-Agudo, J.M. Molecule-based magnets formed by bimetallic three-dimensional oxalate networks and chiral tris(bipyridyl) complex cations. The series [ZII(bpy)$_3$][ClO$_4$][MIICrIII(ox)$_3$] (ZII = Ru, Fe, Co, and Ni; MII = Mn, Fe, Co, Ni, Cu, and Zn; ox = oxalate dianion). *Inorg. Chem.* **2001**, *40*, 113–120. [CrossRef] [PubMed]

111. Abrahams, B.F.; Hudson, T.A.; McCormick, L.J.; Robson, R. Coordination polymers of 2,5-dihydroxybenzoquinone and chloranilic acid with the (10,3)-atopology. *Crys. Growth Des.* **2011**, *11*, 2717–2720. [CrossRef]

112. Frenzer, W.; Wartchow, R.; Bode, H. Crystal structure of disilver 2,5-dichloro-[1,4]benzoquinone-3,6-diolate, Ag$_2$(C$_6$O$_4$Cl$_2$). *Z. Kristallogr. Cryst. Mater.* **1997**, *212*, 237. [CrossRef]

113. Junggeburth, S.C.; Diehl, L.; Werner, S.; Duppel, V.; Sigle, W.; Lotsch, B.V. Ultrathin 2D coordination polymer nanosheets by surfactant-mediated synthesis. *J. Am. Chem. Soc.* **2013**, *135*, 6157–6164. [CrossRef] [PubMed]

114. Saines, P.J.; Tan, J.-C.; Yeung, H.H.-M.; Barton, P.T.; Cheetham, A.K. Layered inorganic-organic frameworks based on the 2,2-dimethylsuccinate ligand: Structural diversity and its effect on nanosheet exfoliation and magnetic properties. *Dalton Trans.* **2012**, *41*, 8585–8593. [CrossRef] [PubMed]

115. Atzori, M.; Marchiò, L.; Clérac, R.; Serpe, A.; Deplano, P.; Avarvari, N.; Mercuri, M.L. Hydrogen-bonded supramolecular architectures based on tris(hydranilato)metallate(III) (M = Fe, Cr) metallotectons. *Cryst. Growth Des.* **2014**, *14*, 5938–5948. [CrossRef]

116. Min, K.S.; Rhinegold, A.L.; Miller, J.S. Tris(chloranilato)ferrate(iii) anionic building block containing the (dihydroxo)oxodiiron(III) dimer cation: Synthesis and characterization of [(Tpa)(OH)Fe(III)OFe(III) (OH)(Tpa)][Fe(CA)$_3$]$_{0.5}$(BF$_4$)$_{0.5,1.5}$MeOH,H$_2$O [Tpa = tris(2-pyridylmethyl)amine; CA = chloranilate]. *J. Am. Chem. Soc.* **2006**, *128*, 40–41. [PubMed]

117. Hazell, A.; Jensen, K.B.; McKenzie, C.J.; Toftlund, H. Synthesis and reactivity of (.Mu.-oxo)diiron(III) complexes of tris(2-pyridylmethyl)amine. X-ray crystal structures of [Tpa(OH)feofe(H$_2$O)tpa](ClO$_4$)$_3$ and [Tpa(Cl)FeOFe(Cl)Tpa](ClO$_4$)$_2$. *Inorg. Chem.* **1994**, *33*, 3127–3134. [CrossRef]

118. Atzori, M.; Artizzu, F.; Sessini, E.; Marchio, L.; Loche, D.; Serpe, A.; Deplano, P.; Concas, G.; Pop, F.; Avarvari, N.; et al. Halogen-bonding in a new family of tris(haloanilato)metallate(III) magnetic molecular building blocks. *Dalton Transactions* **2014**, *43*, 7006–7019. [CrossRef] [PubMed]

119. Atzori, M.; Artizzu, F.; Marchiò, L.; Loche, D.; Caneschi, A.; Serpe, A.; Deplano, P.; Avarvari, N.; Mercuri, M.L. Switching-on luminescence in anilate-based molecular materials. *Dalton Trans.* **2015**, *44*, 15727–16178. [CrossRef] [PubMed]

120. Benmansour, S.; Valles-Garcia, C.; Gomez-Claramunt, P.; Minguez Espallargas, G.; Gomez-Garcia, C.J. 2d and 3d anilato-based heterometallic M(I)M(III) lattices: The missing link. *Inorg. Chem.* **2015**, *54*, 5410–5418. [CrossRef] [PubMed]

121. Mercuri, M.; Deplano, P.; Serpe, A.; Artizzu, F. Multifunctional materials of interest in molecular electronics. In *Multifunctional Molecular Materials*; Pan Stanford Publishing: Boca Raton, FL, USA, 2013; pp. 219–280.

122. Kurmoo, M.; Graham, A.W.; Day, P.; Coles, S.J.; Hursthouse, M.B.; Caulfield, J.L.; Singleton, J.; Pratt, F.L.; Hayes, W. Superconducting and semiconducting magnetic charge transfer salts: (BEDT-TTF)$_4$AFe (C$_2$O$_4$)$_3$.C6H5CN (A = H$_2$O, K, NH4). *J. Am. Chem. Soc.* **1995**, *117*, 12209–12217. [CrossRef]

123. Coronado, E.; Day, P. Magnetic molecular conductors. *Chem. Rev.* **2004**, *104*, 5419–5448. [CrossRef] [PubMed]

124. Atzori, M.; Benmansour, S.; Minguez Espallargas, G.; Clemente-Leon, M.; Abherve, A.; Gomez-Claramunt, P.; Coronado, E.; Artizzu, F.; Sessini, E.; Deplano, P.; et al. A Family of layered chiral porous magnets exhibiting tunable ordering temperatures. *Inorg. Chem.* **2013**, *52*, 10031–10040. [CrossRef] [PubMed]

125. Kherfi, H.; Hamadène, M.; Guehria-Laïdoudi, A.; Dahaoui, S.; Lecomte, C. Synthesis, structure and thermal behavior of oxalato-bridged Rb$^+$ and H$_3$O$^+$ extended frameworks with different dimensionalities. *Materials* **2010**, *3*, 1281. [CrossRef]

126. Cañadillas-Delgado, L.; Fabelo, O.; Rodríguez-Velamazán, J.A.; Lemée-Cailleau, M.-H.; Mason, S.A.; Pardo, E.; Lloret, F.; Zhao, J.-P.; Bu, X.-H.; Simonet, V.; et al. The role of order–disorder transitions in the quest for molecular multiferroics: Structural and magnetic neutron studies of a mixed valence Iron(II)–Iron(III) formate framework. *J. Am. Chem. Soc.* **2012**, *134*, 19772–19781. [CrossRef] [PubMed]

127. Kobayashi, H.; Cui, H.; Kobayashi, A. Organic metals and superconductors based on bets (bets = bis(ethylenedithio)tetraselenafulvalene). *Chem. Rev.* **2004**, *104*, 5265–5288. [CrossRef] [PubMed]

128. Enoki, T.; Miyazaki, A. Magnetic ttf-based charge-transfer complexes. *Chem. Rev.* **2004**, *104*, 5449–5478. [CrossRef] [PubMed]

129. Coronado, E.; Giménez-Saiz, C.; Gómez-García, C.J. Recent advances in polyoxometalate-containing molecular conductors. *Coord. Chem. Rev.* **2005**, *249*, 1776–1796. [CrossRef]

130. Schlueter, J.A.; Geiser, U.; Whited, M.A.; Drichko, N.; Salameh, B.; Petukhov, K.; Dressel, M. Two alternating BEDT-TTF packing motifs in α-κ-(BEDT-TTF)$_2$Hg(SCN)$_3$. *Dalton Trans.* **2007**, 2580–2588. [CrossRef] [PubMed]

131. Rashid, S.; Turner, S.S.; Day, P.; Howard, J.A.K.; Guionneau, P.; McInnes, E.J.L.; Mabbs, F.E.; Clark, R.J.H.; Firth, S.; Biggs, T. New superconducting charge-transfer salts (BEDT-TTF)$_4$[A·M(C$_2$O$_4$)$_3$]·C$_6$H$_5$NO$_2$ (A = H$_3$O or NH$_4$, M = Cr or Fe, BEDT-TTF = bis(ethylenedithio)tetrathiafulvalene). *J. Mater. Chem.* **2001**, *11*, 2095–2101. [CrossRef]

132. Martin, L.; Turner, S.S.; Day, P.; Mabbs, F.E.; McInnes, E.J.L. New molecular superconductor containing paramagnetic Chromium(III) ions. *Chem. Commun.* **1997**, 1367–1368. [CrossRef]

133. Uji, S.; Shinagawa, H.; Terashima, T.; Yakabe, T.; Terai, Y.; Tokumoto, M.; Kobayashi, A.; Tanaka, H.; Kobayashi, H. Magnetic-field-induced superconductivity in a two-dimensional organic conductor. *Nature* **2001**, *410*, 908–910. [CrossRef] [PubMed]

134. Fujiwara, H.; Fujiwara, E.; Nakazawa, Y.; Narymbetov, B.Z.; Kato, K.; Kobayashi, H.; Kobayashi, A.; Tokumoto, M.; Cassoux, P. A novel antiferromagnetic organic superconductor κ-(BETS)$_2$FeBr$_4$ [where BETS = bis(ethylenedithio)tetraselenafulvalene]. *J. Am. Chem. Soc.* **2001**, *123*, 306–314. [CrossRef] [PubMed]

135. Day, P.; Kurmoo, M.; Mallah, T.; Marsden, I.R.; Friend, R.H.; Pratt, F.L.; Hayes, W.; Chasseau, D.; Gaultier, J. Structure and properties of tris[bis(ethylenedithio)tetrathiafulvalenium]tetrachlorocopper(II) hydrate, (BEDT-TTF)$_3$CuCl$_4$·H2O: First evidence for coexistence of localized and conduction electrons in a metallic charge-transfer salt. *J. Am. Chem. Soc.* **1992**, *114*, 10722–10729. [CrossRef]

136. Martin, L.; Day, P.; Clegg, W.; Harrington, R.W.; Horton, P.N.; Bingham, A.; Hursthouse, M.B.; McMillan, P.; Firth, S. Multi-layered molecular charge-transfer salts containing alkali metal ions. *J. Mater. Chem.* **2007**, *17*, 3324–3329. [CrossRef]

137. Coronado, E.; Curreli, S.; Giménez-Saiz, C.; Gómez-García, C.J.; Alberola, A. Radical salts of bis(ethylenediseleno)tetrathiafulvalene with paramagnetic tris(oxalato)metalate anions. *Inorg. Chem.* **2006**, *45*, 10815–10824. [CrossRef] [PubMed]

138. Coronado, E.; Curreli, S.; Giménez-Saiz, C.; Gómez-García, C.J. The series of molecular conductors and superconductors ET$_4$[AFe(C$_2$O$_4$)$_3$]·PhX (ET = bis(ethylenedithio)tetrathiafulvalene; (C$_2$O$_4$)$^{2-}$ = oxalate; A+ = H$_3$O+, K+; X = F, Cl, Br, and I): Influence of the halobenzene guest molecules on the crystal structure and superconducting properties. *Inorg. Chem.* **2012**, *51*, 1111–1126. [PubMed]

139. Coronado, E.; Curreli, S.; Gimenez-Saiz, C.; Gomez-Garcia, C.J. A novel paramagnetic molecular superconductor formed by bis(ethylenedithio)tetrathiafulvalene, tris(oxalato)ferrate(III) anions and bromobenzene as guest molecule: $Et_4[(H_3O)Fe(C_2O_4)_3]\cdot C_6H_5Br$. *J. Mater. Chem.* **2005**, *15*, 1429–1436. [CrossRef]

140. Fourmigué, M.; Batail, P. Activation of hydrogen- and halogen-bonding interactions in tetrathiafulvalene-based crystalline molecular conductors. *Chem. Rev.* **2004**, *104*, 5379–5418. [CrossRef] [PubMed]

141. Coronado, E.; Curreli, S.; Giménez-Saiz, C.; Gómez-García, C.J.; Deplano, P.; Mercuri, M.L.; Serpe, A.; Pilia, L.; Faulmann, C.; Canadell, E. New BEDT-TTF/$[Fe(C_5O_5)_3]_3$- hybrid system: Synthesis, crystal structure, and physical properties of a chirality-induced α phase and a novel magnetic molecular metal. *Inorg. Chem.* **2007**, *46*, 4446–4457. [CrossRef] [PubMed]

142. Gomez-Garcia, C.J.; Coronado, E.; Curreli, S.; Gimenez-Saiz, C.; Deplano, P.; Mercuri, M.L.; Pilia, L.; Serpe, A.; Faulmann, C.; Canadell, E. A chirality-induced alpha phase and a novel molecular magnetic metal in the BEDT-TTF/tris(croconate)Ferrate(III) hybrid molecular system. *Chem. Commun.* **2006**, 4931–4933. [CrossRef] [PubMed]

143. Avarvari, N.; Wallis, J.D. Strategies towards chiral molecular conductors. *J. Mater. Chem.* **2009**, *19*, 4061. [CrossRef]

144. Pop, F.; Auban-Senzier, P.; Canadell, E.; Rikken, G.L.J.A.; Avarvari, N. Electrical magnetochiral anisotropy in a bulk chiral molecular conductor. *Nat. Commun.* **2014**, *5*, 3757. [CrossRef] [PubMed]

145. Rikken, G.L.J.A.; Fölling, J.; Wyder, P. Electrical magnetochiral anisotropy. *Phys. Rev. Lett.* **2001**, *87*, 236602. [CrossRef] [PubMed]

146. Krstic, V.; Roth, S.; Burghard, M.; Kern, K.; Rikken, G.L.J.A. Magneto-chiral anisotropy in charge transport through single-walled carbon nanotubes. *J. Chem. Phys.* **2002**, *117*, 11315. [CrossRef]

147. De Martino, A.; Egger, R.; Tsvelik, A.M. Nonlinear magnetotransport in interacting chiral nanotubes. *Phys. Rev. Lett.* **2006**, *97*, 076402. [CrossRef] [PubMed]

148. Rethore, C.; Fourmigue, M.; Avarvari, N. Tetrathiafulvalene based phosphino-oxazolines: A new family of redox active chiral ligands. *Chem. Commun.* **2004**, 1384–1385. [CrossRef] [PubMed]

149. Réthoré, C.; Avarvari, N.; Canadell, E.; Auban-Senzier, P.; Fourmigué, M. Chiral molecular metals: Syntheses, structures, and properties of the AsF6- salts of Racemic ($\pm$)-, (R)-, and (S)-tetrathiafulvalene—Oxazoline derivatives. *J. Am. Chem. Soc.* **2005**, *127*, 5748–5749. [CrossRef] [PubMed]

150. Madalan, A.M.; Rethore, C.; Fourmigue, M.; Canadell, E.; Lopes, E.B.; Almeida, M.; Auban-Senzier, P.; Avarvari, N. Order versus disorder in chiral tetrathiafulvalene-oxazoline radical-cation salts: Structural and theoretical investigations and physical properties. *Chemistry* **2010**, *16*, 528–537. [CrossRef] [PubMed]

151. Pop, F.; Auban-Senzier, P.; Frąckowiak, A.; Ptaszyński, K.; Olejniczak, I.; Wallis, J.D.; Canadell, E.; Avarvari, N. Chirality driven metallic versus semiconducting behavior in a complete series of radical cation salts based on dimethyl-ethylenedithio-tetrathiafulvalene (DM-EDT-TTF). *J. Am. Chem. Soc.* **2013**, *135*, 17176–17186. [CrossRef] [PubMed]

152. Karrer, A.; Wallis, J.D.; Dunitz, J.D.; Hilti, B.; Mayer, C.W.; Bürkle, M.; Pfeiffer, J. Structures and electrical properties of some new organic conductors derived from the donor molecule tmet (s,s,s,s,-bis (dimethylethylenedithio) tetrathiafulvalene). *Helvetica Chim. Acta* **1987**, *70*, 942–953. [CrossRef]

153. Wallis, J.D.; Karrer, A.; Dunitz, J.D. Chiral metals? A chiral substrate for organic conductors and superconductors. *Helvetica Chim. Acta* **1986**, *69*, 69–70. [CrossRef]

154. Pop, F.; Laroussi, S.; Cauchy, T.; Gomez-Garcia, C.J.; Wallis, J.D.; Avarvari, N. Tetramethyl-bis(ethylenedithio)-tetrathiafulvalene (TM-BEDT-TTF) revisited: Crystal structures, chiroptical properties, theoretical calculations, and a complete series of conducting radical cation salts. *Chirality* **2013**, *25*, 466–474. [CrossRef] [PubMed]

155. Galán-Mascarós, J.R.; Coronado, E.; Goddard, P.A.; Singleton, J.; Coldea, A.I.; Wallis, J.D.; Coles, S.J.; Alberola, A. A chiral ferromagnetic molecular metal. *J. Am. Chem. Soc.* **2010**, *132*, 9271–9273. [CrossRef] [PubMed]

156. Madalan, A.M.; Canadell, E.; Auban-Senzier, P.; Branzea, D.; Avarvari, N.; Andruh, M. Conducting mixed-valence salt of bis(ethylenedithio)tetrathiafulvalene (BEDT-TTF) with the paramagnetic heteroleptic anion $[Cr^{III}(oxalate)_2(2,2'$-bipyridine)]. *New J. Chem.* **2008**, *32*, 333–339. [CrossRef]

157. Martin, L.; Day, P.; Horton, P.; Nakatsuji, S.I.; Yamada, J.I.; Akutsu, H. Chiral conducting salts of BEDT-TTF containing a single enantiomer of tris(oxalato)chromate(III) crystallised from a chiral solvent. *J. Mater. Chem.* **2010**, *20*, 2738–2742. [CrossRef]

158. Pop, F.; Allain, M.; Auban-Senzier, P.; Martínez-Lillo, J.; Lloret, F.; Julve, M.; Canadell, E.; Avarvari, N. Enantiopure conducting salts of dimethylbis(ethylenedithio)tetrathiafulvalene (DM-BEDT-TTF) with the hexachlororhenate(IV) anion. *Eur. J. Inorg. Chem.* **2014**, *2014*, 3855–3862. [CrossRef]

159. Coronado, E.; Minguez Espallargas, G. Dynamic Magnetic MOFs. *Chem. Soc. Rev.* **2013**, *42*, 1525–1539. [CrossRef] [PubMed]

160. Gütlich, P.; Goodwin, H.A. *Spin Crossover in Transition Metal Compounds i*; Springer: Berlin, Germany, 2004; Volume 233.

161. Halcrow, M.A. *Spin-Crossover Materials: Properties and Applications*; Wiley: New York, NY, USA, 2013.

162. Min, K.S.; DiPasquale, A.G.; Golen, J.A.; Rheingold, A.L.; Miller, J.S. Synthesis, structure, and magnetic properties of valence ambiguous dinuclear antiferromagnetically coupled cobalt and ferromagnetically coupled iron complexes containing the chloranilate(2−) and the significantly stronger coupling chloranilate(3−) radical trianion. *J. Am. Chem. Soc.* **2007**, *129*, 2360–2368. [PubMed]

163. Atzori, M.; Pop, F.; Auban-Senzier, P.; Gomez-Garcia, C.J.; Canadell, E.; Artizzu, F.; Serpe, A.; Deplano, P.; Avarvari, N.; Mercuri, M.L. Structural diversity and physical properties of paramagnetic molecular conductors based on bis(ethylenedithio)tetrathiafulvalene (BEDT-TTF) and the tris(chloranilato)FerrateIII) complex. *Inorg. Chem.* **2014**, *53*, 7028–7039. [CrossRef] [PubMed]

164. Takehiko, M. Structural genealogy of BEDT-TTF-based organic conductors i. Parallel molecule s: B and β″ phases. *Bull. Chem. Soc. Jpn.* **1998**, *71*, 2509–2526.

165. Benmansour, S.; Coronado, E.; Giménez-Saiz, C.; Gómez-García, C.J.; Rößer, C. Metallic charge-transfer salts of bis(ethylenedithio)tetrathiafulvalene with paramagnetic tetrachloro(oxalato)rhenate(IV) and tris(chloranilato)ferrate(III) anions. *Eur. J. Inorg. Chem.* **2014**, 3949–3959. [CrossRef]

166. Atzori, M.; Pop, F.; Auban-Senzier, P.; Clerac, R.; Canadell, E.; Mercuri, M.L.; Avarvari, N. Complete series of chiral paramagnetic molecular conductors based on tetramethyl-bis(ethylenedithio)-tetrathiafulvalene (TM-BEDT-TTF) and chloranilate-bridged heterobimetallic honeycomb layers. *Inorg. Chem.* **2015**, *54*, 3643–3653. [CrossRef] [PubMed]

167. Abherve, A.; Clemente-Leon, M.; Coronado, E.; Gomez-Garcia, C.J.; Verneret, M. One-dimensional and two-dimensional anilate-based magnets with inserted spin-crossover complexes. *Inorg. Chem.* **2014**, *53*, 12014–12026. [CrossRef] [PubMed]

168. Clemente-Leo, M.; Coronado, E.; Martí-Gastaldoza, C.; Romero, F.M. Multifunctionality in hybrid magnetic materials based on bimetallic oxalate complexes. *Chem. Soc. Rev.* **2011**, *40*, 473–497. [CrossRef] [PubMed]

169. Boča, R. Zero-field splitting in metal complexes. *Coord. Chem. Rev.* **2004**, *248*, 757–815. [CrossRef]

170. Teppei, Y.; Shota, M.; Hiroshi, K. Structures and proton conductivity of one-dimensional M(dhbq)·nH2O (M = Mg, Mn, Co, Ni, and Zn, H2(dhbq) = 2,5-dihydroxy-1,4-benzoquinone) promoted by connected hydrogen-bond networks with absorbed water. *Bull. Chem. Soc. Jpn.* **2010**, *83*, 42–48.

171. Abhervé, A.; Mañas-Valero, S.; Clemente-León, M.; Coronado, E. Graphene related magnetic materials: Micromechanical exfoliation of 2D layered magnets based on bimetallic anilate complexes with inserted [FeIII(acac2-trien)]+ and [FeIII(sal2-trien)]+ molecules. *Chem. Sci.* **2015**, *6*, 4665–4673. [CrossRef]

172. Castellanos-Gomez, A.; Buscema, M.; Molenaar, R.; Singh, V.; Janssen, L.; van der Zant, H.S.J.; Steele, G.A. Deterministic transfer of two-dimensional materials by all-dry viscoelastic stamping. *2D Mater.* **2014**, *1*, 011002. [CrossRef]

173. Jiang, Y.; Gao, J.; Guo, W.; Jiang, L. Mechanical exfoliation of track-etched two-dimensional layered materials for the fabrication of ultrathin nanopores. *Chem. Commun.* **2014**, *50*, 14149–14152. [CrossRef] [PubMed]

174. Li, H.; Wu, J.; Yin, Z.; Zhang, H. Preparation and applications of mechanically exfoliated single-layer and multilayer MoS2 and WSe2 nanosheets. *Acc. Chem. Res.* **2014**, *47*, 1067–1075. [CrossRef] [PubMed]

175. Li, P.-Z.; Maeda, Y.; Xu, Q. Top-down fabrication of crystalline metal-organic framework nanosheets. *Chem. Commun.* **2011**, *47*, 8436–8438. [CrossRef] [PubMed]

176. Gallego, A.; Hermosa, C.; Castillo, O.; Berlanga, I.; Gómez-García, C.J.; Mateo-Martí, E.; Martínez, J.I.; Flores, F.; Gómez-Navarro, C.; Gómez-Herrero, J.; et al. Solvent-induced delamination of a multifunctional two dimensional coordination polymer. *Adv. Mater.* **2013**, *25*, 2141–2146. [CrossRef] [PubMed]

177. Amo-Ochoa, P.; Welte, L.; Gonzalez-Prieto, R.; Sanz Miguel, P.J.; Gomez-Garcia, C.J.; Mateo-Marti, E.; Delgado, S.; Gomez-Herrero, J.; Zamora, F. Single layers of a multifunctional laminar Cu(I,II) coordination polymer. *Chem. Commun.* **2010**, *46*, 3262–3264. [CrossRef] [PubMed]

178. Beldon, P.J.; Tominaka, S.; Singh, P.; Saha Dasgupta, T.; Bithell, E.G.; Cheetham, A.K. Layered structures and nanosheets of pyrimidinethiolate coordination polymers. *Chem. Commun.* **2014**, *50*, 3955–3957. [CrossRef] [PubMed]

179. Saines, P.J.; Steinmann, M.; Tan, J.-C.; Yeung, H.H.M.; Li, W.; Barton, P.T.; Cheetham, A.K. Isomer-directed structural diversity and its effect on the nanosheet exfoliation and magnetic properties of 2,3-dimethylsuccinate hybrid frameworks. *Inorg. Chem.* **2012**, *51*, 11198–11209. [CrossRef] [PubMed]

180. Tan, J.-C.; Saines, P.J.; Bithell, E.G.; Cheetham, A.K. Hybrid nanosheets of an Inorganic–Organic framework material: Facile synthesis, structure, and elastic properties. *ACS Nano* **2012**, *6*, 615–621. [CrossRef] [PubMed]

181. Kumagai, H.; Kawata, S.; Kitagawa, S. Fabrication of infinite two-dimensional sheets of tetragonal metal(II) lattices X-ray crystal structures and magnetic properties of [M(CA)(pyz)]n (M^{2+} = Mn^{2+} and Co^{2+};H_2CA=chloranilic acid; pyz=pyrazine). *Inorg. Chim. Acta* **2002**, *337*, 387–392. [CrossRef]

182. Nielsen, R.B.; Kongshaug, K.O.; Fjellvag, H. Delamination, synthesis, crystal structure and thermal properties of the layered metal-organic compound Zn($C_{12}H_{14}O_4$). *J. Mater. Chem.* **2008**, *18*, 1002–1007. [CrossRef]

183. Nagayoshi, K.; Kabir, M.K.; Tobita, H.; Honda, K.; Kawahara, M.; Katada, M.; Adachi, K.; Nishikawa, H.; Ikemoto, I.; Kumagai, H.; et al. Design of novel inorganic—Organic hybrid materials based on iron-chloranilate mononuclear complexes: Characteristics of hydrogen-bond-supported layers toward the intercalation of guests. *J. Am. Chem. Soc.* **2003**, *125*, 221–232. [CrossRef] [PubMed]

184. Darago, L.E.; Aubrey, M.L.; Yu, C.J.; Gonzalez, M.I.; Long, J.R. Electronic conductivity, ferrimagnetic ordering, and reductive insertion mediated by organic mixed-valence in a ferric semiquinoid metal—Organic framework. *J. Am. Chem. Soc.* **2015**, *137*, 15703–15711. [CrossRef] [PubMed]

185. Kawata, S.; Kitagawa, S.; Kumagai, H.; Ishiyama, T.; Honda, K.; Tobita, H.; Adachi, K.; Katada, M. Novel intercalation host system based on transition metal (Fe^{2+}, Co^{2+}, Mn^{2+})—Chloranilate coordination polymers. Single crystal structures and properties. *Chem. Mater.* **1998**, *10*, 3902–3912. [CrossRef]

186. Wrobleski, J.T.; Brown, D.B. Synthesis, magnetic susceptibility, and moessbauer spectra of Iron(III) dimers and Iron(II) polymers containing 2,5-dihydroxy-1,4-benzoquinones. *Inorg. Chem.* **1979**, *18*, 498–504. [CrossRef]

187. Wrobleski, J.T.; Brown, D.B. Synthesis, magnetic susceptibility, and spectroscopic properties of single- and mixed-valence iron oxalate, squarate, and dihydroxybenzoquinone coordination polymers. *Inorg. Chem.* **1979**, *18*, 2738–2749. [CrossRef]

188. Clemente-León, M.; Coronado, E.; Gómez-García, C.J.; Soriano-Portillo, A. Increasing the ordering temperatures in oxalate-based 3d chiral magnets: The series [Ir(ppy)2(bpy)][MIIMIII(ox)3]·0.5H2O (MIIMIII = MnCr, FeCr, CoCr, NiCr, ZnCr, MnFe, FeFe); bpy = 2,2′-bipyridine; ppy = 2-phenylpyridine; ox = oxalate dianion). *Inorg. Chem.* **2006**, *45*, 5653–5660. [CrossRef] [PubMed]

189. Coronado, E.; Galán-Mascarós, J.R.; Gómez-García, C.J.; Martínez-Ferrero, E.; Almeida, M.; Waerenborgh, J.C. Oxalate-based 3D chiral magnets: The series [ZII(bpy)3][ClO4][MIIFeIII(ox)3] (ZII = Fe, Ru; MII = Mn, Fe; bpy = 2,2′-bipyridine; ox = oxalate dianion). *Eur. J. Inorg. Chem.* **2005**, *2005*, 2064–2070. [CrossRef]

190. Decurtins, S.; Schmalle, H.W.; Schneuwly, P.; Ensling, J.; Guetlich, P. A concept for the synthesis of 3-dimensional homo- and bimetallic oxalate-bridged networks [M2(ox)3]n. Structural, moessbauer, and magnetic studies in the field of molecular-based magnets. *J. Am. Chem. Soc.* **1994**, *116*, 9521–9528. [CrossRef]

191. Shaikh, N.; Goswami, S.; Panja, A.; Wang, X.-Y.; Gao, S.; Butcher, R.J.; Banerjee, P. New route to the mixed valence semiquinone-catecholate based mononuclear FeIII and catecholate based dinuclear Mn^{III} complexes: First experimental evidence of valence tautomerism in an iron complex. *Inorg. Chem.* **2004**, *43*, 5908–5918. [CrossRef] [PubMed]

192. D'Alessandro, D.M.; Keene, F.R. Current trends and future challenges in the experimental, theoretical and computational analysis of intervalence charge transfer (IVCT) transitions. *Chem. Soc. Rev.* **2006**, *35*, 424–440. [CrossRef] [PubMed]

193. D'Alessandro, D.M.; Keene, F.R. Intervalence charge transfer (IVCT) in trinuclear and tetranuclear complexes of Iron, Ruthenium, and Osmium. *Chem. Rev.* **2006**, *106*, 2270–2298. [CrossRef] [PubMed]

194. Demadis, K.D.; Hartshorn, C.M.; Meyer, T.J. The localized-to-delocalized transition in mixed-valence chemistry. *Chem. Rev.* **2001**, *101*, 2655–2686. [CrossRef] [PubMed]

195. Hankache, J.; Wenger, O.S. Organic mixed valence. *Chem. Rev.* **2011**, *111*, 5138–5178. [CrossRef] [PubMed]

196. Miller, J.S. Magnetically ordered molecule-based materials. *Chem. Soc. Rev.* **2011**, *40*, 3266–3296. [CrossRef] [PubMed]

197. Ward, M.D. A dinuclear Ruthenium(II) complex with the dianion of 2,5-dihydroxy-1,4-benzoquinone as bridging ligand. Redox, spectroscopic, and mixed-valence properties. *Inorg. Chem.* **1996**, *35*, 1712–1714. [CrossRef] [PubMed]

198. Miyazaki, A.; Yamazaki, H.; Aimatsu, M.; Enoki, T.; Watanabe, R.; Ogura, E.; Kuwatani, Y.; Iyoda, M. Crystal structure and physical properties of conducting molecular antiferromagnets with a halogen-substituted donor: $(EDO-TTFBr_2)_2FeX_4$ (X = Cl, Br). *Inorg. Chem.* **2007**, *46*, 3353–3366. [CrossRef] [PubMed]

199. Clérac, R.; O'Kane, S.; Cowen, J.; Ouyang, X.; Heintz, R.; Zhao, H.; Bazile, M.J.; Dunbar, K.R. Glassy magnets composed of metals coordinated to 7,7,8,8-tetracyanoquinodimethane: $M(tcnq)_2$ (M = Mn, Fe, Co, Ni). *Chem. Mater.* **2003**, *15*, 1840–1850. [CrossRef]

200. Jeon, I.-R.; Negru, B.; Van Duyne, R.P.; Harris, T.D. A 2D semiquinone radical-containing microporous magnet with solvent-induced switching from Tc = 26 to 80 k. *J. Am. Chem. Soc.* **2015**, *137*, 15699–15702. [CrossRef] [PubMed]

201. Zeng, M.-H.; Yin, Z.; Tan, Y.-X.; Zhang, W.-X.; He, Y.-P.; Kurmoo, M. Nanoporous Cobalt(II) mof exhibiting four magnetic ground states and changes in gas sorption upon post-synthetic modification. *J. Am. Chem. Soc.* **2014**, *136*, 4680–4688. [CrossRef] [PubMed]

202. Sun, L.; Hendon, C.H.; Minier, M.A.; Walsh, A.; Dincă, M. Million-fold electrical conductivity enhancement in $Fe_2(DEBDC)$ versus $Mn_2(DEBDC)$ (E = S, O). *J. Am. Chem. Soc.* **2015**, *137*, 6164–6167. [CrossRef] [PubMed]

203. Murdock, C.R.; Hughes, B.C.; Lu, Z.; Jenkins, D.M. Approaches for synthesizing breathing MOFs by exploiting dimensional rigidity. *Coord. Chem. Rev.* **2014**, *258–259*, 119–136. [CrossRef]

204. Ferey, G.; Serre, C. Large breathing effects in three-dimensional porous hybrid matter: Facts, analyses, rules and consequences. *Chem. Soc. Rev.* **2009**, *38*, 1380–1399. [CrossRef] [PubMed]

205. Nakabayashi, K.; Ohkoshi, S.-I. Monometallic lanthanoid assembly showing ferromagnetism with a curie temperature of 11 k. *Inorg. Chem.* **2009**, *48*, 8647–8649. [CrossRef] [PubMed]

206. Przychodzeń, P.; Pełka, R.; Lewiński, K.; Supel, J.; Rams, M.; Tomala, K.; Sieklucka, B. Tuning of magnetic properties of polynuclear lanthanide(III)—Octacyanotungstate(V) systems: Determination of ligand-field parameters and exchange interaction. *Inorg. Chem.* **2007**, *46*, 8924–8938. [CrossRef] [PubMed]

207. Ishikawa, N.; Sugita, M.; Wernsdorfer, W. Nuclear spin driven quantum tunneling of magnetization in a new lanthanide single-molecule magnet: Bis(phthalocyaninato)Holmium anion. *J. Am. Chem. Soc.* **2005**, *127*, 3650–3651. [CrossRef] [PubMed]

magnetochemistry

MDPI

Article

Synthesis and Characterization of Ethylenedithio-MPTTF-PTM Radical Dyad as a Potential Neutral Radical Conductor

Manuel Souto [1], Dan Bendixen [2], Morten Jensen [2], Valentín Díez-Cabanes [3], Jérôme Cornil [3], Jan O. Jeppesen [2], Imma Ratera [1,*], Concepció Rovira [1] and Jaume Veciana [1,*]

[1] Department of Molecular Nanoscience and Organic Materials, Institut de Ciència de Materials de Barcelona (ICMAB-CSIC)/CIBER-BBN, Campus de la UAB, 08193 Bellaterra, Spain; msouto@icmab.com (M.S.); cun@icmab.es (C.R.)

[2] Department of Physics, Chemistry and Pharmacy, University of Southern Denmark, 5230 Odense, Denmark; danbendixen88@hotmail.com (D.B.); mortenj@sdu.dk (M.J.); joj@sdu.dk (J.O.J.)

[3] Laboratory for the Chemistry of Novel Materials, Université de Mons, 7000 Mons, Belgium; valentin.diez@hotmail.com (V.D.-C.); jerome.cornil@umons.uc.be (J.C.)

* Correspondence: iratera@icmab.es (I.R.); vecianaj@icmab.es (J.V.); Tel.: +34-935-801-853 (I.R. & J.V.)

Academic Editor: Manuel Almeida

Received: 3 November 2016; Accepted: 9 December 2016; Published: 16 December 2016

Abstract: During the last years there has been a high interest in the development of new purely-organic single-component conductors. Very recently, we have reported a new neutral radical conductor based on the perchlorotriphenylmethyl (PTM) radical moiety linked to a monopyrrolo-tetrathiafulvalene (MPTTF) unit by a π-conjugated bridge (**1**) that behaves as a semiconductor under high pressure. With the aim of developing a new material with improved conducting properties, we have designed and synthesized the radical dyad **2** which was functionalized with an ethylenedithio (EDT) group in order to improve the intermolecular interactions of the tetrathiafulvalene (TTF) subunits. The physical properties of the new radical dyad **2** were studied in detail in solution to further analyze its electronic structure.

Keywords: neutral organic radical; perchlorotriphenylmethyl radical; tetrathiafulvalene; donor-acceptor

1. Introduction

During the past few years there has been a growing interest in the development of purely-organic single-component conductors [1,2]. The interest of such systems is that they offer the possibility of three-dimensional electronic structure, because of the absence of any counterion [3]. However, realization of purely-organic single-component conductors is a major challenge since organic materials are normally insulating.

One interesting possibility to achieve this goal is to use neutral organic radicals as single-component conductors. Although most neutral radicals behave as a Mott insulator, high conductivity can be attained if the electronic bandwidth W, which is directly related to the intermolecular interactions, is maximized and the intra-site Coulomb repulsion energy U is minimized. In this sense, most of these systems are based on delocalized and planar radicals in order to increase the W/U ratio, such as spirobisphenalenyl [4–6] and thiazolyl radicals [7–9] reported by Haddon and Oakley, respectively. In addition, Mori and coworkers have also recently reported a purely-organic single-component conductor by utilizing strong hydrogen-bonding interactions between tetrathiafulvalene (TTF)-based electron-donor molecules [10,11].

In this direction, we have very recently reported the organic donor-acceptor (D-A) dyad **1** based on the non-planar and spin localized radical perchlorotriphenylmethyl (PTM) radical linked to a monopyrrolo-tetrathiafulvalene (MPTTF) unit that exhibited semiconducting behavior with high conductivity upon application of high pressure [12–15]. The origin of such conducting behavior was attributed to the increased electronic bandwidth W, thanks to short intermolecular interactions between the MPTTF and PTM subunits, and the decreasing of the Coulomb repulsion energy U due to the charge reorganization occurring at high pressure. The intramolecular charge transfer between the TTF and PTM moieties, as demonstrated by the formation of charge-separated zwitterionic species in solution when using polar solvents [12,13], is an optimal condition for obtaining neutral radical conductors since the Coulomb repulsion energy U can be minimized.

To design and obtain new neutral radicals based on this family of compounds exhibiting higher conductivity, one synthetic approach could be the functionalization of the MPTTF unit with an ethylenedithio (EDT) group in order to increase the intermolecular interactions as it has been reported for several organic superconductors [16]. Thus, in this article we report the synthesis and characterization of the D-A radical dyad **2** (EDT-MPTTF-PTM) which contains the same molecular structure as **1** but is functionalized with an EDT group in order to improve the self-assembly between the radical molecules (Scheme 1). Subsequently, we evaluate the physical properties in solution of the radical dyad **2** that could be a neutral radical conductor in the solid state.

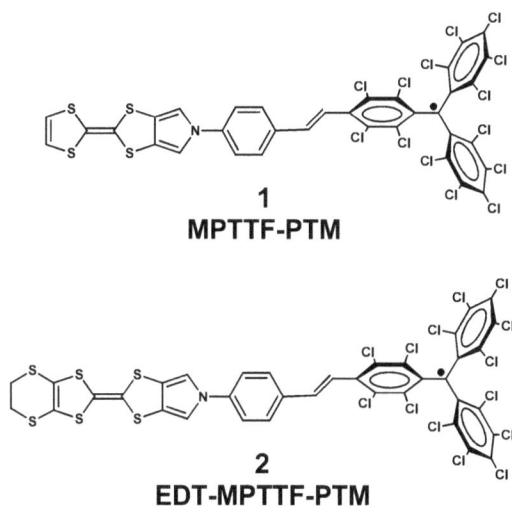

1
MPTTF-PTM

2
EDT-MPTTF-PTM

Scheme 1. Molecular structures of the monopyrrolo-tetrathiafulvalene (MPTTF)-perchlorotriphenylmethyl (PTM) (**1**); and ethylenedithio (EDT)-MPTTF-PTM (**2**) radical dyads.

2. Results and Discussion

2.1. Synthesis

The strategy to synthesize dyad **2** was based on the Wittig–Horner coupling between the phosphonated PTM derivative **8** [17] and the EDT-MPTTF aldehyde **7** functionalized with an ethylenedithio group.

The synthesis of the aldehyde precursor **7** was carried out as outline in Scheme 2. First, the cross-coupling of the ethylendithio-thione (**3**) [18] and 5-tosyl-(1,3)-dithiolo[4,5-*c*]pyrrole-2-one (**4**) [19] in anhydrous $(EtO)_3P$ gave **5** in 82% yield after column chromatography. Subsequently, the tosyl protecting group on derivative **5** was removed using NaOMe in a THF/MeOH mixture to obtain compound **6** as a yellow solid which was used in the next reaction without further purification. Finally, by mixing derivative **6**, 4-bromobenzaldehyde, an excess of cuprous iodide (CuI), and *trans*-1,2-diaminocyclohexane gave compound **7** in 41% overall yield (from compound **5**).

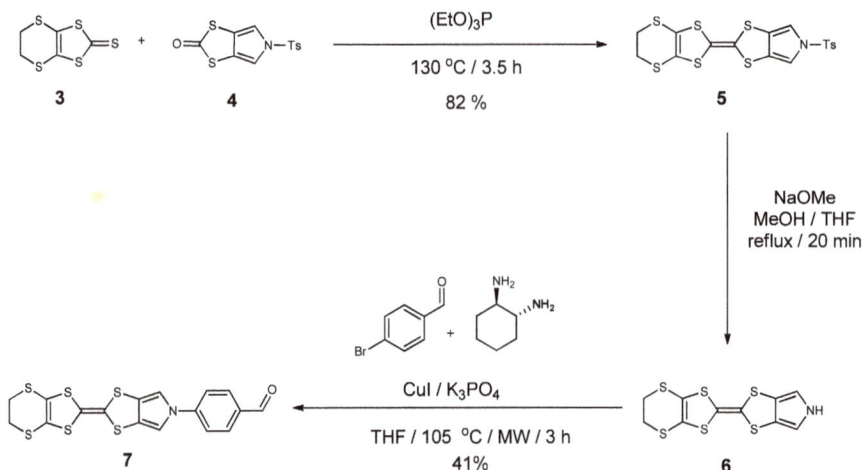

Scheme 2. Synthesis of compound **7**. Ts = tosyl.

Radical dyad **2** was synthesized in three steps, as summarized in Scheme 3. First, compound **9** was obtained in a 52% yield through a Horner–Wadsworth–Emmons reaction of aldehyde **7** and the phosphonated PTM derivative **8** and was fully characterized (Figures S1–S3) [17]. Subsequent deprotonation and oxidation of the generated carbanion with silver nitrate provided radical dyad **2** (69%) as a black solid.

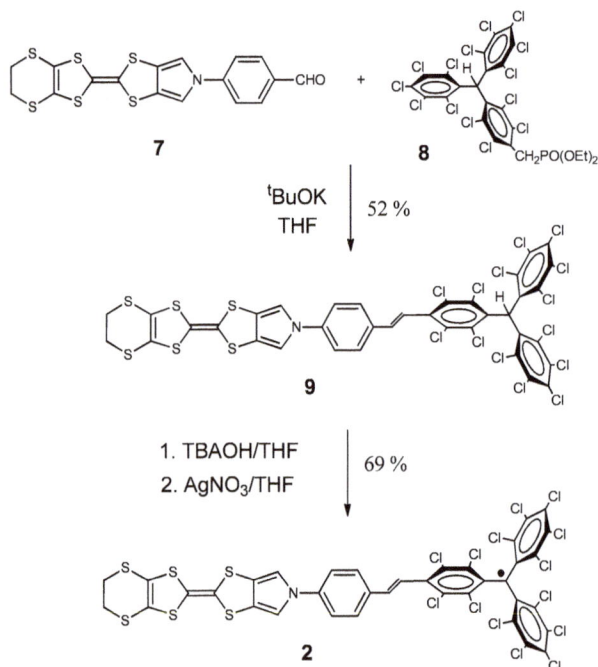

Scheme 3. Synthesis of radical dyad **2**.

2.2. Characterization

2.2.1. Electrochemical Properties

Figure 1 displays the cyclic voltammograms (CV) of radical dyads **1** and **2** recorded in CH_2Cl_2 at room temperature. Both dyads exhibit three reversible waves assigned to the following redox pairs: $PTM^-/PTM^{\cdot}$ ($E^{red}_{1/2}$), $MPTTF/MPTTF^{+\cdot}$ ($E^{ox1}_{1/2}$), and $MPTTF^{+\cdot}/MPTTF^{2+}$ ($E^{ox2}_{1/2}$). The difference between the reduction and the first oxidation potential, ($\Delta E^{(1)}$), is directly related to the intrasite Coulomb repulsion energy U of the neutral radical compound and should be minimized for obtaining optimal molecular conductors [13]. As it transpires from Table 1, $\Delta E^{(1)}$ was smaller for radical dyad **1** indicating that the repulsion energy U was slightly increased in radical dyad **2**. On the other hand, the difference between the first and second oxidation potential, ($\Delta E^{(2)}$), slightly decreases in the case of radical dyad **2**.

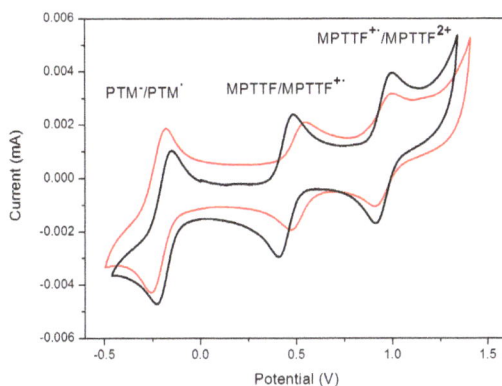

Figure 1. Cyclic voltammograms of dyads **1** (black line) and **2** (red line) in CH_2Cl_2 versus Ag/AgCl with Bu_4NPF_6 (0.1 M) as electrolyte at 300 K under argon at scan rate of 0.1 V/s.

Table 1. Electrochemical data of radical dyads **1** and **2**.

Dyad	$E^{red}_{1/2}$ (V)	$E^{ox1}_{1/2}$ (V)	$E^{ox2}_{1/2}$ (V)	$\Delta E^{(1)}$ (V) [a]	$\Delta E^{(2)}$ (V) [b]
1	−0.19	+0.45	+0.95	+0.64	+0.50
2	−0.22	+0.51	+0.95	+0.73	+0.44

[a] $\Delta E^{(1)} = E^{ox1}_{1/2} - E^{red}_{1/2}$; [b] $\Delta E^{(2)} = E^{ox2}_{1/2} - E^{ox1}_{1/2}$.

2.2.2. Optical Properties

The UV-vis-near infrared (NIR) spectra of dyads **1** and **2** were recorded in CH_2Cl_2 at 300 K (Figure 2) and show similar bands. The intense peak at 385 nm corresponds to the radical chromophore of the PTM subunit and the shoulders appearing around 450 and 550 nm are assigned to the electronic conjugation of the unpaired electron into the conjugated π-framework [12]. The lowest energy band, which appears around 800 nm, is ascribed to the intramolecular charge transfer (ICT) occurring between the electron-donor TTF and electron-acceptor PTM units. This band appears at similar energies for both dyads indicating that compounds **1** and **2** exhibit a similar degree of charge transfer.

On the other hand, we have observed a bistable behavior in solution, between neutral and zwitterionic species, for similar TTF-PTM dyads that can be switched by tuning the polarity of the solvents [12,20,21]. In order to evaluate this bistability phenomenon in solution for this new system, we recorded the optical spectra of radical dyad **2** in three solvents with different polarities (CH_2Cl_2, Me_2CO and DMF) (Figure 3). As we can observe, dyad **2** was present in the neutral state in CH_2Cl_2 and Me_2CO since we only observed the band at 387 nm attributed to the radical neutral species. Thus, the bistability of dyad **2** was only observed in the most polar solvent (i.e., DMF), as demonstrated by the presence of the intense band at 512 nm attributed to the anionic form of the PTM unit due to the formation of zwitterionic species. Moreover, solution of dyad **2** in DMF was ESR silent suggesting that the zwitterionic species formed diamagnetic dimers as it has been observed for similar TTF-PTM dyads [12,14,20]. Most of the absorption bands related to the TTF radical cation dimer are overlapped by the intense absorption band related to the PTM anion as it has been previously observed (Figure S4) [22]. This behavior was similar for dyad **1** [12] indicating that dyads **1** and **2** exhibit similar electronic properties. DFT calculations confirmed that frontier molecular orbitals were similar for dyads **1** and **2** (Supplementary Materials Figure S4). This bistable behavior in solution suggests that a charge transfer between the TTF and PTM units in dyad **2** can be promoted upon application of an external stimulus (i.e., solvent polarity, high pressure, temperature, etc.). This charge reorganization is an optimal condition for obtaining neutral radical conductors since the Coulomb repulsion energy U can be minimized [13].

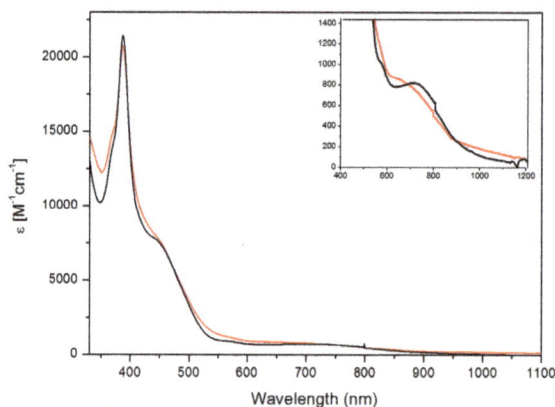

Figure 2. UV-vis-near infrared (NIR) spectra recorded in CH_2Cl_2 at 300 K of a 0.05 mM solution of dyads **1** (black line) and **2** (red line). The inset shows the low-energy range of the absorption spectra of **1** and **2**.

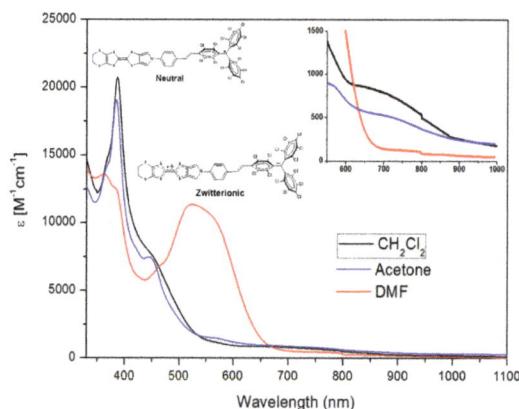

Figure 3. UV-vis-NIR spectra of radical dyad **2** recorded in CH_2Cl_2 (black line), Me_2CO (blue line) and DMF (red line) at 300 K. Molecular structures of the possible neutral and zwitterionic states for radical dyad **2**. The inset shows the low-energy range of the absorption spectra.

2.2.3. Magnetic Properties

Magnetic properties were studied in solution by Electron Spin Resonance (ESR) spectroscopy. The ESR spectrum of radical dyad **2** recorded in CH_2Cl_2 (0.05 mM) at room temperature (Figure 4) showed two main lines with a *g*-value of 2.0025 corresponding to the hyperfine coupling of the unpaired electron with one ^1H atom of the vinylene bridge. The isotropic hyperfine coupling constants with such a proton showed the typical values for vinyelene-substituted PTM radicals (a_H = 1.9 Hz; ΔH_{pp} = 1.1 Hz). The few satellite weak lines are attributed to the hyperfine coupling of the unpaired electron with naturally abundant ^{13}C isotopes at the α and aromatic positions.

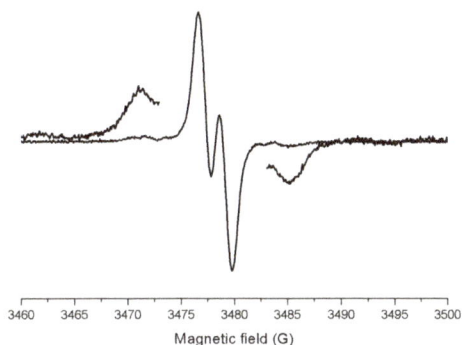

Figure 4. Electron Spin Resonance (ESR) spectrum recorded of a 0.05 mM solution of radical dyad **2** in CH_2Cl_2 at 300 K.

On the other hand, we generated **2**$^{\bullet+}$ by chemical oxidation with $FeCl_3$ (Figure S5) and recorded ESR spectra at different temperatures (220–300 K) (Figure 5). The ESR spectrum at room temperature consists of two separated groups of lines (a single line and two symmetrical lines) with g values of 2.0056 and 2.0029 attributed to the open-shell TTF$^{\bullet+}$ and PTM$^{\bullet}$ radical subunits, respectively. When temperature is lowered, the lines associated with both TTF and PTM units progressively decreased in intensity and the intensity of the TTF line disappeared completely at 220 K. These changes were completely reversible with temperature and this behavior was consistent with the dimerization of **2**$^{\bullet+}$ at low temperature as it has already been observed for other TTF-PTM oxidized dyads [22].

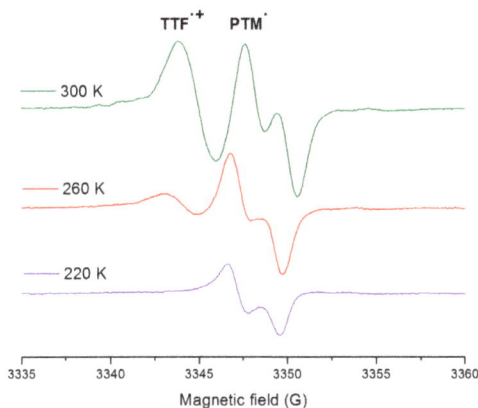

Figure 5. ESR spectra of **2**$^{\bullet+}$ recorded in THF (0.05 mM) at 300 (green line), 260 (red line) and 220 K (purple line).

As noted previously for similar systems [22], the observation that the intensity of the ESR signal decreases when decreasing the temperature suggests that in the dimer $[2^{\bullet+}]_2$ the two TTF$^{\bullet+}$ subunits are magnetically coupled in a singlet (ESR-silent) ground state and, in addition, are able to transmit an antiferromagnetic coupling between the two unpaired spins of the PTM subunits. Such behavior indicates that there is a great intermolecular interaction between the dyads **2** which is an important prerequisite for obtaining optimal molecular conductors in the solid state.

Concerning the properties in solid state of radical dyad **2**, the external and additional sulfur atoms as well as the ethylene groups could play a key role in obtaining an optimal crystal packing since it is well known that S···S and C–H···S interactions are very useful as secondary-interactions to achieve supramolecular assemblies of TTF derivatives [16]. Therefore, we expect that intermolecular interactions in dyad **2** will be increased in comparison to dyad **1** and that the electronic bandwidth W will be enhanced. On the other hand, although the estimated Coulomb repulsion energy U of **2** seems to increase in comparison with **1** from the CV, we expect that charge transfer occurring between the TTF and PTM subunits, as it has been observed in solution, will help to decrease such repulsion energy.

In view of the promising conducting properties that could exhibit crystals of dyad **2**, several efforts for crystallizing the neutral compound has been realized by diffusion and slow evaporation methods using different solvents (hexane, toluene, CH_2Cl_2, Et_2O ...) and conditions (temperature, concentration ...). Unfortunately, up to now, it has not been possible to obtain crystals of high quality which has prevented the determination of its X-ray crystal structure. Moreover, in view of the several reported examples of magnetic conductors and superconductors based on BEDT-TTF charge transfer salts [16], it is also planned to obtain radical ion-salts based on dyad **2** by electrocrystallization in the future. Thus, we will continue our efforts for crystallizing such compounds that could have a potential use in spintronics if they are able to combine both conductivity and magnetism in a cooperative manner, as the spin-polarized donor radicals reported by Sugawara and coworkers [23].

3. Materials and Methods

3.1. General Methods for Synthesis and Characterization

All reagents and solvents employed for the syntheses were of high purity grade and were purchased from Sigma-Aldrich Co. (St Louis, MO, USA), Merck (Darmstadt, Germany), and SDS, except compound **3** [18] and **4** [19] which were synthesized according to literature procedures Anhydrous solvents were used in the chemical reactions and for recording the cyclic voltammograms. ^{1}H NMR spectra were recorded using a Bruker Avance 250, 400, or 500 instruments (Bruker, Billerica, MA, USA) and Me$_4$Si was used an internal standard. Infrared spectra were recorded with Spectrum One FT-IR Spectroscopy instrument (Perkin Elmer, Waltham, MA, USA) and UV/Vis/NIR spectra were measured using Cary 5000E Varian (Agilent, Santa Clara, CA, USA). ESR spectra were performed with a Bruker ESP 300 E (Bruker, Billerica, MA, USA) equipped with a rectangular cavity T102 that works with an X-band (9.5 GHz). The solutions were degassed by argon bubbling before the measurements. LDI/TOF MS were recorded in a Bruker Ultraflex LDI-TOF spectrometer (Bruker, Billerica, MA, USA). Cyclic voltammetry measurements were obtained with a potentiostat 263a (Ametek, Berwyn, PA, USA) from EG&G Princeton Applied Research in a standard 3 electrodes cell. The IR-NIR spectra have been collected with a Bruker FT-IR IFS-66 spectrometer (Perkin Elmer, Waltham, MA, USA) equipped with a Hyperion microscope. The spectral resolution is about 2 cm^{-1} for both spectrometers. The manipulation of the radicals in solution was performed under red light.

3.2. Synthesis and Characterization

Compound **1** was synthesized as reported in reference [12].

Compound **5**: Anhydrous (EtO)$_3$P (40 mL) was heated to 130 °C before the thione **3** (1.10 g, 4.90 mmol) and the ketone **4** (0.99 g, 3.18 mmol) were added. After stirring the reaction mixture at 130 °C for 10 min, an additional portion of **3** (1.08 g, 4.81 mmol) was added, then the reaction mixture was heated at 130 °C for 3.5 h. The reaction mixture was cooled to room temperature, whereupon cold MeOH (50 mL, −18 °C) was added. Leaving the mixture overnight at −18 °C produced a yellow precipitate, which was filtered off and washed with cold MeOH (3 × 20 mL, −18 °C). The yellow solid was dissolved in CH_2Cl_2 (500 mL) and Celite 545 (40 mL) was added, before the solvent was removed in vacuo. The resulting residue was purified by column chromatography (silica gel: 1. CH_2Cl_2/petroleum ether (b.p. 60–80 °C) 1:1, 2. CH_2Cl_2) and the broad yellow band was collected and concentrated to the give the title compound **5** as a yellow powder containing traces of grease. Yield (1.27 g, 82%). Characterization: M.p. > 250 °C. ^{1}H-NMR ($CDCl_3$): δ = 2.41 (s, 3H), 3.29 (s, 4H), 6.93 (s, 2H), 7.30 (d, 2H, J = 8.3 Hz), 7.72 (d, 2H, J = 8.3 Hz). MS (ESI): m/z = 486 (M^+). HiRes-FT ESI-MS: Calculated m/z = 486.8991; Found m/z = 486.8987. IR (KBr): v/cm^{-1} = 3125; 2919; 1595; 1371; 1225; 1188; 1171; 1090; 1054.

Compound **7**: A solution of the MPTTF compound **5** (0.390 g, 0.800 mmol) in a mixture of anhydrous THF (50 mL) and anhydrous MeOH (50 mL) was degassed (N_2, 20 min.), before a solution of NaOMe in MeOH (25% w/w, 1.85 mL, 8.1 mmol) was added. The reaction mixture was heated under reflux for 20 min and then allowed to cool down to room temperature. The solvent was removed in vacuo and the residue was dissolved in CH_2Cl_2 (500 mL). This mixture was washed with H_2O (3 × 200 mL) and dried ($MgSO_4$) before the solvent was removed in vacuo to produce a yellow solid containing **6**, which was used in the next reaction without further purification. The crude product containing the MPTTF compound **6**, 4-brombenzaldehyd (0.681 g, 3.68 mmol), CuI (0.315 g, 1.65 mmol), and K_3PO_4 (0.548 g, 2.58 mmol) were dissolved in anhydrous THF (20 mL) and degassed (N_2, 20 min) before ± trans 1,2 diaminocyclohexane (0.40 mL, 0.38 g, 3.3 mmol) was added. The reaction mixture was heated to 105 °C under microwave (MW) irradiation for 3 h and then cooled to room temperature before the solvent was removed in vacuo. The crude product was dissolved in CH_2Cl_2 (500 mL) and washed with H_2O (3 × 250 mL). The aqueous phases were combined and extracted with CH_2Cl_2 (16 × 250 mL). The combined organic phases were dried ($MgSO_4$) and Celite 545 (20 mL) was added, before the solvent was removed in vacuo. The resulting residue was purified by column chromatography (silica gel: 1. EtOAc, 2. CH_2Cl_2) and the broad yellow band was collected and concentrated to give a yellow powder, which was recrystallized from PhMe (400 mL) and washed with cold PhMe (0 °C, 50 mL) providing the title compound **7** as an orange powder. Yield (0.143 g, 41%). Characterization: M.p. > 250 °C. ^{1}H NMR ((CD_3)$_2$SO): δ = 7.66 (s, 2H), 7.77 (s, 2H), 7.98 (s, 2H), 9.97 (s, 1H, CHO), the signal form the four CH_2 protons are obscured under the water signal present in the NMR solvent. ^{1}H NMR ($CDCl_3$): δ = 3.31 (bs, 4H), 7.00 (bs, 2H); 7.44 (bs, 2H), 7.94 (bs, 2H), 9.98 (s, 1H). MS (ESI): m/z = 436(M^+). HiRes-FT ESI-MS: Calculated m/z = 436.9165; Found m/z = 436.9140. IR (KBr): v/cm^{-1} = 2921; 2844; 1690; 1600; 1584; 1519; 1492; 1391; 1311; 1172.

Compound **9**: A solution of the phosphonated PTM derivative **8** [17] (219 mg. 0.25 mmol) was dissolved in anhydrous THF (50 mL) under strict inert conditions and cooled down to −78 °C. Potassium *tert*-butoxide (84 mg, 0.75 mmol) was added and the reaction mixture was stirred for 20 min to form the yellow-orange ylide. Subsequently, compound **7** (120 mg, 0.27 mmol) was added and the reaction mixture was allowed up to room temperature and then stirred for 3 days. Then the reaction mixture was extracted with CH_2Cl_2 (150 mL) washed with H_2O (75 mL), and dried ($MgSO_4$) before the solvents were evaporated under reduced pressure. Finally, the product was purified by column chromatography (silica gel: ether/hexane 1:1) to obtain compound **9** as a light yellow powder. Yield (150 mg, 52%). Characterization: ^{1}H NMR (400 MHz, CD_3)$_2$SO): δ = 7.75 (d, 2H, J = 8.6 Hz), 7.58 (d, 2H, J = 8.6 Hz), 7.55 (s, 2H), 7.22 (d, 1H, J = 16.6 Hz), 7.13 (d, 1H, J = 16.6 Hz); 6.94 (s, 1H); 3.41 (s, 4H). FT-IR: v/cm^{-1} = 2922 (w), 2851 (w), 2743 (w), 1603 (m, CH=CH), 1519 (s), 1486 (w), 1461 (w), 1365 (w), 1337 (w), 1311 (s), 1298 (s), 1187 (m), 1138 (m), 1038 (m), 967 (m), 934 (m), 875 (w), 808 (s), 748 (m), 719 (w), 685 (w), 669 (s). LDI-TOF (positive mode): m/z (amu/e^-): 1158.944 ($M^{+\cdot}$); (negative mode): m/z (amu/e^-): 1158.845 ($M^{-\cdot}$). Cyclic voltammetry (Bu_4NPF_6 0.1 M in CH_2Cl_2 as electrolyte): $E^{ox1}_{\frac{1}{2}}$ = +0.425 V; $E^{ox2}_{\frac{1}{2}}$ = +0.925 V. M.p.: 220 °C.

Compound **2**: Tetrabutylammonium hydroxide (1.0 M in methanol, 60 µL, 0.060 mmol) was added to a solution of compound **9** (50 mg, 0.043 mmol) in distilled THF (20 mL) and the purple solution was stirred for 2 h. Then $AgNO_3$ (12 mg, 0.071 mmol) dissolved in MeCN (10 mL) was added and the reaction mixture was stirred for additional 60 min. The solution changes from purple to dark brown and a precipitate of silver (Ag^0) was formed. The reaction mixture was subsequently filtered and the filtrate was evaporated under reduced pressure. Finally, the product was purified by flash column chromatography (silica gel: CH_2Cl_2/hexane 1:1) affording the radical dyad **2** as a dark reddish-brown powder. Yield (35 mg, 69%). Characterization: FT-IR: v/cm^{-1} = 2952 (w), 2918 (w), 2853 (w), 1602 (m, CH=CH), 1519 (s), 1484 (w), 1461 (w), 1430 (w), 1380 (w), 1361 (w), 1333 (m), 1309 (s), 1298 (s), 1257 (m), 1228 (w), 1183 (w), 1155 (w), 1118 (w), 1038 (m), 966 (w), 935 (m), 859 (w), 814 (s), 766 (w), 752 (w), 737 (m), 707 (m), 666 (w), 651 (s). UV-VIS-NIR (CH_2Cl_2, λ_{max}/nm, $\varepsilon/M^{-1}\cdot cm^{-1}$): 325 (22269), 370 (22658), 386 (31503), 437 (12011). LDI-TOF (positive mode): m/z (amu/e$^-$): 1158.396 (M$^+$). Cyclic voltammetry (Bu$_4$NPF$_6$ 0.1 M in CH_2Cl_2 as electrolyte): $E_{\frac{1}{2}}^1$=−0.220 V; $E_{\frac{1}{2}}^2$= +0.510 V; $E_{\frac{1}{2}}^3$= +0.953 V. ESR (CH_2Cl_2): g = 2.0025; a$_H$ = 1.9 Hz; ΔH_{pp} = 1.1 Hz. M.p.: 220 °C.

4. Conclusions

In summary, we have reported the synthesis and characterization of the new dyad **2** based on the EDT-MPTTF moiety as electron donor linked to a PTM radical as electron acceptor through a phenyl-vinyelene bridge. This new dyad exhibits optimal electronic properties in solution and have the potential to give rise to new radical conductors in the solid state. Neutral and charge transfer salt crystals based on this system could have a potential use in spintronics if they are able to combine both conductivity and magnetism in a cooperative way.

Supplementary Materials: The following are available online at www.mdpi.com/2312-7481/2/4/46/s1, Figure S1: ^{1}H NMR spectrum of **9**, Figure S2: IR spectrum of **2** and **9**, Figure S3: CVs of **2** and **9**, Figure S4: UV-vis-NIR of **2**$^{·+}$ in THF. Figure S5: Frontier molecular orbital energy diagrams (left) and topologies (right) for the dyads **1** (black) and **2** (red). The two series of lines represent the orbitals associated to the spin up (α) and spin down (β), respectively.

Acknowledgments: This work was supported by the EU ITN iSwitch 642196 and "Nano2Fun" 607721 DGI grant (BeWell; CTQ2013-40480-R), the Networking Research Center on Bioengineering, Biomaterials, and Nanomedicine (CIBER-BBN), and the Generalitat de Catalunya (grant 2014-SGR-17). ICMAB acknowledges support from the Spanish Ministry of Economy and Competitiveness, through the "Severo Ochoa" Programme for Centres of Excellence in R&D (SEV-2015-0496). In Denmark, this work was supported by the Danish Council for Independent Research | Natural Sciences (#11-106744). M.S. is grateful to Spanish Ministerio de Educación, Cultura y Deporte for a FPU grant. We thank Vega Lloveras for ESR spectroscopy and Amable Bernabé for MALDI spectroscopy.

Author Contributions: M.S., D.B. and M.J. synthesized and characterized the compounds; V.D.-C. and J.C. performed the theoretical calculations; J.O.J., I.R., C.R and J.V. conceived and designed the experiments.

Conflicts of Interest: The authors declare no conflict of interest.

References

1. Haddon, R.C. Design of organic metals and superconductors. *Nature* **1975**, *256*, 394–396. [CrossRef]
2. Cordes, A.W.; Haddon, R.C.; Oakley, R.T. Molecular conductors from neutral heterocyclic π-radicals. *Adv. Mater.* **1994**, *6*, 798–802. [CrossRef]
3. Kobayashi, A.; Fujiwara, E.; Kobayashi, H. Single-Component Molecular Metals with Extended-TTF Dithiolate Ligands. *Chem. Rev.* **2004**, *104*, 5243–5264. [PubMed]
4. Pal, S.K.; Itkis, M.E.; Tham, F.S.; Reed, R.W.; Oakley, R.T.; Haddon, R.C. Resonating valence-bond ground state in a phenalenyl-based neutral radical conductor. *Science* **2005**, *309*, 281–284. [CrossRef] [PubMed]
5. Mandal, S.K.; Samanta, S.; Itkis, M.E.; Jensen, D.W.; Reed, R.W.; Oakley, R.T.; Tham, F.S.; Donnadieu, B.; Haddon, R.C. Resonating valence bond ground state in oxygen-functionalized phenalenyl-based neutral radical molecular conductors. *J. Am. Chem. Soc.* **2006**, *128*, 1982–1994. [CrossRef] [PubMed]
6. Pal, S.K.; Itkis, M.E.; Tham, F.S.; Reed, R.W.; Oakley, R.T.; Haddon, R.C. Trisphenalenyl-based neutral radical molecular conductor. *J. Am. Chem. Soc.* **2008**, *130*, 3942–3951. [CrossRef] [PubMed]

7. Tenn, N.; Bellec, N.; Jeannin, O.; Piekara-sady, L.; Auban-senzier, P. A Single-Component Molecular Metal Based on a Thiazole Dithiolate Gold Complex. *J. Am. Chem. Soc.* **2009**, *131*, 16961–16967. [CrossRef] [PubMed]

8. Yu, X.; Mailman, A.; Lekin, K.; Assoud, A.; Robertson, C.M.; Noll, B.C.; Campana, C.F.; Howard, J.A.K.; Dube, P.A.; Oakley, R.T. Semiquinone-bridged bisdithiazolyl radicals as neutral radical conductors. *J. Am. Chem. Soc.* **2012**, *134*, 2264–2275. [CrossRef] [PubMed]

9. Leitch, A.A.; Lekin, K.; Winter, S.M.; Downie, L.E.; Tsuruda, H.; Tse, J.S.; Mito, M.; Desgreniers, S.; Dube, P.A.; Zhang, S.; et al. From magnets to metals: The response of tetragonal bisdiselenazolyl radicals to pressure. *J. Am. Chem. Soc.* **2011**, *133*, 6050–6060. [CrossRef] [PubMed]

10. Isono, T.; Kamo, H.; Ueda, A.; Takahashi, K.; Nakao, A.; Kumai, R.; Nakao, H.; Kobayashi, K.; Murakami, Y.; Mori, H. Hydrogen bond-promoted metallic state in a purely organic single-component conductor under pressure. *Nat. Commun.* **2013**, *4*, 1344–1346. [CrossRef] [PubMed]

11. Ueda, A.; Yamada, S.; Isono, T.; Kamo, H.; Nakao, A.; Kumai, R.; Nakao, H.; Murakami, Y.; Yamamoto, K.; Nishio, Y.; et al. Hydrogen-Bond-Dynamics-Based Switching of Conductivity and Magnetism: A Phase Transition Caused by Deuterium and Electron Transfer in a Hydrogen-Bonded Purely Organic Conductor Crystal. *J. Am. Chem. Soc.* **2014**, *136*, 12184–12192. [CrossRef] [PubMed]

12. Souto, M.; Solano, M.V.; Jensen, M.; Bendixen, D.; Delchiaro, F.; Girlando, A.; Painelli, A.; Jeppesen, J.O.; Rovira, C.; Ratera, I.; et al. Self-Assembled Architectures with Segregated Donor and Acceptor Units of a Dyad Based on a Monopyrrolo-Annulated TTF-PTM Radical. *Chem. Eur. J.* **2015**, *21*, 8816–8825. [CrossRef] [PubMed]

13. Souto, M.; Cui, H.; Peña-Álvarez, M.; Baonza, V.G.; Jeschke, H.O.; Tomic, M.; Valent, R.; Blasi, D.; Ratera, I.; Rovira, C.; et al. Pressure-Induced Conductivity in a Neutral Nonplanar Spin-Localized Radical. *J. Am. Chem. Soc.* **2016**, *138*, 11517–11525. [CrossRef] [PubMed]

14. Souto, M.; Rovira, C.; Ratera, I.; Veciana, J. TTF–PTM dyads: From switched molecular self assembly in solution to radical conductors in solid state. *CrystEngComm* **2016**. [CrossRef]

15. Pop, F.; Avarvari, N. Covalent non-fused tetrathiafulvalene-acceptor systems. *Chem. Commun.* **2016**, *52*, 7906–7927. [CrossRef] [PubMed]

16. Rovira, C. Bis(ethylenethio)tetrathiafulvalene (BET-TTF) and related dissymmetrical electron donors: From the molecule to functional molecular materials and devices (OFETs). *Chem. Rev.* **2004**, *104*, 5289–5317. [CrossRef] [PubMed]

17. Rovira, C.; Ruiz-Molina, D.; Elsner, O.; Vidal-Gancedo, J.; Bonvoisin, J.; Launay, J.P.; Veciana, J. Influence of topology on the long-range electron-transfer phenomenon. *Chem. Eur. J.* **2001**, *7*, 240–250. [CrossRef]

18. Simonsen, K.B.; Svenstrup, N.; Lau, J.; Simonsen, O.; Mørk, P.; Kristensen, G.J.; Becher, J. Sequential Functionalisation of Bis-Protected Tetrathiafulvalene-dithiolates. *Synthesis (Stuttg.)* **1996**, *1996*, 407–418. [CrossRef]

19. O'Driscoll, L.J.; Andersen, S.S.; Solano, M.V.; Bendixen, D.; Jensen, M.; Duedal, T.; Lycoops, J.; Van Der Pol, C.; Sørensen, R.E.; Larsen, K.R.; et al. Advances in the synthesis of functionalised pyrrolotetrathiafulvalenes. *Beilstein J. Org. Chem.* **2015**, *11*, 1112–1122. [CrossRef] [PubMed]

20. Guasch, J.; Grisanti, L.; Lloveras, V.; Vidal-Gancedo, J.; Souto, M.; Morales, D.C.; Vilaseca, M.; Sissa, C.; Painelli, A.; Ratera, I.; et al. Induced self-assembly of a tetrathiafulvalene-based open-shell dyad through intramolecular electron transfer. *Angew. Chem. Int. Ed.* **2012**, *51*, 11024–11028. [CrossRef] [PubMed]

21. Souto, M.; Guasch, J.; Lloveras, V.; Mayorga, P.; López Navarrete, J.T.; Casado, J.; Ratera, I.; Rovira, C.; Painelli, A.; Veciana, J. Thermomagnetic molecular system based on TTF-PTM radical: Switching the spin and charge delocalization. *J. Phys. Chem. Lett.* **2013**, *4*, 2721–2726. [CrossRef]

22. Guasch, J.; Grisanti, L.; Souto, M.; Lloveras, V.; Vidal-Gancedo, J.; Ratera, I.; Painelli, A.; Rovira, C.; Veciana, J. Intra- and intermolecular charge transfer in aggregates of tetrathiafulvalene-triphenylmethyl radical derivatives in solution. *J. Am. Chem. Soc.* **2013**, *135*, 6958–6967. [CrossRef] [PubMed]

23. Sugawara, T.; Komatsu, H.; Suzuki, K. Interplay between magnetism and conductivity derived from spin-polarized donor radicals. *Chem. Soc. Rev.* **2011**, *40*, 3105–3118. [CrossRef] [PubMed]

MDPI

Article

New Ethylenedithio-TTF Containing a 2,2,5,5-Tetramethylpyrrolin-1-yloxyl Radical through a Vinylene Spacer and Its FeCl₄⁻ Salt—Synthesis, Physical Properties and Crystal Structure Analyses

Kazuki Horikiri and Hideki Fujiwara *

Department of Chemistry, Graduate School of Science, Osaka Prefecture University, Gakuen-cho, Naka-ku, Sakai 599-8531, Japan; k_horikiri@c.s.osakafu-u.ac.jp
* Correspondence: hfuji@c.s.osakafu-u.ac.jp; Tel.: +81-72-254-9818

Academic Editor: Manuel Almeida
Received: 13 January 2017; Accepted: 6 February 2017; Published: 9 February 2017

Abstract: To develop novel magnetic conductors exhibiting conducting/magnetic bifunctionalities and peculiar responses to applied magnetic fields, we synthesized new EDT-TTF (ethylenedithio-tetrathiafulvalene) donor containing a 2,2,5,5-tetramethylpyrrolin-1-yloxyl radical through a π-conjugated vinylene spacer 1 and examined its electronic and crystal structures, and physical properties. We also prepared its cation radical salts by an electrochemical oxidation method and successfully cleared the crystal structures and magnetic properties of the cation radical salts, 1·FeCl₄ and 1·GaCl₄. These salts have strongly dimerized one-dimensional arrays of the fully oxidized donor molecules, giving rise to the formation of spin-singlet state of the π cation radical spins in the dimer. On the other hand, the FeCl₄⁻ anion locates on the side of the dimers with very short S-Cl contacts and mediates very strong π-d interaction between the donor and anion moieties, resulting in the antiferromagnetic behavior of the Weiss temperature of $\theta = -3.9$ K through its d-π-d interaction.

Keywords: TTF; molecular conductors; magnetic properties; X-ray crystal structure analyses; stable organic radicals; 2,2,5,5-tetramethylpyrrolin-1-yloxyl radical

1. Introduction

In the field of molecular conductors, much interest has been focused on the development of magnetic conductors, which simultaneously exhibit conducting properties of organic layers of π-electron donors and magnetic properties of inorganic layers of magnetic transition metal anions. In such conductors, the conducting properties of organic layers can be controlled by the application of external magnetic fields and several peculiar physical phenomena such as field-induced superconductivities in the λ- and κ-type BETS salts with FeX_4^- anions (BETS = bis(ethylenedithio)tetraselenafulvalene; X = Cl and Br) [1–3] and anomalies of magnetoresistances corresponding to the spin-flop transitions of magnetic Fe^{3+} spins [4,5] have been yielded through the π-d interaction between the π-electrons of the organic layers and the magnetic d spins. On the other hand, the studies on molecular conductors using donor molecules substituted with stable organic radicals such as 2,2,6,6-tetramethylpiperidin-1-yloxyl (TEMPO) and nitronyl nitroxide radicals have been performed by several research groups [6–9] because cation radical salts of such stable radical-containing donors are expected to have strong intramolecular magnetic interactions between their π-cation radical spins and stable radicals and to show outstanding responses to the application of external magnetic fields [10]. Among them, we have also reported several tetrathiafulvalene (TTF)-based donor molecules containing stable organic radical parts such as TEMPO and 2,2,5,5-tetramethylpyrrolidin-1-yloxy (PROXYL) radicals [11–13], and discovered highly conducting cation radical salts using bis-fused TTF (TTP = 2,5-bis(1,3-dithiol-2-ylidene)-1,3,4,6-tetrathiapentalene) donors and a PROXYL radical [14–16]. Furthermore, we have developed new TTF and TTP donors containing a 2,2,5,5-tetramethylpyrrolin-1-yloxyl radical [17] because this radical part has an unsaturated 3-pyrroline ring with a C=C bond and smaller steric hindrance in comparison to the PROXYL radical having a saturated 3-pyrrolidine ring. However, conducting cation radical salts of these donors containing a 2,2,5,5-tetramethylpyrrolin-1-yloxyl radical could not be obtained probably due to the steric bulkiness of the radical part that is connected almost orthogonally to the ethylenedithio bridge of these donor molecules [17]. Therefore, to minimize the steric bulkiness of this radical part, we designed a new ethylenedithio-tetrathiafulvalene (EDT-TTF) molecule containing a 2,2,5,5-tetramethylpyrrolin-1-yloxyl radical through a π-conjugated vinylene spacer **1** because its π-conjugated hexatriene chain will ensure the coplanarity between the TTF and stable radical parts. In this paper, we will report the synthesis, electronic and crystal structures, and physical properties of new molecule **1**. Furthermore, we will discuss the detail of X-ray single crystal structure analyses and magnetic properties of the FeCl$_4^-$ and GaCl$_4^-$ salts of molecule **1** prepared by an electrochemical oxidation method. Because the FeCl$_4^-$ salt contains three kinds of spins, namely, the π-cation radical spins of the donor parts, the stable organic radicals and the Fe^{3+} d spins, the magnetic interactions between these spins are of special interest.

2. Results and Discussion

2.1. Synthesis of Donor **1**

The EDT-TTF molecule containing a 2,2,5,5-tetramethylpyrrolin-1-yloxyl radical **1** was prepared as described in Scheme 1. The Wittig reagent of stable radical part **4** was prepared by the reported method from the corresponding bromomethyl derivative **3** [18], and was reacted successively with t-BuOK and formyl-substituted EDT-TTF **2** [19] in dry benzene at room temperature. After the column-chromatographic separation of the resultant mixture, donor molecule **1** having a vinylene spacer was obtained as air-stable orange microcrystals in 53% yield.

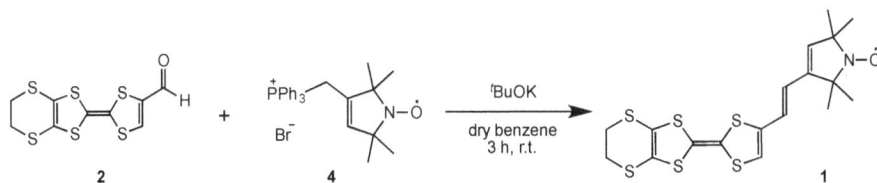

Scheme 1. Synthesis of donor **1**.

2.2. Electrochemical Properties and DFT Calculation of Donor **1**

Electrochemical properties of **1** were investigated by cyclic voltammetry technique. Cyclic voltammograms were measured in benzonitrile at 25 °C using tetra-*n*-butylammonium hexafluorophosphate as a supporting electrolyte. The obtained redox potentials of donor **1** are summarized in Table 1 together with those of EDT-TTF and 2,2,5,5-tetramethylpyrrolin-1-yloxyl radical derivative **5** [20] measured under the identical conditions. Molecule **1** showed three pairs of one-electron reversible redox waves (E_1, E_2 and E_3). The E_1 and E_2 values (+0.50 V and +0.87 V vs. Ag/AgCl) are almost the same to those of EDT-TTF (+0.48 V and +0.88 V), suggesting that the first and second oxidations occur at the EDT-TTF part. Similarly, the third oxidation (E_3) occurs at the stable radical part because the E_3 value of **1** (+0.96 V) is almost the same as that of **5** (+0.99 V). These results suggest that the HOMO orbital of donor **1** mainly locates on the EDT-TTF part.

A molecular orbital calculation of donor **1** was performed on the basis of the DFT theory at UB3LYP/6-31G(d, p) level using GAUSSIAN 09 package [21]. Figure 1 shows the molecular orbitals of donor **1** below the LUMO+2 level. The highest occupied two orbitals (120α and 119β: −4.76 eV) originate from the EDT-TTF part and the SOMO located under the HOMO orbitals (119α: −5.00 eV) localizes at the stable radical part. These results correspond to the above-mentioned electrochemical studies and indicate that the generated cation radical spin that exists on the EDT-TTF part and the localized stable radical spin can coexist in its cation radical salts prepared by an electrochemical oxidation method.

Table 1. Redox potentials [1] of **1**, ethylenedithio-tetrathiafulvalene (EDT-TTF) and stable organic radical **5**.

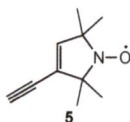

5

Compound	E_1	E_2	E_3	E_2-E_1
1	+0.50	+0.87	+0.96	0.37
EDT-TTF	+0.48	+0.88	-	0.40
5	-	-	+0.99	-

[1] V vs. Ag/AgCl, 0.1 mol·L^{-1} n-Bu$_4$NPF$_6$ in benzonitrile at 25 °C, Pt electrodes, scan rate of 50 mV·s^{-1}. The potentials were corrected with Ferrocene; E(Fc/Fc+) = +0.48 V vs. Ag/AgCl.

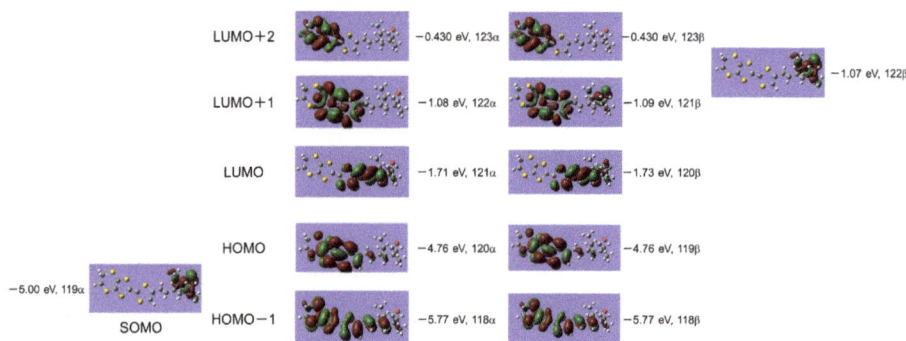

Figure 1. Molecular orbitals of donor **1** calculated on the basis of the DFT method at UB3LYP/6-31G (d, p) level.

2.3. Crystal Structure Analysis and Physical Properties of Donor 1

X-ray crystal structure analysis was performed on an orange platelike single crystal of **1**, which was obtained by recrystallization from dichloromethane/n-hexane. Figure 2 shows the ORTEP drawings of the molecular structure of **1**. This crystal belongs to the monoclinic $P2_1/c$ space group and one crystallographically independent molecule exists in the unit cell. As shown in Figure 2b, the TTF part adopts a boat-form conformation as is often observed in the neutral TTF derivatives. The vinylene spacer part has a *trans*-conformation and the vinylene spacer and pyrrolin-1-yloxyl parts show high planarity. This molecule has a slightly twisted molecular structure with a dihedral angle of 26° between the EDT-TTF moiety and radical moiety. As shown in Figure 3, the molecules are dimerized in a head-to-tail manner and form a so-called "κ type" molecular arrangement. The shortest S–S contacts are 3.61 Å in the dimer and 3.90 Å between the dimers. Because the shortest distances between the oxygen atoms of the stable radicals are 6.10, 6.17, and 6.43 Å, the intermolecular interaction between the radical parts seems to be very weak in the neutral crystal.

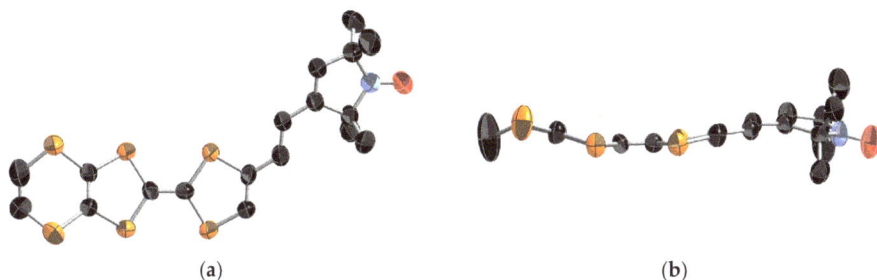

(a)   (b)

Figure 2. ORTEP drawings of the molecular structure of neutral donor **1** (**a**) Top view and (**b**) side view. The hydrogen atoms are omitted for clarity.

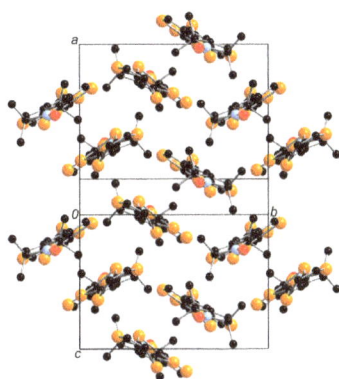

Figure 3. Crystal structure of neutral donor **1**. The hydrogen atoms are omitted for clarity.

Magnetic susceptibilities of neutral donor **1** were measured with a SQUID magnetometer (MPMS-XL, Quantum Design Inc., San Diego, CA, USA) under the applied field of 10 kOe in the temperature range of 2–300 K using a powder sample of **1**. The temperature dependence of magnetic susceptibilities can be fitted by a Curie-Weiss law with a Curie constant of 0.365 emu·K·mol^{-1} that corresponds to the one $S = 1/2$ spin of the stable radical with a g-value of 2.00 per molecule and small Weiss temperature of $\theta = -1.5$ K, suggesting the existence of very weak antiferromagnetic interaction between the stable radicals (See Figure S1a).

2.4. Crystal Structure Analyses and Magnetic Properties of the Cation Rarical Salts of Donor 1

X-ray crystal structure analyses were performed on needle-like single crystals of the FeCl$_4^-$ and GaCl$_4^-$ salts of donor **1**, **1**·FeCl$_4$ and **1**·GaCl$_4$ obtained by galvanostatic oxidation in the mixture of dry 1,2-dichloroethane and dry ethanol ($v/v = 1:9$) using the tetraethylammonium salts of the counteranions as supporting electrolytes. Because these two crystals belong to the monoclinic $P2_1/n$ space group and are completely isostructural to each other, only the structure of **1**·FeCl$_4$ will be discussed in detail in this section. In the unit cell, one donor **1** and one anion moiety are crystallographically independent, indicating that the donor:anion ratio of these crystals is 1:1 and each of the donor moieties is in the monocation radical state **1**$^{+\cdot}$. The ORTEP drawings of the molecular structure and the crystal structures of **1**·FeCl$_4$ are shown in Figures 4 and 5, respectively. As shown in Figure 4, the molecular structure of the donor moiety of **1**·FeCl$_4$ is similar to that of the neutral one (Figure 2), but is more planar with a dihedral angle of 15° between the EDT-TTF moiety and radical moiety than that of the neutral one (26°). Furthermore, the TTF framework has quite high planarity with a mean deviation from the least-squares plane of 0.027 Å due to its +1 oxidized state. As shown in the crystal structures (Figure 5), two donor molecules form a dimer in a head-to-tail manner with a very short interplanar distance of 3.41 Å in the dimer (interaction a1; the shortest S–S contact is 3.44 Å). The dimers construct a one-dimensional stacking structure along the *a*-axis with a longer interplanar distance of 3.63 Å between the dimers (interaction a2; the shortest S–S contact is 3.93 Å). Overlap integrals between the donor moieties along these stackings, which are calculated by the extended Hückel method [22], are $a1 = 41.5 \times 10^{-3}$ and $a2 = 0.29 \times 10^{-3}$, indicating its quite strongly dimerized one-dimensional electronic structure. On the other hand, the anion moiety locates on the side of the dimers with very short S–Cl contacts of 3.34 Å (interaction I) and 3.59 Å (interaction II), suggesting very strong π-d interaction between the donor and anion moieties. These anions form a uniform one-dimensional array with a relatively long Cl–Cl contact of 3.90 Å (interaction dd1) along the *a*-axis, suggesting weak direct intermolecular interaction between the anions. Furthermore, the oxygen atom of the stable radical part has a short O–S contact of 2.94 Å with the neighboring EDT-TTF moiety as shown in Figure 5a.

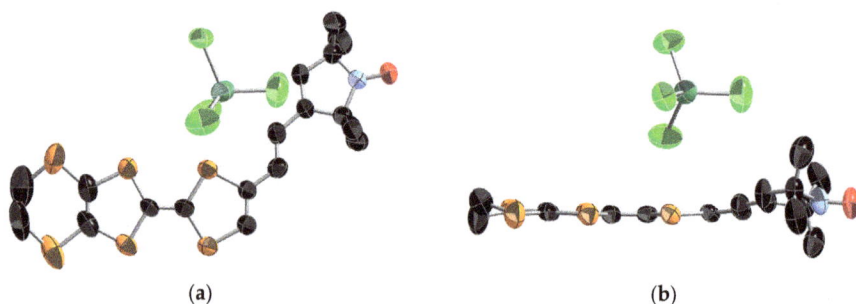

Figure 4. ORTEP drawings of the molecular structure of **1**·FeCl$_4$ salt (**a**) Top view and (**b**) side view. The hydrogen atoms are omitted for clarity.

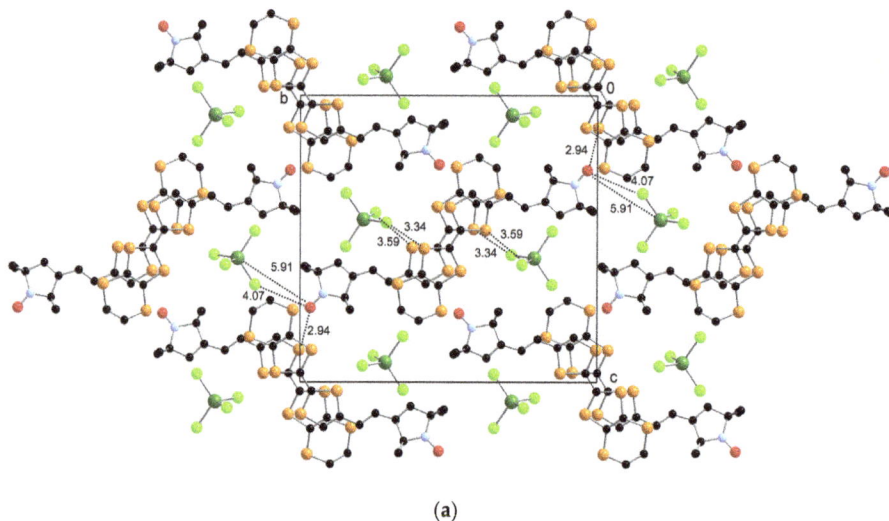

(a)

Figure 5. *Cont.*

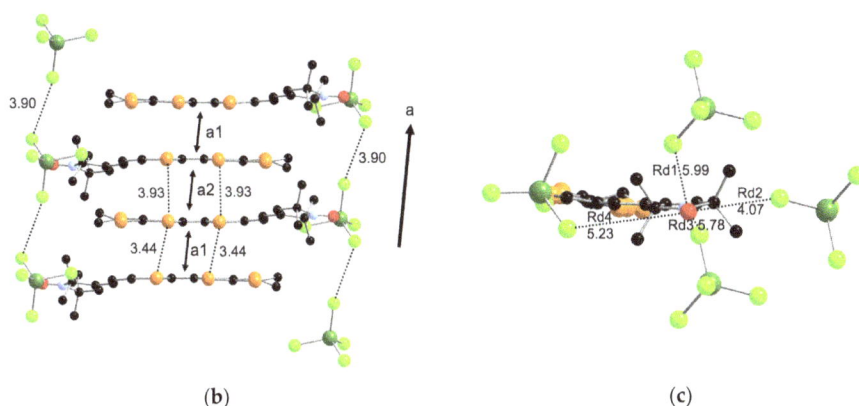

Figure 5. (a) Crystal structure of **1**·FeCl$_4$ projected onto the *bc*-plane; (b) Stacking structure of **1**·FeCl$_4$ along the *a*-axis; (c) Interactions around the stable radical. The hydrogen atoms are omitted for clarity. Short intermolecular contacts (Å) are indicated as dotted lines. Overlap integrals: *a*1 = 41.5 × 10^{-3} and *a*2 = 0.29 × 10^{-3}.

Paramagnetic susceptibilities (χ_p) of **1**·FeCl$_4$ and **1**·GaCl$_4$ were measured under the applied field of 10 kOe in the temperature range of 2–300 K. The temperature dependence of χ_p values of the GaCl$_4^-$ salt obeyed a Curie law with a Curie constant of 0.371 emu·K·mol^{-1} that is very close to the calculated value for one $S = 1/2$ spin system per 1:1 salt (0.375 emu·K·mol^{-1}) (See Figure S1c). This result suggests that two π cation radical spins on two EDT-TTF moieties in one dimer are paired to each other due to the strong intradimer interaction *a*1, resulting in the formation of a nonmagnetic spin-singlet state, and only the contribution from the stable radical part can be observed. On the other hand, the FeCl$_4^-$ salt indicated a Curie-Weiss fitting of the temperature dependence of χ_p values with a Curie constant of 4.89 emu·K·mol^{-1} that corresponds to the sum of the calculated contributions from the high-spin Fe^{3+} spin ($S = 5/2$; 4.375 emu·K·mol^{-1}) and the stable radical ($S = 1/2$; 0.375 emu·K·mol^{-1}) (See Figure S1b). The obtained Weiss temperature of −3.9 K suggests the existence of antiferromagnetic interaction between these spins. Due to the strongly dimerized structure of fully oxidized EDT-TTF moieties, these two salts showed insulating conducting behaviors.

Magnetic exchange interactions ($J_{\pi\pi}$, J_{dd}, $J_{\pi d}$ and J_{Rd}) between these paramagnetic moieties (the π cation radical on the EDT-TTF (π), the stable radical (R) and the magnetic FeCl$_4^-$ anion (d)) are estimated from the overlap integrals calculated by the extended Hückel method [23,24]. The π-π interactions in the dimer (*a*1 = 41.5 × 10^{-3}) and between the dimers (*a*2 = 0.29 × 10^{-3}) correspond to $J_{\pi\pi1}$ = 3996 K and $J_{\pi\pi2}$ = 0.19 K, respectively, confirming the nonmagnetic state of the donor parts due to the spin-singlet formation caused by the strong dimerization. The calculated direct d-d interaction between the anions along the *a*-axis is a very small value of J_{dd1} = 0.09 K due to its long Cl–Cl distance of 3.9 Å in comparison to the sum of the van der Waals radii of chlorine atoms (3.6 Å). On the other hand, the π-d interactions between the EDT-TTF moiety and the FeCl$_4^-$ anion are very large values of J_I = 7.7 K and J_{II} = 19.6 K reflecting very short S–Cl contacts of 3.34 Å (interaction I) and 3.59 Å (interaction II) between them mentioned above, while the interactions between the stable radical R and the anion moiety are estimated to be small values of J_{Rd1} = 0.30 K, J_{Rd2} = 0.32 K, J_{Rd3} = 0.03 K, and J_{Rd4} = 0.77 K using the SOMO orbital localized on the stable radical (See Figure 5c). Such short S–Cl contacts can mediate the antiferromagnetic interaction between the donor and anion moieties as reported in (TTF)$_3$ [(Cl)(Mo$_6$Cl$_{14}$)] complex in which a charge-enhanced S(δ^+)–Cl(δ^-) intermolecular interaction (3.229 Å) plays an important role to cause antiferromagnetic ordering of the complex [25]. Although the short O–S contact of 2.94 Å between the oxygen atom of the stable radical part and the neighboring EDT-TTF moiety might mediate the magnetic interaction between the stable radical parts, quite small

Weiss temperature of 0.03 K of the $GaCl_4^-$ salt suggests that the magnetic exchange interaction $J_{R\pi}$ between the radical part and the donor part is negligible. The overall J_{dd}, $J_{\pi d}$ and J_{Rd} values of this salt are obtained as the mean-field sum; $J_{dd} = 2J_{dd1} = 0.18$ K, $J_{\pi d} = J_I + J_{II} = 27.3$ K and $J_{Rd} = J_{Rd1} + J_{Rd2} + J_{Rd3} + J_{Rd4} = 1.42$ K, respectively, suggesting that the π-d interaction $J_{\pi d}$ seems to be dominant in the magnetic properties of this salt. Because the d-d interaction J_{dd} between the $FeCl_4^-$ anion is a very small value of 0.18 K compared to the Weiss temperature of -3.9 K, the strong π-d interactions (I and II) and strong π-π interaction in the dimer (a1) will mediate an indirect antiferromagnetic d-d interaction of θ = -3.9 K through its d-π-d interaction in addition to the small contribution by the magnetic interaction between the stable radical and the anions, $J_{Rd} = 1.39$ K.

3. Materials and Methods

General Remarks: Benzene was distilled under nitrogen atmosphere over calcium hydride. 1, 2-Dichloroethane was distilled under nitrogen atmosphere over P_2O_5. Other chemical reagents were purchased and used without further purification. High-resolution mass spectra (HRMS) using FAB⁺ method was measured using a JEOL JMS-700 mass spectrometer (JEOL Ltd., Akishima, Tokyo, Japan). IR spectra were recorded on KBr pellets using a JASCO FT/IR-4100 spectrometer (JASCO Corp., Hachioji, Tokyo, Japan). Cyclic voltammograms were measured using a BAS Electrochemical Analyzer Model 612B (BAS Inc., Sumida-ku, Tokyo, Japan). ESR spectrum of the benzene solution of **1** was measured at room temperature with a JEOL JES-RE1X X-band ESR spectrometer (JEOL Ltd., Akishima, Tokyo, Japan).

Synthesis of 1: A solid mixture of *t*-BuOK (70 mg, 0. 62 mmol) and **4** [18] (180 mg, 0.37 mmol) was suspended in 30 mL of dry benzene under nitrogen atmosphere and stirring for 2 h at room temperature. Then, **2** [19] (100 mg, 0.31 mmol) was added and the reaction mixture was further stirred for 3 h. After the solvent was evaporated in vacuo, the crude mixture was purified by column-chromatography on silica gel with dichloromethane (R_f = 0.58) as an eluent. Further purification by recrystallization with dichloromethane/*n*-hexane gave orange microcrystals of **1** (76 mg, 0.17 mol, 53%). m.p. 198–200 °C (dec.); HRMS FAB⁺ (Matrix = 3-Nitrobenzyl alcohol) ($C_{16}H_{18}NOS_4$): Found 368.0291; Calcd. 368.0271; IR (KBr) v1151, 2359, 2972, 3747 cm⁻¹; ESR (in benzene solution at r.t.) g = 2.0042, a_N = 1.44 mT

Preparation of cation radical salts of 1: The $FeCl_4^-$ and $GaCl_4^-$ salts of **1** were prepared as black needle-like crystals by a galvanostatic (I = 0.4 μA) oxidation using a conventional H-type electrocrystallization cell in the presence of **1** (5.0 mg) and the corresponding tetraethylammonium salts of the anions (100 mg) as a supporting electrolyte under nitrogen atmosphere in the mixture of dry 1, 2-dichloroethane and dry ethanol (10 ml, v/v = 1:9) at 16 °C for a few weeks.

X-ray data collection and reduction for the single crystalline samples: X-ray diffraction data were collected for the single crystal of neutral donor **1** on a Rigaku AFC-7 Mercury CCD diffractometer (Rigaku Corp., Akishima, Tokyo, Japan) with a graphite monochromated Mo-Kα radiation (λ = 0.7107 Å) and for the single crystals of **1**·$FeCl_4$ and **1**·$GaCl_4$ on a Rigaku AFC-8 Mercury CCD diffractometer (Rigaku Corp., Akishima, Tokyo, Japan) with confocal X-ray mirror system [Mo-Kα radiation (λ = 0.71075 Å)] and a rotating anode generator (0.8 kW). Lorentz and polarization corrections were applied. The structures were solved by a direct method (SIR92) [26], expanded (DIRDIF94) [27] and refined on F with full-matrix least-squares analysis. The non-hydrogen atoms were refined anisotropically. Hydrogen atoms were refined using the riding model. All the calculations were performed using the CrystalStructure crystallographic software package of the Molecular Structure Corporation [28]. Crystal data and structure refinement parameters are given in Table 2. CCDC-1527054 (**1**), 1527055 (**1**·$FeCl_4$) and 1527056 (**1**·$GaCl_4$) contains the supplementary crystallographic data for this paper. These data can be obtained free of charge from The Cambridge Crystallographic Data Centre via www.ccdc.cam.ac.uk/data_request/cif.

Magnetic property measurements: Magnetic susceptibilities were measured in the temperatures range of 2–300 K under the applied field of 10 kOe with a SQUID magnetometer (MPMS-XL, Quantum Design Inc., San Diego, CA, USA). Paramagnetic susceptibilities (χ_p) were obtained by subtracting the diamagnetic contribution estimated using Pascal's constants [29] from the observed magnetic susceptibilities.

Table 2. Crystallographic data for **1**, **1**·FeCl$_4$ and **1**·GaCl$_4$.

Crystal Data	1	1·FeCl$_4$	1·GaCl$_4$
Temperature/K	293.3	293.3	293.1
Chemical Formula	C$_{18}$H$_{20}$NOS$_6$	C$_{18}$H$_{20}$NOS$_6$FeCl$_4$	C$_{18}$H$_{20}$NOS$_6$GaCl$_4$
Formula weight	458.72	656.38	670.25
Crystal color, habit	orange, platelet	black, platelet	black, platelet
Dimensions, mm	0.50 × 0.40 × 0.10	0.10 × 0.05 × 0.04	0.10 × 0.10 × 0.08
Crystal system	Monoclinic	Monoclinic	Monoclinic
a/Å	15.172(9)	7.209(3)	7.242(2)
b/Å	12.274(8)	20.106(7)	20.166(5)
c/Å	11.458(7)	18.831(7)	18.930(5)
β/°	93.166(9)	95.504(11)	95.571(6)
V/Å^3	2130(3)	2717(2)	2751.7(11)
Space group, Z	$P2_1/c$, 4	$P2_1/n$, 4	$P2_1/n$, 4
$D_{calc.}$/g·cm^{-3}	1.430	1.605	1.618
μ/cm^{-1}	6.50	14.217	18.561
F_{000}	956.00	1332.00	1352.00
$2\theta_{max}$/°	61.7	61.1	60.8
Reflections collected	18831	29638	29387
Independent reflections	5971 (R_{int} = 0.0749)	7008 (R_{int} = 0.0758)	7104 (R_{int} = 0.0474)
Reflections used	1879 ($I > 2.50\sigma(I)$)	2308 ($I > 2.50\sigma(I)$)	3520 ($I > 3.00\sigma(I)$)
Number of variables	255	280	280
GOF on F	1.087	1.034	1.077
R_1	0.0688 ($I > 2.50\sigma(I)$)	0.0704 ($I > 2.50\sigma(I)$)	0.0787 ($I > 3.00\sigma(I)$)
wR	0.0738 ($I > 2.50\sigma(I)$)	0.0818 ($I > 2.50\sigma(I)$)	0.0733($I > 3.00\sigma(I)$)

4. Conclusions

We synthesized new EDT-TTF donor containing a 2,2,5,5-tetramethylpyrrolin-1-yloxyl radical through a π-conjugated vinylene spacer **1** and examined its electronic and crystal structures, and physical properties. We also prepared its cation radical salts by an electrochemical oxidation method and successfully cleared the crystal structures and magnetic properties of the cation radical salts, **1**·FeCl$_4$ and **1**·GaCl$_4$. These salts have the strongly dimerized one-dimensional arrays of the fully oxidized donor molecules, giving rise to the formation of spin-singlet state of the π cation radical spins in the dimer. On the other hand, the FeCl$_4^-$ anion locates on the side of the dimers with very short S-Cl contacts and mediates very strong π-d interaction between the donor and anion moieties, resulting in the antiferromagnetic behavior of θ = −3.9 K through its d-π-d interaction. The new findings in this paper will lead to the future construction of organic magnetic conductors based on the stable radical-containing donors. The preparation of new cation radical salts of donor **1** and its analogues, especially partially oxidized conducting materials, are now in progress to realize magnetic-conducting bifunctional materials.

Supplementary Materials: The following are available online at www.mdpi.com/2312-7481/3/1/8/s1, Figure S1: The temperature dependences of magnetic susceptibilities of neutral donor **1** (a), **1**·FeCl$_4$ (b) and **1**·GaCl$_4$ (c) measured in the temperatures range of 2–300 K under the applied field of 10 kOe. Calcd. lines indicate Curie-Weiss fittings of the data.

Acknowledgments: This work was financially supported in part by Grants-in-Aid for Scientific Research (No. 15K05483) from the Ministry of Education, Culture, Sports, Science and Technology of Japan. This work was also supported by the Institute for Molecular Science, Okazaki, Japan for the X-ray diffraction measurements on a Rigaku AFC-8 Mercury CCD diffractometer.

Author Contributions: H.F. conceived and designed the experiments; K.H. performed the synthetic and magnetic susceptibility experiments; K.H. and H.F. analyzed the X-ray data; K.H. and H.F. wrote the paper.

Conflicts of Interest: The authors declare no conflict of interest. The founding sponsor had no role in the design of the study; in the collection, analyses, or interpretation of data; in the writing of the manuscript, and in the decision to publish the results.

References

1. Uji, S.; Shinagawa, H.; Terashima, T.; Yakabe, T.; Terai, Y.; Tokumoto, M.; Kobayashi, A.; Tanaka, H.; Kobayashi, H. Magnetic-field-induced superconductivity in a two-dimensional organic conductor. *Nature* **2001**, *410*, 908–910. [CrossRef] [PubMed]

2. Fujiwara, H.; Fujiwara, E.; Nakazawa, Y.; Narymbetov, B.Z.; Kato, K.; Kobayashi, H.; Kobayashi, A.; Tokumoto, M.; Cassoux, P. A novel antiferromagnetic organic superconductor κ-(BETS)$_2$FeBr$_4$ [where BETS = bis(ethylenedithio)tetraselenafulvalene]. *J. Am. Chem. Soc.* **2001**, *123*, 306–314. [CrossRef] [PubMed]

3. Kobayashi, H.; Cui, H.B.; Kobayashi, A. Organic metals and superconductors based on BETS (BETS = bis(ethylenedithio)tetraselenafulvalene). *Chem. Rev.* **2004**, *104*, 5265–5288. [CrossRef] [PubMed]

4. Hayashi, T.; Xiao, X.; Fujiwara, H.; Sugimoto, T.; Nakazumi, H.; Noguchi, S.; Fujimoto, T.; Yasuzuka, S.; Yoshino, H.; Murata, K.; et al. A metallic (EDT-DSDTFVSDS)$_2$FeBr$_4$ salt: Antiferromagnetic ordering of d spins of FeBr$_4{}^-$ ions and anomalous magnetoresistance due to preferential π-d interaction. *J. Am. Chem. Soc.* **2006**, *128*, 11746–11747. [CrossRef] [PubMed]

5. Maesato, M.; Kawashima, T.; Furushima, Y.; Saito, G.; Kitagawa, H.; Shirahata, T.; Kibune, M.; Imakubo, T. Spin-Flop Switching and Memory in a Molecular Conductor. *J. Am. Chem. Soc.* **2012**, *134*, 17452–17455.

6. Sugano, T.; Fukasawa, T.; Kinoshita, M. Magnetic interactions among unpaired electrons in charge-transfer complexes of organic donors having a neutral radical. *Synth. Met.* **1991**, *43*, 3281–3284. [CrossRef]

7. Kumai, R.; Matsushita, M.M.; Izuoka, A.; Sugawara, T. Intramolecular Exchange Interaction in a Novel Cross-Conjugated Spin System Composed of π-Ion Radical and Nitronyl Nitroxide. *J. Am. Chem. Soc.* **1994**, *116*, 4523–4524. [CrossRef]

8. Komatsu, H.; Matsushita, M.M.; Yamamura, S.; Sugawara, Y.; Suzuki, K.; Sugawara, T. Influence of Magnetic Field upon the Conductance of a Unicomponent Crystal of a Tetrathiafulvalene-Based Nitronyl Nitroxide. *J. Am. Chem. Soc.* **2010**, *132*, 4528–4529. [CrossRef] [PubMed]

9. Souto, M.; Solano, M.V.; Jensen, M.; Bendixen, D.; Delchiaro, F.; Girlando, A.; Painelli, A.; Jeppesen, J.O.; Rovira, C.; Ratera, I.; et al. Self-Assembled Architectures with Segregated Donor and Acceptor Units of a Dyad Based on a Monopyrrolo-Annulated TTF-PTM Radical. *Chem. Eur. J.* **2015**, *21*, 8816–8825. [CrossRef] [PubMed]

10. Matsushita, M.M.; Kawakami, H.; Sugawara, T.; Ogata, M. Molecule-based system with coexisting conductivity and magnetism and without magnetic inorganic ions. *Phys. Rev. B* **2008**, *77*, 195208. [CrossRef]

11. Fujiwara, H.; Kobayashi, H. New pi-extended organic donor containing a stable TEMPO radical as a candidate for conducting magnetic multifunctional materials. *Chem. Commun.* **1999**, 2417–2418. [CrossRef]

12. Fujiwara, H.; Fujiwara, E.; Kobayashi, H. Novel π-electron donors for magnetic conductors containing a PROXYL radical. *Chem. Lett.* **2002**, 1048–1049. [CrossRef]

13. Fujiwara, H.; Fujiwara, E.; Kobayashi, H. Synthesis, structure and physical properties of donors containing a PROXYL radical. *Synth. Met.* **2003**, *135*, 533–534. [CrossRef]

14. Fujiwara, H.; Lee, H.J.; Kobayashi, H.; Fujiwara, E.; Kobayashi, A. A novel TTP donor containing a PROXYL radical for magnetic molecular conductors. *Chem. Lett.* **2003**, *32*, 482–483. [CrossRef]

15. Fujiwara, H.; Lee, H.J.; Cui, H.B.; Kobayashi, H.; Fujiwara, E.; Kobayashi, A. Synthesis, structure, and physical properties of a new organic conductor based on a π-extended donor containing a stable 2,2,5,5-tetramethyl-1-pyrrolidinyloxy radical. *Adv. Mater.* **2004**, *16*, 1765–1769. [CrossRef]

16. Otsubo, S.; Cui, H.B.; Lee, H.J.; Fujiwara, H.; Takahashi, K.; Okano, Y.; Kobayashi, H. A magnetic organic conductor based on a π donor with a stable radical and a magnetic anion—A step to magnetic organic metals with two kinds of localized spin systems. *Chem. Lett.* **2006**, *35*, 130–131. [CrossRef]

17. Fujiwara, E.; Aonuma, S.; Fujiwara, H.; Sugimoto, T.; Misaki, Y. New π-electron donors with a 2,2,5,5-tetramethylpyrrolin-1-yloxyl radical designed for magnetic molecular conductors. *Chem. Lett.* **2008**, *37*, 84–85. [CrossRef]

18. Krishna, M.C.; DeGraff, W.; Hankovszky, O.H.; Sár, C.P.; Kálai, T.; Jekő, J.; Russo, A.; Mitchell, J.B.; Hideg, K. Studies of Structure—Activity Relationship of Nitroxide Free Radicals and Their Precursors as Modifiers Against Oxidative Damage. *J. Med. Chem.* **1998**, *41*, 3477–3492. [CrossRef] [PubMed]

19. Garin, J.; Orduna, J.; Uriel, S.; Moore, A.J.; Bryce, M.R.; Wegener, S.; Yufit, D.S.; Howard, J.A.K. Improved Syntheses of Carboxytetrathiafulvalene, Formyltetrathiafulvalene and (Hydroxymethyl)Tetrathiafulvalene —Versatile Building-Blocks for New Functionalized Tetrathiafulvalene Derivatives. *Synthesis* **1994**, *1994*, 489–493. [CrossRef]

20. Kálai, T.; Balog, M.; Jekö, J.; Hideg, K. Synthesis and Reactions of a Symmetric Paramagnetic Pyrrolidine Diene. *Synthesis* **1999**, *1999*, 973–980. [CrossRef]

21. Frisch, M.J.; Trucks, G.W.; Schlegel, H.B.; Scuseria, G.E.; Robb, M.A.; Cheeseman, J.R.; Scalmani, G.; Barone, V.; Mennucci, B.; Petersson, G.A.; et al. *Gaussian 09, Revision B.01*; Gaussian, Inc.: Wallingford, CT, USA, 2010.

22. Mori, T.; Kobayashi, A.; Sasaki, Y.; Kobayashi, H.; Saito, G.; Inokuchi, H. The Intermolecular Interaction of Tetrathiafulvalene and Bis(ethylenedithio)tetrathiafulvalene in Organic Metals. Calculation of Orbital Overlaps and Models of Energy-band Structures. *Bull. Chem. Soc. Jpn.* **1984**, *57*, 627–633. [CrossRef]

23. Mori, T.; Katsuhara, M. Estimation of πd-Interactions in Organic Conductors Including Magnetic Anions. *J. Phys. Soc. Jpn.* **2002**, *71*, 826–844. [CrossRef]

24. Mori, T.; Katsuhara, M.; Akutsu, H.; Kikuchi, K.; Yamada, J.; Fujiwara, H.; Matsumoto, T.; Sugimoto, T. Estimation of π d-interactions in magnetic molecular conductors. *Polyhedron* **2005**, *24*, 2315–2320. [CrossRef]

25. Batail, P.; Livage, C.; Parkin, S.S.P.; Coulon, C.; Martin, J.D.; Canadell, E. Antiperovskite Structure with Ternary Tetrathiafulvalenium Salts: Construction, Distortion, and Antiferromagnetic Ordering. *Angew. Chem. Int. Ed. Engl.* **1991**, *30*, 1498–1500. [CrossRef]

26. Altomare, A.; Cascarano, G.; Giacovazzo, C.; Guagliardi, A.; Burla, M.C.; Polidori, G.; Camalli, M. SIR92—A program for automatic solution of crystal structures by direct methods. *J. Appl. Crystallogr.* **1994**, *27*, 435. [CrossRef]

27. Beurskens, P.T.A.; Beurskens, G.; Bosman, W.P.; de Gelder, D.; Israel, R.; Smith, J.M.M. *Technical Report of the Crystallography Laboratory*; University of Nijmegen: Nijmegen, The Netherlands, 1994.

28. *CrystalStructure, 4.0*; Crystal Structure Analysis Package; Rigaku Corp.: Akishima, Tokyo, Japan, 2000–2010.

29. König, E. *Landolt Bornstein, Group II: Atomic and Molecular Physics, Vol. 2, Magnetic Properties of Coordination and Organometallic Transition Metal Compounds*; Spring: Berlin, Germany, 1966.

magnetochemistry

MDPI

Article

New Dmit-Based Organic Magnetic Conductors (PO-CONH-C$_2$H$_4$N(CH$_3$)$_3$)[M(dmit)$_2$]$_2$ (M = Ni, Pd) Including an Organic Cation Derived from a 2,2,5,5-Tetramethyl-3-pyrrolin-1-oxyl (PO) Radical

Hiroki Akutsu [1,*], Scott S. Turner [2] and Yasuhiro Nakazawa [1]

[1] Department of Chemistry, Graduate School of Science, Osaka University, 1-1 Machikaneyama, Toyonaka, Osaka 560-0043, Japan; nakazawa@chem.sci.osaka-u.ac.jp

[2] Department of Chemistry, University of Surrey, Guildford, Surrey GU2 7XH, UK; s.s.turner@surrey.ac.uk

* Correspondence: akutsu@chem.sci.osaka-u.ac.jp; Tel.: +81-6-6850-5399

Academic Editor: Manuel Almeida

Received: 11 January 2017; Accepted: 20 February 2017; Published: 25 February 2017

Abstract: We have prepared two dmit-based salts with a stable organic radical-substituted ammonium cation, (PO-CONH-C$_2$H$_4$N(CH$_3$)$_3$)[Ni(dmit)$_2$]$_2$·CH$_3$CN and (PO-CONH-C$_2$H$_4$N(CH$_3$)$_3$)[Pd(dmit)$_2$]$_2$ where PO is 2,2,5,5-Tetramethyl-3-pyrrolin-1-oxyl and dmit is 2-Thioxo-1,3-dithiol-4,5-dithiolate. The salts are not isostructural but have similar structural features in the anion and cation packing arrangements. The acceptor layers of both salts consist of tetramers, which gather to form 2D conducting layers. Magnetic susceptibility measurements indicate that the Ni salt is a Mott insulator and the Pd salt is a band insulator, which has been confirmed by band structure calculations. The cationic layers of both salts have a previously unreported polar structure, in which the cation dipoles order as ↗↘↗↘ along the acceptors stacking direction to provide dipole moments. The dipole moments of nearest neighbor cation layers are inverted in both salts, indicating no net dipole moments for the whole crystals. The magnetic network of the [Ni(dmit)$_2$] layer of the Ni salt is two-dimensional so that the magnetic susceptibility would be expected to obey the 1D or 2D Heisenberg model that has a broad maximum around $T \approx \theta$. However, the magnetic susceptibility after subtraction of the contribution from the PO radical has no broad maximum. Instead, it shows Curie–Weiss behavior with $C = 0.378$ emu·K/mol and $\theta = -35.8$ K. The magnetic susceptibility of the Pd salt obeys a Curie–Weiss model with $C = 0.329$ emu·K·mol^{-1} and $\theta = -0.88$ K.

Keywords: 1D Heisenberg chains; dmit; organic stable radicals

1. Introduction

Over the past two decades, organic magnetic conductors that combine electrical conductivity with magnetic moments have attracted great interest [1–3]. This is due to reports of unique and interesting properties that emerge from the interplay between itinerant and localized electrons. In particular, λ-(BETS)$_2$FeCl$_4$ (BETS = bis(ethylenedithio)tetraselenafulvalene) and its derivatives show unique physical properties such as a transition to antiferromagnetic order coupled to a metal-insulator transition, colossal magnetoresistance, field-induced superconductivity and a superconductor-to-insulator transition [1]. In fact, many inorganic magnetic anions have already been used as counteranions in organic conductors [2]. Alternatively, researchers have introduced organic free radicals as the source of magnetism [4]. However, thus far, no salts have shown significant magnetic interactions between the conducting electrons and the magnetic moments on the incorporated radicals apart from Sugawara's radical-substituted TTF-based salts [5,6] (TTF = tetrathiafulvalene).

Over the past 15 years, we have prepared several purely organic magnetic conductors, which consist of cationic organic donors (e.g., TTF, bis(ethylenedithio)tetrathiafulvalene (BEDT-TTF)) with monoanionic sulfo ($-SO_3^-$) derivatives of organic free radicals (2,2,6,6-tetramethylpiperidin-1-oxyl (TEMPO) [7–12], 2,2,5,5-tetramethylpyrrolidin-1-oxyl (PROXYL) [13] or 2,2,5,5-Tetramethyl-3-pyrrolin-1-oxyl (PO) [14–18]). Of the salt that we have obtained, none showed a significant interplay between the conducting and localized electrons. This was despite some having evidence of short contacts between the moieties with conducting and those with localized electrons, through short S(BEDT-TTF)···O(Spin centers of the free radicals) distances.

Recently, we have noted that some of these salts have unique structural features, which we have classified into two types, I and II (see Figure 1). Almost all donor-anion type salts consist of 1D or 2D conducting layers interleaved by insulating counterions and/or incorporated neutral molecules. Type I and II salts have structurally unusual anionic layers, in which each individual anisotropic anion aligns in the same orientation as the other anions in the same layer. The result is that each layer has a dipole moment. In Type I salts, such as κ-β″-(BEDT-TTF)$_2$**A1** [15], (TTF)$_3$(**A2**)$_2$ [16] and α′-α′-(BEDT-TTF)$_2$**A3**·H$_2$O [18] (see Scheme 1 for molecular structures), as illustrated in Figure 1a, the dipole moments of successive anionic layers oppose each other, so there is no net dipole moment. As a consequence, Type I salts have two crystallographically independent cation layers (A and B), one of which (A in Figure 1a) is surrounded by the negative ends of the anionic layer's dipole and the other (B) is bordered by the positive ends of the dipole moments. We also observed self-doping in the Type I salts, where the A layers were more positively charged than B layers. By contrast, a Type II salt, exemplified by α-(BEDT-TTF)$_2$**A4**·3H$_2$O [17], has a net dipole moment for the whole crystal because all polarized anionic layers are oriented in almost the same direction as shown in Figure 1b. The donor layers of Type II salts also have a dipole moment, which cancels that of the anionic layers. At present, we do not understand the reason why these specific salts possess polarized anionic layers, although it is noted that Type I and II salts all contain sulfo derivatives of the PO radical.

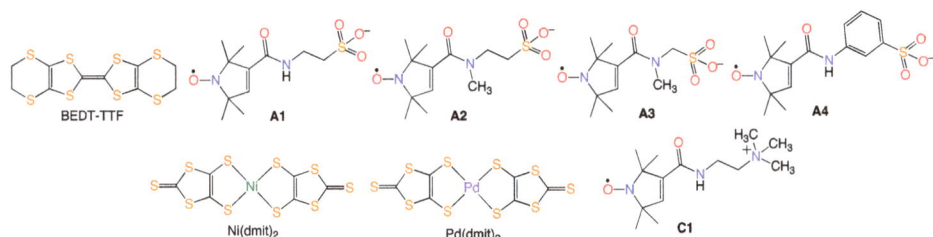

Scheme 1. Molecular structures where the prefixes **A** and **C** stand for anion and cations, respectively.

There are many types of organic conductors, although salts with a conducting donor and counter-anion are the most widely researched, and it is clear that this type has the greatest number of examples of metals and superconductors. This is the reason why we previously concentrated on the synthesis of this type of purely organic magnetic conductors. The second most widely researched type of salt is that with a conducting acceptor and counter-cation, often made with acceptors [M(dmit)$_2$] (where M = Ni, Pd, Pt, etc.). This type has also provided many metals and superconductors [19–21]. Here, we report organic magnetic conductors consisting of [M(dmit)$_2$] (M = Ni and Pd) and a new organic cation bearing PO radical, PO–CONH–C$_2$H$_4$N(CH$_3$)$_3^+$ (**C1**$^+$), the molecular shape of which is similar to that of **A1**. The resultant salts have a new type of structural feature, Type III as shown in Figure 1c.

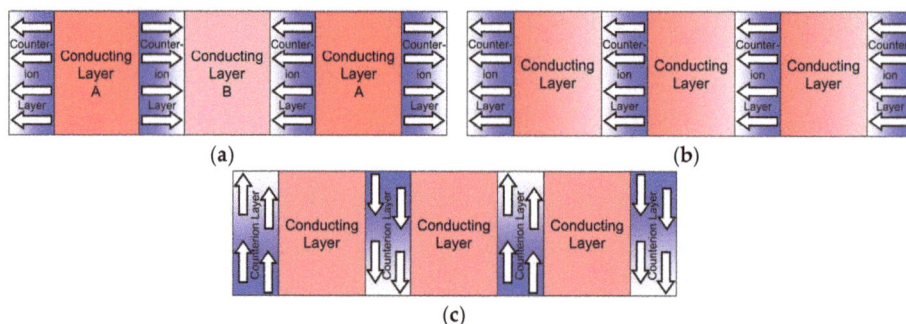

(a) (b)

(c)

Figure 1. Schematic diagrams of the crystal structures of (**a**) Type I; (**b**) II and (**c**) III salts where the electrical dipoles of the counterions are indicated by arrows, conducting layers are shown as red squares and counterion layers are shown as blue rectangles.

2. Results and Discussion

2.1. (PO-CONH-C₂H₄N(CH₃)₃)Cl (*C1Cl*, 1)

2.1. (PO-CONH-C_2H_4N(CH$_3$)$_3$)Cl (C1Cl, 1)

Yellow prisms of **1** are deliquescent on exposure to the air so a crystal, suitable for X-ray analysis, was sealed in a capillary tube. X-ray data was collected at 200 K. Figure 2 shows the crystal structure of **1** and indicates the short O···O contact between spin centres. The contacts are approximately 0.8 Å longer than the van der Waals distance (3.04 Å), indicating that the interaction between radicals is weak. Temperature-dependent magnetic susceptibility of a powdered sample of **1** obeys the Curie law with $C = 0.377$ emu·K·mol^{-1} and $\theta = -2.1$ K.

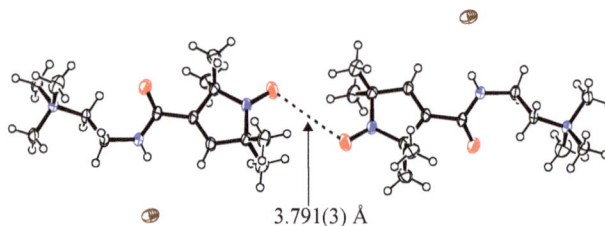

3.791(3) Å

Figure 2. Crystal structure of (PO-CONH-C_2H_4N(CH$_3$)$_3$)Cl (**1**).

2.2. (PO-CONH-C₂H₄N(CH₃)₃)[Ni(dmit)₂]₂·CH₃CN ((C1)[Ni(dmit)₂]₂·CH₃CN, 2)

2.2. (PO-CONH-C_2H_4N(CH$_3$)$_3$)[Ni(dmit)$_2$]$_2$·CH$_3$CN ((C1)[Ni(dmit)$_2$]$_2$·CH$_3$CN, 2)

The asymmetric unit consists of two [Ni(dmit)₂] molecules, a **C1** cation and an incorporated acetonitrile molecule. Figure 3a shows the crystal structure of **2**. The structure has alternating acceptor and cation/acetonitrile layers propagating along the *b*-axis. The structure of the two-dimensional conducting acceptor layer is shown in Figure 3b. Dotted lines indicate short S···S contacts (<3.70 Å). A greater number of short contacts are observed between A–A′ and A–B than between B–B′ along the stacking direction (// *c*), suggesting the formation of B′–A′–A–B tetramers (see also Figure 3c). To confirm tetramer formation, the transfer integrals were calculated using Prof. Takehiko Mori's tight-binding band structure calculation program [22]. The values of *p*1 (A–B), *p*2 (A–A′) and *p*3 (B–B′) (see Figure 3b) are 3.00, 0.87 and -4.54×10^{-3} eV, respectively. The B–B′ interaction, which has fewer short S···S contacts, is approximately five times stronger than that of A–A′, suggesting the formation of a tetramer designated by the A–B–B′–A′. The formula charge of [Ni(dmit)₂] is -0.5 so that the charge of the tetramer is -2. This suggests that the salt is either a Mott insulator for which the tetramer would have two spins or a band insulator for which the tetramer consists of spin dimers. The band structure calculation [22] of **2** has been performed. Figure 4 shows the band dispersions (left) and Fermi surfaces (right) of **2**. There is a mid-gap between the upper and lower bands, which is the so-called "Mott gap", suggesting that the salt is a Mott insulator. Temperature-dependent electrical resistivity of **2** indicates that the salt is a semiconductor with $\rho_{RT} = 3.4 \ \Omega \cdot cm$ and $E_a = 0.032$ eV. We will further discuss whether the salt is a Mott or band insulator below.

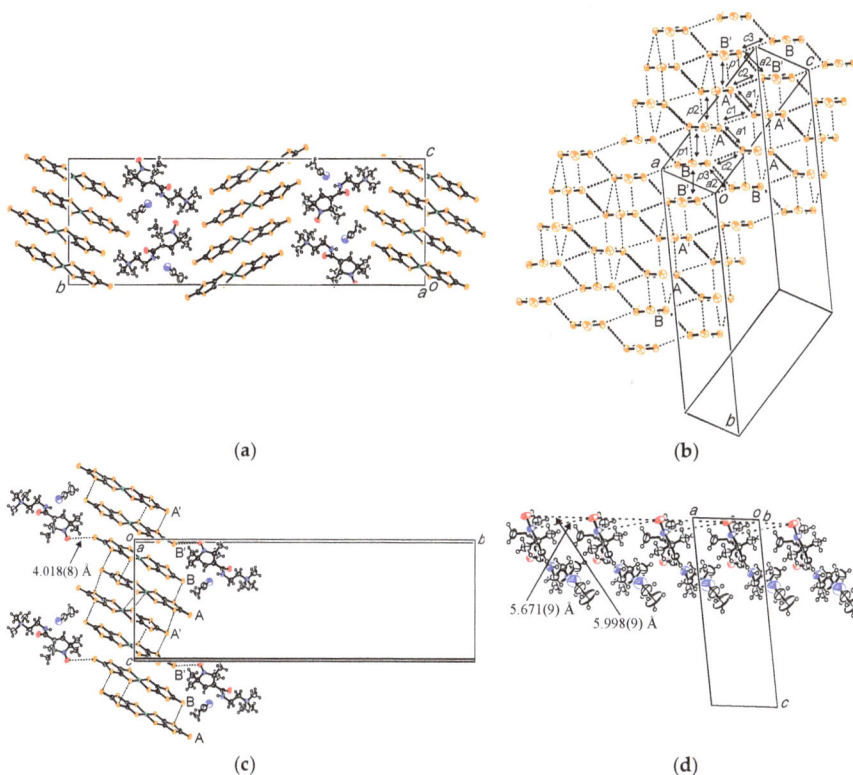

(a)

(b)

(c)

(d)

Figure 3. (**a**) crystal structure of **2**; (**b**) [Ni(dmit)$_2$] arrangement in conducting layers of **2**, where dotted lines indicate shorter S$\cdots$S contacts than the van der Walls distance (3.70 Å); (**c**) interactions between [Ni(dmit)$_2$] and **C1** in **2**, where dotted lines indicate short S$\cdots$S, S$\cdots$O and Ni$\cdots$S (<3.48 Å) contacts and (**d**) 1D magnetic chain of **C1** in (**C1**)[Ni(dmit)$_2$]$_2\cdot$CH$_3$CN (**2**).

The salt **2** also exhibits magnetically important interactions. Each **C1** cation stacks along the *a*-axis with short contacts between spin centres (N$\cdots$O and O$\cdots$O) to form a 1D magnetic chain as shown in Figure 3d. The contacts have longer distances than the van der Waals contacts, but contacts shorter than 6 Å can be significant for magnetic interactions. The spin center O atoms are also close to the outer S atoms of the [Ni(dmit)$_2$] network that forms a 2D conducting sheet as shown in Figure 3c. Thus, the 1D chains of the PO radicals and the 2D layers of the [Ni(dmit)$_2$] molecules interact to form a 2D magnetic system. Magnetic susceptibility measurements of **2** are shown in Figure 5a as a χT-T plot. The right vertical axis indicates $S = 1/2$ spin concentration. The figure indicates that there are approximately two spins at room temperature, suggesting that both the PO radicals of **C1** and also a [Ni(dmit)$_2$]$_2{}^-$ dimer have one spin. This clearly indicates that the salt **2** is a Mott insulator. The χT value decreases with decreasing temperature, indicating that, at very low temperatures, the magnetic contribution of the [Ni(dmit)$_2$]$_2{}^-$ dimers is negligible and that of the PO radicals is dominant. We can estimate the contribution of the PO radicals ($\chi_{radical}$) by fitting to a Curie–Weiss expression in the very low temperature region (2–4 K), yielding $C = 0.313$ emu·K·mol^{-1} and $\theta = +0.08$ K. The negligibly small Weiss constant suggests that the O···O short contacts between PO radicals (Figure 3d) do not significantly affect the magnetism. To obtain the magnetic contribution of the [Ni(dmit)$_2$] layers, we subtracted the Curie term ($\chi_{radical}$) from the total data (χ_{total}). The resultant $\chi_{[Ni(dmit)_2]}$-T curve is shown in Figure 5b, which can be modeled by the Curie–Weiss law with $C = 0.378$ emu·K·mol^{-1} and $\theta = -35.8$ K (solid line in Figure 5b). The C value is close to that for $S = 1/2$ spin (0.375 emu·K·mol^{-1}), suggesting that the [Ni(dmit)$_2$]$_2^-$ dimer has one spin. The θ value indicates that the spins interact with each other much more strongly than with the spins on **C1**. Organic charge-transfer salts in the Mott insulating state usually have a broad maximums in their χ-T plots at around $T = \theta$ as a result of their low dimensionality (1D or 2D Heisenberg systems). However, Figure 5b has no broad maximum and the χ value increases with decreasing temperature. A possible explanation is that **2** may have a 3D magnetic network, although this is not consistent with the structural data from X-ray analysis.

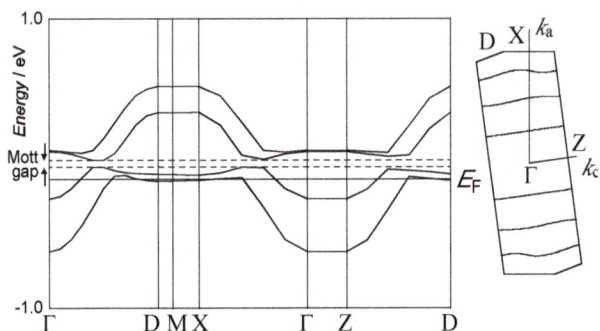

Figure 4. Band dispersions (**left**) and Fermi surfaces (**right**) of **2**. Transfer integrals of $p1$, $p2$, $p3$, $a1$, $a2$, $c1$, $c2$ and $c3$ (see Figure 3b) are 3.00, 0.87, −4.54, −4.14, 21.37, −1.29, −5.94 and -13.89×10^{-3} eV, respectively.

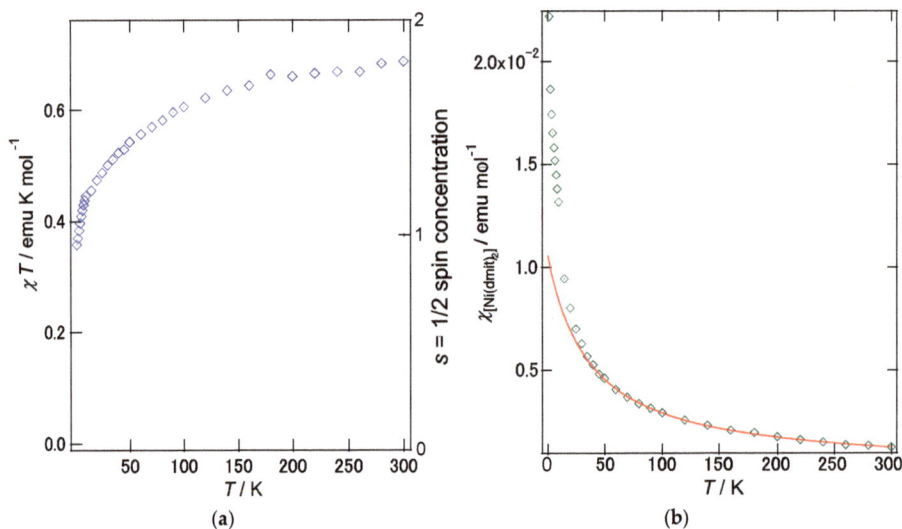

Figure 5. (a) χT-T plots of **2**. (b) $\chi_{[Ni(dmit)_2]}$-T plots of **2** where $\chi_{[Ni(dmit)_2]} = \chi_{total} - \chi_{radical}$. See text for explanation.

Finally, we discuss the dipole structure of the salt **2**. The salt crystallizes in the monoclinic space group $P2_1/c$, which is centrosymmetric and, therefore, there is no net dipole moment. However, each cation layer has a dipole moment (see Figure 3a). We calculate the dipole moment of **C1** using MOPAC7 [23] to be 23.6 debye. There are two **C1** molecules, which are related by a c glide operation, the symmetry operation partially canceling the dipole moments. The dipole moment of the two molecules is calculated by MOPAC7 [23] to be 19.2 debye (D) per one molecule, the direction of the dipole's vector being parallel to the c-axis. The dipole moments of nearest neighbor cation layers are crystallographically inverted. Thus, the salt **2** belongs to Type III as shown in Figure 1c.

2.3. (PO-CONH-C$_2$H$_4$N(CH$_3$)$_3$)[Pd(dmit)$_2$]$_2$ ((**C1**)[Pd(dmit)$_2$]$_2$, **3**)

Crystals **2** and **3** are not isostructural but they do have similar structural features. The asymmetric unit of **3** contains two [Pd(dmit)$_2$] molecules and a **C1** cation but does not have any incorporated neutral solvent molecules. The structure consists of alternating acceptor and cation layers propagating along the b-axis as shown in Figure 6a. Figure 6b shows a conducting layer consisting of [Pd(dmit)$_2$] molecules. The pattern of S$\cdots$S short contacts indicates that the acceptor layers of compound **3** consist of B–A–A$'$–B$'$ tetramers (see also Figure 6a). The transfer integral calculations [22] for $p1$ (A–B) = 44.16, $p2$ (A–A$'$) = 31.80 and $p3$ (B–B$'$) = 2.39×10^{-3} eV, which also suggest the formation of a B–A–A$'$–B$'$ tetramer. The formula charge of [Pd(dmit)$_2$] is -0.5, indicating a charge per tetramer of -2. This, again, suggests that the salt is a Mott insulator or a band insulator. The ratio of $p2$ (A–A$'$)/$p3$ (B–B$'$) of 13.3 indicates that the tetramer in **3** is more isolated than that of **2**, suggesting the formation of a spin dimer and that the salt is a band insulator. If the salt is a Mott insulator like **2**, the salt would be a paramagnet. Figure 7 shows a variable temperature magnetic susceptibility and, unlike for compound **2**, the data can be fitted well by a Curie–Weiss model with C = 0.329 emu·K·mol^{-1}, θ = -0.88 K and an additional temperature independent term a = 3.1×10^{-5} emu·mol^{-1}. The C value is close to 0.375 emu·K·mol^{-1} for 100% of $S = 1/2$ spin, which suggests that the free radical spins on **C1** dominate the Curie–Weiss term. The a value is much less than 2×10^{-4} emu·mol^{-1}, so that the contribution from conduction electrons is negligible (normally, a would be about 2–6 $\times 10^{-4}$ emu·mol^{-1} if the salt is a Pauli paramagnet). These results suggest that compound **3** is a band insulator.

Although no Ni$\cdots$Ni short contacts were observed in **2** (all >4 Å), short Pd$\cdots$Pd intra-tetramer contacts in **3** were observed, but no inter-tetramer Pd$\cdots$Pd contacts are present (Figure 6a). Therefore, the intra-tetramer interactions in **3** are stronger than those in **2**. Band dispersions (Figure 8 left) and a Brillouin zone (Figure 8 right) of **3** show no Fermi surfaces, which also points towards a band insulator description. The temperature-dependent electrical resistivity of **3** indicates that the salt **3** is a semiconductor with ρ_{RT} = 0.81 Ω·cm and E_a = 0.17 eV. The E_a value of **3** is 5.3 times larger than that of **2**. The **C1** cations in **3** form a 1D magnetic chain (Figure 6c); however, the shortest N$\cdots$O contact of **3** is slightly shorter than 6 Å, indicating that the magnetic interactions are weak. This is consistent with the fact that salt **3** shows a small θ value. Three short S$\cdots$O contacts between **C1** and [Pd(dmit)$_2$] were observed as shown in Figure 6a. This indicates that the 2D [Pd(dmit)$_2$] layers are connected by a PO radical to form a 3D magnetic network. However, the S$\cdots$O contacts, ranging from 5.647 to 5.874 Å (Figure 6a) are much longer than those in compound **2** (4.018(8) Å), suggesting that the magnetic interactions within **3** are much weaker than those of **2**, which is again consistent with a small θ value of **3**.

The dipole moment of **C1** in the salt **3** was calculated using MOPAC7 [23]. The value of 22.3 D is slightly smaller than that of **2**. Two **C1** molecules in the unit cell are related by n glide operation, which partially cancels their dipole moments. The dipole moment of two molecules is 13.6 D per one molecule, the vector of which is parallel to the n (=$a + c$) axis. In addition, [Pd(dmit)$_2$] stacks along the n-axis. The dipole moments of nearest neighbor cation layers are opposite. Thus, salt **3** has a structural feature of Type III as shown in Figure 1c.

(a)

(b)

N···O = 5.965(16) Å

(c)

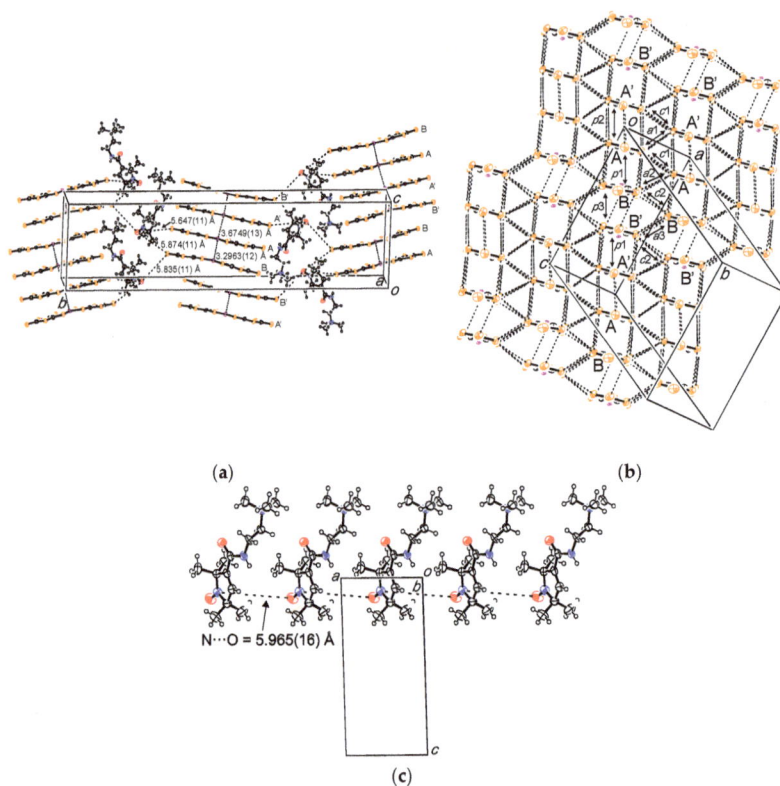

Figure 6. (a) crystal structure of **3** where dotted lines indicate the Pd···Pd and S···O short contacts; (b) the molecular arrangement in a [Pd(dmit)$_2$] conducting layer of **3** where dotted lines indicate the short S···S distances (<3.70 Å) and (c) the 1D magnetic chain of **C1** in (**C1**)[Pd(dmit)$_2$]$_2$ (**3**).

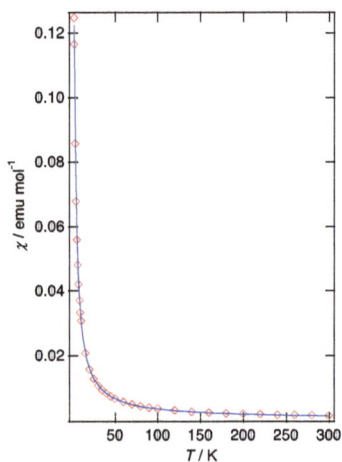

Figure 7. Temperature dependence of the magnetic susceptibility of **3**. The solid line is calculated on the basis of a Curie–Weiss model.

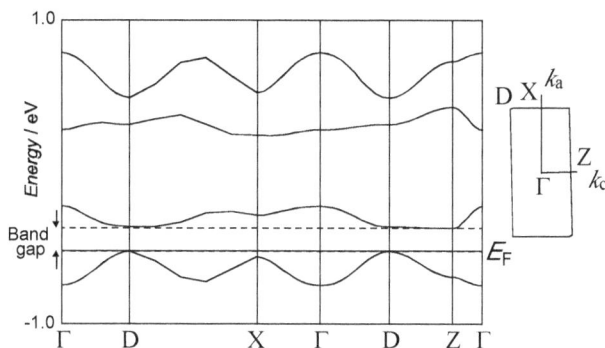

Figure 8. Band dispersions (**left**) and a Brillouin zone (**right**) of **3**. Transfer integrals of $p1$, $p2$, $p3$, $a1$, $a2$, $a3$, $c1$ and $c2$ (see Figure 6b) are 44.16, 31.80, 2.39, 9.47, 4.63, 8.41, -0.60 and -1.61×10^{-3} eV, respectively.

3. Materials and Methods

3-carboxy-2,2,5,5-tetramethyl-3pyrrolin-1-oxyl (PO-COOH) was prepared according to the literature method [24]. **C1**Cl (**1**) was prepared by reacting PO-COOH (1.0 g, 5.4 mmol) with $H_2NC_2H_4N(CH_3)_3Cl \cdot HCl$ (0.95 g, 5.4 mmol) in the presence of N,N'-dicyclohexylcarbodiimide (DCC, 2.7 g, 13 mmol) and 4-dimethylaminopyridine (DMAP, 2.4 g, 20 mmol) in 50 mL of CH_2Cl_2 at room temperature with stirring for two weeks. The resultant solution was filtered, and the filtrate was purified by column chromatography (silica gel, eluents: CH_2Cl_2 then acetone and finally with acetone/methanol = 1:1). Recrystallisation from acetonitrile and ethyl acetate gave hygroscopic yellow rods of **1** (yield 0.63 g (38%); m.p. 209–210 °C). Elemental analysis data were consistent with (**C1**)Cl·0.6H_2O probably because of its strong hygroscopic nature (Anal. Calcd. for $C_{14}H_{27}N_3O_2Cl \cdot 0.6H_2O$: C, 53.27; H, 9.00; N, 13.31. Found: C, 53.31; H, 8.84; N, 13.30). Constant-current (0.5 µA) electrocrystallisation of **1** (10 mg) with $(n\text{-}(C_4H_9)_4P)[Ni(dmit)_2]$ (purchased from Tokyo Chemical Industry Co., Ltd., Tokyo,. Japan (TCI), 10 mg) in acetonitrile (20 mL) in a conventional H-shaped cell yielded black needles of **C1**$[Ni(dmit)_2]_2 \cdot CH_3CN$ (**2**; m.p. >300 °C; Anal. Calcd. for $C_{28}H_{30}N_4O_2S_{20}Ni_2$: C, 27.72; H, 2.49; N, 4.62. Found: C, 27.98; H, 2.50; N, 4.43). Constant-current (1.0 µA) electrocrystallisation of **1** (10 mg) with $(n\text{-}(C_4H_9)_4N)_2[Pd(dmit)_2]$ (purchased from TCI, 10 mg) in acetonitrile (20 mL) in a conventional H-shaped cell provided a thin black elongated plates of **C1**$[Pd(dmit)_2]_2$ (**3**; m.p. > 300 °C; Anal. Calcd. for $C_{26}H_{27}N_3O_2S_{20}Pd_2$: C, 24.63; H, 2.15; N, 3.31. Found: C, 24.50; H, 2.15; N, 3.09).

X-ray diffraction data of **1**, **2** and **3** were collected at 200, 290 and 250 K, respectively, with a Rigaku Rapid II imaging plate system (Rigaku, Tokyo, Japan) with MicroMax-007 HF/VariMax rotating-anode X-ray generator with confocal monochromated MoKα radiation. The crystallographic data of **1**, **2** and **3** are listed in Table 1. Electrical resistance measurements were performed by the conventional four-probe method. Temperature dependence of magnetic susceptibility of a polycrystalline sample from 2–300 K was measured using a Quantum Design MPMS-2S SQUID magnetometer (Quantum Design, San Diego, CA, USA). The magnetic susceptibility data of **2** and **3** were corrected for a contribution from an aluminum foil sample folder and the diamagnetic contributions of the samples were estimated from Pascal's constant.

Table 1. Crystallographic data of **1**, **2** and **3**.

Compound	1	2	3
Composition	**C1**Cl	(**C1**)[Ni(dmit)$_2$]$_2$·CH$_3$CN	(**C1**)[Pd(dmit)$_2$]$_2$
Formula	C$_{14}$H$_{27}$N$_3$O$_2$Cl$_1$	C$_{28}$H$_{30}$O$_2$N$_4$S$_{20}$Ni$_2$	C$_{26}$H$_{27}$O$_2$N$_3$S$_{20}$Pd$_2$
Fw	304.84	1213.17	1267.52
Space Group	$P2_1/n$	$P2_1/c$	$P2_1/n$
a (Å)	6.2618(3)	5.9977(3)	6.4052(3)
b (Å)	11.5017(7)	47.6504(19)	50.0326(18)
c (Å)	23.6111(14)	16.4464(7)	13.5041(5)
β (°)	92.693(7)	97.507(7)	91.407(6)
V (Å^3)	1698.62(16)	4660.0(4)	4326.3(3)
Z	4	4	4
T (K)	200	290	250
d_{calc} (g·cm^{-1})	1.192	1.729	1.946
μ (cm^{-1})	2.304	17.392	18.309
F (000)	660	2472	2528
2θ range (°)	6–55	6–55	6–55
Total ref.	15621	43525	39491
Unique ref.	3844	10638	9854
R_{int}	0.0337	0.0605	0.0607
Parameters	188	505	478
R_1 ($I > 2\sigma(I)$)	0.066	0.071	0.079
wR_2 (all data)	0.178	0.227	0.217
S	0.981	0.915	1.115
$\Delta\rho_{max}$ (e Å^{-3})	0.61	0.81	1.71
$\Delta\rho_{min}$ (e Å^{-3})	−0.32	−0.71	−1.30
CCDC reference	1532085	1532069	1532070

Each dipole moment of **C1** was calculated using MOPAC7 [23]. The molecular geometry of **C1** was observed in the structure of the relevant charge transfer salt (**2** or **3**) and was used without further structural optimization. In addition, we considered the effect of the incorporated acetonitrile molecule for **2**, which itself has a 3.4 D dipole moment. We performed the dipole moment calculation of one **C1** molecule with the one incorporated CH$_3$CN molecule by MOPAC7, giving a value of 23.1 D. The calculation of two **C1** cations with the two incorporated CH$_3$CN molecules by MOPAC2016 [25] was also performed, yielding a value of 19.4 D per one **C1** with one CH$_3$CN. The value is almost the same as for the result without CH$_3$CN (19.2 D). All MOPAC calculations were performed using MOPAC keywords of PM3, 1SCF and PRECISE.

4. Conclusions

We have prepared new acceptor-cation type organic magnetic conductors containing an aminoxyl radical: (PO–CONH–C$_2$H$_4$N(CH$_3$)$_3$)[Ni(dmit)$_2$]$_2$·CH$_3$CN (**2**) and (PO–CONH–C$_2$H$_4$N(CH$_3$)$_3$)[Pd(dmit)$_2$]$_2$ (**3**). Neither salt shows significant magnetic interactions between itinerant and localized electrons. An interesting feature is that they have similar structural features, in that they do not have net dipole moments but possess a periodicity of dipoles that we have termed Type III structures. The magnetic susceptibility measurements and band structure calculations indicate that **2** is a Mott insulator and **3** is a band insulator.

Acknowledgments: The authors thank Shusaku Imajo at Osaka Unversity for help with electrical resistivity measurements. This work was supported by the Murata Science Foundation, Nagaokakyo, Japan.

Author Contributions: H.A. conceived and designed the experiments. H.A. performed the experimental work. All authors contributed to the preparation of the manuscript.

Conflicts of Interest: The authors declare no conflict of interest.

References

1. Kobayashi, H.; Cui, H.; Kobayashi, A. Organic metals and superconductors based on BETS (BETS = bis(ethylenedithio)tetraselenafulvalene. *Chem. Rev.* **2004**, *104*, 5265–5288. [CrossRef] [PubMed]
2. Day, P.; Coronado, E. Magnetic molecular conductors. *Chem. Rev.* **2004**, *104*, 5419–5448.
3. Vyaselev, O.M.; Kartsovnik, M.V.; Biberacher, W.; Zorina, L.V.; Kushch, N.D.; Yagubskii, E.B. Magnetic transformations in the organic conductor κ-(BETS)$_2$Mn[N(CN)$_2$]$_3$[N(CN)$_2$]$_2$ at the metal-insulator transition. *Phys. Rev. B* **2011**, *83*, 094425:1–094425:6. [CrossRef]
4. Nakatsuji, S. Preparations, Reactions, and Properties of Functional Nitroxide Radicals. In *Nitroxide: Application in Chemistry, Biomedicine, and Materials Chemistry*; Likhtenshtein, G.I., Yamauchi, J., Nakatsuji, S., Smirnov, A.I., Tamura, R., Eds.; Wiley-VCH: Weinheim, Germany, 2008; pp. 161–204.
5. Matsushita, M.M.; Kawakami, H.; Kawada, Y.; Sugawara, T. Negative Magneto-resistance Observed on an Ion-radical Salt of a TTF-based Spin-polarized Donor. *Chem. Lett.* **2007**, *36*, 110–111. [CrossRef]
6. Sugawara, T.; Komatsu, H.; Suzuki, K. Interplay between magnetism and conductivity derived from spin-polarized donor radicals. *Chem. Soc. Rev.* **2011**, *40*, 3105–3118. [CrossRef] [PubMed]
7. Akutsu, H.; Yamada, J.; Nakatsuji, S. Preparation and characterization of novel organic radical anions for organic conductors: TEMPO–NHSO$_3^-$ and TEMPO-OSO$_3^-$. *Synth. Met.* **2001**, *120*, 871–872. [CrossRef]
8. Akutsu, H.; Yamada, J.; Nakatsuji, S. A New Organic Anion Consisting of the TEMPO Radical for Organic Charge-Transfer Salts: 2,2,6,6-Tetramethylpiperidinyloxy-4-sulfamate (TEMPO–NHSO$_3^-$). *Chem. Lett.* **2001**, *30*, 208–209. [CrossRef]
9. Akutsu, H.; Yamada, J.; Nakatsuji, S. New BEDT-TTF-based Organic Conductor Including an Organic Anion Derived from the TEMPO Radical, α-(BEDT-TTF)$_3$(TEMPO–NHCOCH$_2$SO$_3$)$_2$·6H$_2$O. *Chem. Lett.* **2003**, *32*, 1118–1119. [CrossRef]
10. Akutsu, H.; Masaki, K.; Mori, K.; Yamada, J.; Nakatsuji, S. New organic free radical anions TEMPO-*A*-CO-(*o*-, *m*-, *p*-)C$_6$H$_4$SO$_3^-$ (*A* = NH, NCH$_3$, O) and their TTF and/or BEDT-TTF salts. *Polyhedron* **2005**, *24*, 2126–2132. [CrossRef]
11. Yamashita, A.; Akutsu, H.; Yamada, J.; Nakatsuji, S. New organic magnetic anions TEMPO–CONA (CH$_2$)$_n$SO$_3^-$ (*n* = 0–3 for *A* = H, *n* = 2 for *A* = CH$_3$) and their TTF, TMTSF and/or BEDT-TTF salts. *Polyhedron* **2005**, *16*, 2796–2802. [CrossRef]
12. Akutsu, H.; Yamada, J.; Nakatsuji, S.; Turner, S.S. A novel BEDT-TTF-based purely organic magnetic conductor, α-(BEDT-TTF)$_2$(TEMPO–N(CH$_3$)COCH$_2$SO$_3$)·3H$_2$O. *Solid State Commun.* **2006**, *140*, 256–260. [CrossRef]
13. Akutsu, H.; Sato, K.; Yamashita, S.; Yamada, J.; Nakatsuji, S.; Turner, S.S. The first organic paramagnetic metal containing the aminoxyl radical. *J. Mater. Chem.* **2008**, *18*, 3313–3315. [CrossRef]
14. Akutsu, H.; Yamashita, S.; Yamada, J.; Nakatsuji, S.; Turner, S.S. Novel Purely Organic Conductor with an Aminoxyl Radical, α-(BEDT-TTF)$_2$(PO–CONHCH$_2$SO$_3$)·2H$_2$O (PO = 2,2,5,5-Tetramethyl-3-pyrrolin-1-oxyl Free Radical). *Chem. Lett.* **2008**, *37*, 882–883. [CrossRef]
15. Akutsu, H.; Yamashita, S.; Yamada, J.; Nakatsuji, S.; Hosokoshi, Y.; Turner, S.S. A Purely Organic Paramagnetic Metal, κ-β''-(BEDT-TTF)$_2$(PO–CONHC$_2$H$_4$SO$_3$), Where PO = 2,2,5,5-Tetramethyl-3-pyrrolin-1-oxyl Free Radical. *Chem. Mater.* **2011**, *23*, 762–764. [CrossRef]
16. Akutsu, H.; Kawamura, A.; Yamada, J.; Nakatsuji, S.; Turner, S.S. Anion polarity-induced dual oxidation states in a dual-layered purely organic paramagnetic charge-transfer salt, (TTF)$_3$(PO–CON(CH$_3$)C$_2$H$_4$SO$_3$)$_2$, where PO = 2,2,5,5-tetramethyl-3-pyrrolin-1-oxyl free radical. *CrystEngComm* **2011**, *13*, 5281–5284. [CrossRef]
17. Akutsu, H.; Ishihara, K.; Yamada, J.; Nakatsuji, S.; Turner, S.S.; Nakazawa, Y. A strongly polarized organic conductor. *CrystEngComm* **2016**, *18*, 8151–8154. [CrossRef]
18. Akutsu, H.; Ishihara, K.; Ito, S.; Nishiyama, F.; Yamada, J.; Nakatsuji, S.; Turner, S.S.; Nakazawa, Y. Anion Polarity-Induced Self-doping in Purely Organic Paramagnetic Conductor, α'-α'-(BEDT-TTF)$_2$ (PO-CONH-*m*-C$_6$H$_4$SO$_3$)·H$_2$O where BEDT-TTF is Bis(ethylenedithio)tetrathiafulvalene and PO is 2,2,5,5-Tetramethyl-3-pyrrolin-1-oxyl Free Radical. *Polyhedron* **2017**, in press. [CrossRef]
19. Kato, R. Conducting Metal Dithiolene Complexes: Structural and Electronic Properties. *Chem. Rev.* **2004**, *104*, 5319–5346. [CrossRef] [PubMed]
20. Kobayashi, H.; Kobayashi, A.; Tajima, H. Studies on Molecular Conductors: From Organic Semiconductors to Molecular Metals and Superconductors. *Chem. Asian J.* **2011**, *6*, 1688–1704. [CrossRef] [PubMed]

21. Cassoux, P.; Valade, L.; Kobayashi, H.; Kobayashi, A.; Clark, R.A.; Underhill, A.E. Molecular metals and superconductors derived from metal complexes of 1,3-dithiol-2-thione-4,5-dithiolate (dmit). *Coord. Chem. Rev.* **1991**, *110*, 115–160. [CrossRef]

22. Mori, T.; Kobayashi, A.; Sasaki, Y.; Kobayashi, H.; Saito, G.; Inokuchi, H. The Intermolecular Interaction of Tetrathiafulvalene and Bis(ethylenedithio)tetrathiafulvalene in Organic Metals. Calculation of Orbital Overlaps and Models of Energy-band Structures. *Bull. Chem. Soc. Jpn.* **1984**, *57*, 627–633. [CrossRef]

23. Stewart, J.J.P. *MOPAC7*, Stewart Computational Chemistry: Colorado Springs, CO, USA, 1993.

24. Rozantsev, E.G. *Free Nitroxide Radicals*; Plenum Press: New York, NY, USA; London, UK, 1970.

25. Stewart, J.J.P. *MOPAC2016*, Stewart Computational Chemistry: Colorado Springs, CO, USA, 2016.

magnetochemistry

MDPI

Article

Mn-Containing Paramagnetic Conductors with Bis(ethylenedithio)tetrathiafulvalene (BEDT-TTF)

Samia Benmansour *, Yolanda Sánchez-Máñez and Carlos J. Gómez-García *

Instituto de Ciencia Molecular (ICMol), Universidad de Valencia, C/Catedrático José Beltrán 2, 46010 Valencia, Spain; yolsanma@gmail.com
* Correspondence: sam.ben@uv.es (S.B.); carlos.gomez@uv.es (C.J.G.-G.);
   Tel.: +34-9-6354-4423 (S.B. & C.J.G.-G.); Fax: +34-9-6354-3273 (S.B. & C.J.G.-G.)

Academic Editor: Manuel Almeida
Received: 24 January 2017; Accepted: 6 February 2017; Published: 9 February 2017

Abstract: Two novel paramagnetic conductors have been prepared with the organic donor bis(ethylenedithio)tetrathiafulvalene (BEDT-TTF = ET) and paramagnetic Mn-containing metallic complexes: κ'-ET$_4$[KMn^{III}(C$_2$O$_4$)$_3$]·PhCN (**1**) and ET[Mn^{II}Cl$_4$]·H$_2$O (**2**). Compound **1** represents the first Mn-containing ET salt of the large Day's series of oxalato-based molecular conductors and superconductors formulated as (ET)$_4$[AM(C$_2$O$_4$)$_3$]·G (A^+ = H$_3$O$^+$, NH$_4{}^+$, K$^+$, ...; M^{III} = Fe, Cr, Al, Co, ...; G = PhCN, PhNO$_2$, PhF, PhCl, PhBr, ...). It crystallizes in the orthorhombic pseudo-κ phase where dimers of ET molecules are surrounded by six isolated ET molecules in the cationic layers. The anionic layers contain the well-known hexagonal honey-comb lattice with Mn(III) and H$_3$O$^+$ ions connected by C$_2$O$_4{}^{2-}$ anions. Compound **2** is one of the very few examples of ET salts containing ET^{2+}. It also presents alternating cationic-anionic layers although the ET molecules lie parallel to the layers instead of the typical almost perpendicular orientation. Both salts are semiconductors with room temperature conductivities of ca. 2×10^{-5} and 8×10^{-5} S/cm and activation energies of 180 and 210 meV, respectively. The magnetic properties are dominated by the paramagnetic contributions of the high spin Mn(III) (S = 2) and Mn(II) (S = 5/2) ions.

Keywords: electro-crystallization; ET-salts; paramagnetic conductors; electrical conductivity; magnetic properties; Mn(II) complexes; Mn(III) complexes

---

1. Introduction

The design and synthesis of multifunctional molecular materials combining electrical and magnetic properties is one of the main challenges in the field of molecular materials [1–4]. An advantage of these materials is that they offer the possibility to study the competition and interplay of these two properties. So far, a large number of molecular materials combining magnetism with conductivity has been obtained. These examples include superconductors with paramagnetic complexes [1,5–9] or with antiferromagnetic lattices [10–15] and ferromagnetic conductors [4,16].

Among the different paramagnetic complexes used to prepare these materials, tris(oxalato)metalate complexes, [M(C$_2$O$_4$)$_3$]$^{n-}$, are, by far, the most used ones. These anions may crystallize as:

(i) Monomers as in (TTF)$_7$[Fe(C$_2$O$_4$)$_3$]$_2$·4H$_2$O [17], (TTF)$_3$[Ru(C$_2$O$_4$)$_3$]·(EtOH)$_{0.5}$·4H$_2$O [18], (BEST)$_4$[M(C$_2$O$_4$)$_3$]·PhCOOH·H$_2$O [19], (BEST)$_4$[M(C$_2$O$_4$)$_3$]·1.5H$_2$O [19], (M = Cr and Fe), (BEST)$_9$[Fe(C$_2$O$_4$)$_3$]$_2$·7H$_2$O [19], (ET)$_2$[Ge(C$_2$O$_4$)$_3$]·PhCN [20], (ET)$_9$Na$_{18}$[M(C$_2$O$_4$)$_3$]$_8$·24H$_2$O (M^{III} = Fe and Cr) [21,22], (ET)$_{12}$[Fe(C$_2$O$_4$)$_3$]$_2$·nH$_2$O [23], (ET)$_5$[Fe(C$_2$O$_4$)$_3$]·CH$_2$Cl$_2$·2H$_2$O [24], (ET)$_5$[Ge(C$_2$O$_4$)$_3$]$_2$ [25] and (ET)$_7$[Ge(C$_2$O$_4$)$_3$](CH$_2$Cl$_2$)$_{0.87}$(H$_2$O)$_{0.09}$ [25]. (TTF = tetrathiafulvalene; BEST = bis(ethylenediseleno)-tetrathiafulvalene; ET = bis(ethylenedithio)tetrathiafulvalene).

(ii) Previously unknown $[M_2(C_2O_4)_5]^{4-}$ dimers (M^{III} = Fe and Cr) with TTF, TMTTF (tetramethyl-tetrathiafulvalene) and ET [17,26].

(iii) Previously unknown $[\{M^{III}(C_2O_4)_3\}_2M^{II}(H_2O)_2]^{4-}$ trimers (M^{III} = Cr and Fe; M^{II} = Mn, Fe, Co, Ni, Cu and Zn) only obtained with TTF [27,28].

(iv) Forming honeycomb-like 2D anionic layers as in the first molecular ferromagnetic metals: $(ET)_3[Mn^{II}Cr^{III}(C_2O_4)_3]$ and $(BETS)_x[Mn^{II}Cr^{III}(C_2O_4)_3]\cdot CH_2Cl_2$ (BETS = bis(ethylenedithio) tetraselenafulvalene; $x \approx 3$) [4,16,29–31], and also in the Day's series of paramagnetic superconductors, metals and semiconductors formulated as $(ET)_4[A^IM^{III}(C_2O_4)_3]\cdot G$ (A^I = H_3O^+, K^+ and NH_4^+; M^{III} = Cr, Fe, Ga, Co, Mn and Al; G = PhCN, $PhNO_2$, py, $PhCl_2$, PhF, PhCl, PhBr, $PhCOCH_3$, $PhCH_2OHCH_3$, Me_2NCHO, CH_2Cl_2, $PhN(CH_3)CHO$, $PhCH_2CN$, …) [1]. This series constitutes, by far, the largest family of paramagnetic superconductors, metals and semiconductors prepared to date. In this series, we can distinguish three different crystal structures: (i) a $C2/c$ (#15) monoclinic β'' phase (Table 1); (ii) an orthorhombic *Pbcn* (#60) pseudo-κ phase (Table 2) and (iii) a triclinic *P*1 (#1) or *P*-1 (#2) $\alpha\beta''$ or α-pseudo-κ phase (Table 3). Besides these three 4:1 series, there is a fourth series with 3:1 cation:anion stoichiometry with either triclinic *P*1 (#1), monoclinic $P2_1$ (#4) and $P2_1/c$ (#14) or orthorhombic $P2_12_12_1$ (#19) crystal structures (Table 4). The main difference between these four series lies in the disposition of the organic molecules in the cationic layers. The monoclinic $C2/c$ (#15) β'' phase presents parallel ET molecules, the orthorhombic *Pbcn* (#60) pseudo-κ phase contains ET dimers surrounded by six monomers, the triclinic phase presents a mixture of alternating θ and β'' (or θ and pseudo-κ) layers, and, finally, the 3:1 salts present alternating tilted dimers and monomers. These structural differences lead to different physical properties: the triclinic and orthorhombic phases are semiconductors (Tables 2–4), whereas the monoclinic salts are metallic or even superconductors (Table 1).

Table 1. Structural and electrical properties of the monoclinic $(ET)_4[A^IM^{III}(C_2O_4)_3]\cdot G$ salts.

CCDC Code	M^{III}	A^I	G	Packing	SG	Elect. Prop.	Ref.
ZIGYET	Fe	H_3O^+	PhCN	β''	$C2/c$	T_c = 7.0–8.5 K	[6,7,32–34]
KILFOB/GOC/GUI/HAP	Fe	H_3O^+	$C_5H_5N_{(1-x)}/PhCN_x$	β''	$C2/c$	T_c = F(x)	[33]
BEMPEO/QAL	Fe	H_3O^+	C_5H_5N	β''	$C2/c$	T_{MI} = 116 K	[33,35]
ECOPIV	Fe	H_3O^+/NH_4^+	$PhNO_2$	β''	$C2/c$	T_c = 6.2 K	[36]
COQNEB	Fe	H_3O^+	$PhNO_2$	β''	$C2/c$	Semicond	[37,38]
PONMEL	Fe	H_3O^+	$PhCl_2$	β''	$C2/c$	T_{MI} = 3.0 K, Metal > 1.5 K	[34,39]
SAPWEM	Fe	H_3O^+	PhBr	β''	$C2/c$	T_c = 4.0 K	[40]
UMACEQ	Fe	NH_4^+	DMF	β''	$C2/c$	Metal > 4 K	[41]
UJOXEX	Fe	H_3O^+	PhF	β''	$C2/c$	T_c = 1.0 K	[34,42,43]
UJOXAT	Fe	H_3O^+	PhCl	β''	$C2/c$	Metal > 0.4 K	[34,42–45]
UJOXIB	Fe	H_3O^+	PhF/PhCN	β''	$C2/c$	T_c = 6.0 K	[34,42]
UJOXOH	Fe	H_3O^+	$PhCl_2$/PhCN	β''	$C2/c$	T_c = 7.2 K	[34]
UJOYAU	Fe	H_3O^+	PhCl/PhCN	β''	$C2/c$	T_c = 6.0 K	[34,42]
UJOYEY	Fe	H_3O^+	PhBr/PhCN	β''	$C2/c$	T_c = 4.2 K	[34,42]
QAXSIT	Fe	K^+	PhI	β''	$C2/c$	E_a = 64 meV	[43]
-	Fe	K^+	PhCl	β''	-	Semicond.	[46]
-	Fe	Rb^+	C_5H_5N	β''	-	Metal > 4.2 K	[44]
JUPGUW01	Cr	H_3O^+	PhCN	β''	$C2/c$	T_c = 5.5–6.0 K	[32,47]
MEQZIR	Cr	H_3O^+	CH_2Cl_2	β''	$C2/c$	T_{MI} = 150 K	[48]
ECOPUH	Cr	H_3O^+/NH_4^+	$PhNO_2$	β''	$C2/c$	T_c = 5.8 K	[36]
-	Cr	H_3O^+	PhBr	β''	$C2/c$	T_c = 1.5 K	[45]
-	Cr	H_3O^+	PhCl	β''	$C2/c$	T_{MI} = 130 K	[45]
UMACAM	Cr	K^+/NH_4^+	DMF	β''	$C2/c$	Metal > 4 K	[41]
UMACIU	Cr	K^+	DMF	β''	$C2/c$	Metal > 4 K	[41]
HUNQIQ	Ga	H_3O^+	C_5H_5N	β''	$C2/c$	$T_c \approx 2$ K	[49]
HUNQUC	Ga	H_3O^+	$PhNO_2$	β''	$C2/c$	T_c = 7.5 K	[49]
HOBROH	Ga	H_3O^+/K^+	PhBr	β''	$C2/c$	metal > 0.5 K	[50]
UDETUU	Ru	H_3O^+/K^+	PhCN	β''	$C2/c$	T_c = 6.3 K	[51]
YUYTUJ	Fe	H_3O^+	2-Cl-Py	β''	$C2/c$	T_c = 4.0 K	[52]
YUYVEV	Fe	H_3O^+	2-Br-py	β''	$C2/c$	T_c = 4.3 K	[52]
YUYVOF	Fe	H_3O^+	3-Cl-py	β''	$C2/c$	metal > 0.5 K	[52]
YUYVUL	Fe	H_3O^+	3-Br-py	β''	$C2/c$	metal > 0.5 K	[52]
DUDWOQ	Fe	$Li^+ + H_2O$	EtOH	η (α'')	$P2_1/n$	E_a = 80 meV	[53]
-	Mn	H_3O^+	PhBr	β''	$C2/c$	T_c = 2.0 K	[43]

T_c = superconducting temperature; T_{MI} = metal insulator temperature; E_a = activation energy.

Table 2. Structural and electrical properties of the orthorhombic $(ET)_4[A^I M^{III}(C_2O_4)_3] \cdot G$ salts.

CCDC Code	M^{III}	A^I	G	ET Packing	Space Group	Electrical Properties	Ref.
UJOXUN	Fe	H_3O^+	PhF/PhCN	pseudo-κ	*Pbcn*	Semiconductor	[34]
ZIWNEY	Fe	NH_4^+	PhCN	pseudo-κ	*Pbcn*	$E_a = 140$ meV	[7,32]
ZIWNIC	Fe	K^+	PhCN	pseudo-κ	*Pbcn*	$E_a = 141$ meV	[7]
JUPGUW	Cr	H_3O^+	PhCN	pseudo-κ	*Pbcn*	$E_a = 153$ meV	[32,47]
QIWMOY	Co	NH_4^+	PhCN	pseudo-κ	*Pbcn*	$E_a = 225$ meV	[32]
QIWMUE	Al	NH_4^+	PhCN	pseudo-κ	*Pbcn*	$E_a = 222$ meV	[32]
UDETOO	Ru	H_3O^+/K^+	PhCN	pseudo-κ	*Pbcn*	-	[51]
1	Mn	K^+	PhCN	pseudo-κ	*Pbcn*	$E_a = 180$ meV	this work

E_a = activation energy.

Table 3. Structural and electrical properties of the triclinic $(ET)_4[A^I M^{III}(C_2O_4)_3] \cdot G$ salts.

CCDC Code	M^{III}	A^I	G	ET Packing	Space Group	Electrical Properties	Ref.
TANDIX	Fe	H_3O^+	PhBr$_2$	α + κ	*P*-1	Metal > 0.4 K	[34,54]
HOBRIB	Ga	H_3O^+/K^+	PhBr$_2$	α + κ	*P*-1	metal > 0.5 K	[50]
ARABEA	Fe	NH_4^+	PhCOCH$_3$	α + β″	*P*-1	No supercond	[55]
CILDIL	Fe	NH_4^+	R/S-Ph-CH$_2$OHCH$_3$	α + β″	*P*-1	$T_{MI} = 170$ K	[56]
NIPTEM	Fe	NH_4^+	S-PhCH$_2$OHCH$_3$	α + β″	*P*-1	$T_{MI} = 150$ K	[56]
AQUZUH	Ga	NH_4^+	PhN(Me)CHO	α + β″	*P*-1	Semicond	[55]
ARABAW	Ga	NH_4^+	PhCH$_2$CN	α + β″	*P*-1	Semicond	[55]

T_{MI} = metal insulator temperature.

Table 4. Structural and electrical properties of $(ET)_3[A^I M^{III}(C_2O_4)_3] \cdot G$ salts.

CCDC Code	M^{III}	A^I	G	ET Packing	Space Group	Electrical Properties	Ref.
BOYTIU	Al	Na^+	CH$_3$NO$_2$	dimers + mon.	$P2_1$	$E_a \approx 140$ meV	[57]
XUNXOU01	Cr	Na^+	CH$_3$NO$_2$	dimers + mon.	$P2_1$	$E_a = 79$ meV	[58]
XUNXOU	Cr	Na^+	CH$_3$NO$_2$	dimers + mon.	$P2_12_12_1$	$E_a = 80$ meV	[58]
-	Cr	NH_4^+	CH$_3$NO$_2$	dimers + mon.	$P2_12_12_1$	$E_a = 80$ meV	[58]
DUXNOA	Cr	Na^+	CH$_2$Cl$_2$	dimers + mon.	$P1$	$E_a = 69$ meV	[22]
DUDWUW	Cr	Li^+	EtOH	dimers + mon.	$P2_1/c$	$E_a = 179$ meV	[53]
-	Fe	Li^+	EtOH	dimers + mon.	$P2_1/c$	$E_a = 126$ meV	[53]
KOGMUG01	Cr	Na^+	CH$_3$CN	dimers + mon.	$P2_1$	$E_a = 79$ meV	[59]
-	Cr	Na^+	DMF	θ-packing	$P1$	$E_a = 43$ meV	[59]
YUCLOZ	Cr	Na^+	EtOH	dimers + mon.	$P1$	no data	[59]

One of the advantages of these series of compounds is the possibility to tune the electrical properties by simply changing the guest solvent molecule (G) located in the centre of the hexagonal cavities formed by the anionic lattice. This guest molecule may interact with the ET molecules, promoting the ordering of the ethylene groups of the ET molecules and, thus, stabilizing the superconductor state [60] as in the case of G = PhCN and PhNO$_2$ [6,7,36,49,61] whose radical salts are superconductors and present the highest T_c's in these series: (T_c = 6.0, 8.5, 5.8, 6.2 and 7.5 K for G/M = PhCN/Cr and PhCN/Fe, PhNO$_2$/Cr, PhNO$_2$/Fe and PhNO$_2$/Ga, respectively, Table 1). For G = pyridine [35,49], dichloromethane [48] or dimethylformamide [41], the disorder remains down to very low temperatures and the salts are not superconductors or present very low T_c. Even more, the mixture of PhCN with other solvents as C$_5$H$_5$N, PhCl$_2$, PhNO$_2$, PhF, PhCl or PhBr changes the ordering effect of the solvent and allows a fine tuning of T_c [33,34].

Although many different guest molecules have been used (see Tables 1–4), the number of trivalent metals used to date is quite limited. Thus, most of the reported salts contain Fe (30 salts), Cr (16 salts) or Ga (6 salts). There are also two reported examples with Ru, two with Al and one with Co. Surprisingly, no radical salts with other trivalent metal ions have been reported to date. In order to investigate the effect of other trivalent metal ions on the final structure and on the physical properties, we have used the $[Mn(C_2O_4)_3]^{3-}$ anion with ET under different synthetic conditions. Here, we present the synthesis, structure, magnetic and electrical properties of the first example of radical salt of the Day's series obtained with Mn(III): κ'-(ET)$_4$[KMn^{III}(C$_2$O$_4$)$_3$]·PhCN (**1**) and of a very original salt obtained with the same $[Mn(C_2O_4)_3]^{3-}$ anion but using different synthetic conditions: (ET)[MnCl$_4$]·H$_2$O (**2**).

2. Results and Discussion

2.1. Syntheses of the Complexes

The synthesis of the two radical salts was performed using the same precursor K$_3$[Mn(C$_2$O$_4$)$_3$] salt (and 18-crown-6 in order to solubilize this salt, see the Experimental section). The main difference is the use of different solvents: a 10:1 (v/v) mixture of PhCN and MeOH for compound **1** and a 10:1 (v/v) mixture of 1,1,2-trichloroethane (TCE) and MeOH for **2**. An additional difference is the use of benzoic acid in the synthesis of compound **1**. Interestingly, benzoic acid does not enter in the structure, but it seems to facilitate the crystallization of the final salt. In fact, attempts to obtain compound **1** without the use of benzoic acid failed. In summary, the use of PhCN gives rise to compound (ET)$_4$[KMn^{III}(C$_2$O$_4$)$_3$]·PhCN (**1**), whereas TCE results in a totally different compound (ET)[MnCl$_4$]·H$_2$O (**2**) with ET^{2+} instead of ET$^{+0.5}$ and with [Mn^{II}Cl$_4$]$^{2-}$ instead of [Mn^{III}(C$_2$O$_4$)$_3$]$^{3-}$. The question is straightforward: why is the solvent so important in the final product? The answer seems to be related with the much lower solubility of the precursor K$_3$[Mn(C$_2$O$_4$)$_3$] in TCE. This lower solubility increases the resistance of the electrochemical cell since the concentration of anions is lower. The higher resistance increases the potential of the source needed to apply the desired constant intensity since the electrochemical synthesis is performed under constant current. The higher voltage results in the cathode in the oxidation of ET to ET^{2+} and in the anode in the reduction of Mn(III) to Mn(II). Additionally, the intensity and time used for compound **2** were higher than for **1** (see experimental section). Finally, the partial decomposition of TCE liberates chloride anions that coordinate to Mn(II) to form the observed [MnCl$_4$]$^{2-}$ anion. Note that the release of chloride anions from the decomposition of chlorinated solvents is quite common in the synthetic conditions of the electrochemical cells and has been observed in other ET salts [62–64].

2.2. Description of the Structures

Structure of (ET)$_4$[KMn^{III}(C$_2$O$_4$)$_3$]·PhCN (**1**). Compound **1** crystallizes in the orthorhombic space group *Pbcn* (Table 5) and is isostructural to those obtained with other trivalent metal ions as Fe, Cr, Co and Al and different monovalent cations as H$_3$O$^+$, NH$_4$$^+$ and K$^+$ (Table 2). Interestingly, there is only one reported example with K$^+$ as monovalent cation and there is no example with Mn(III) as a trivalent cation.

The asymmetric unit contains two independent ET molecules (labelled as A and B) lying on general positions, half $[Mn(C_2O_4)_3]^{3-}$ anion, half benzonitrile molecule and half K$^+$ cation, all lying on a two-fold rotation axis. Figure 1 shows the ellipsoid diagram of the molecules in **1** together with the atom-labelling scheme.

The crystal structure consists of alternating layers of ET molecules adopting the pseudo-κ phase and anionic layers containing $[Mn(C_2O_4)_3]^{3-}$ anions, K$^+$ cations and the guest benzonitrile molecule (Figure 2).

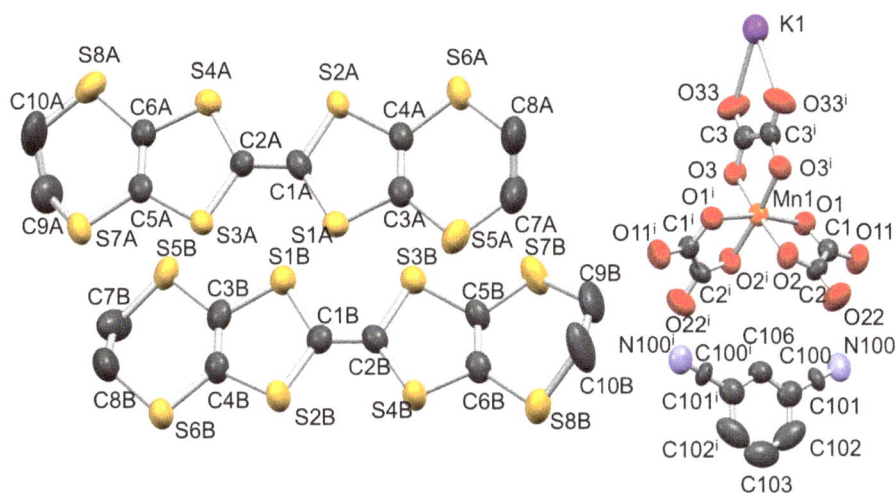

Figure 1. Thermal ellipsoid diagram (at 50% probability) of the molecules in compound **1**. Symmetry code: i = −*x*, *y*, 1/2−*z*.

Table 5. Crystal data and structure refinement of compounds **1** and **2**.

Compound	1	2
Formula	$C_{53}H_{36}KMnNO_{12}S_{32}$	$C_{10}H_{10}MnCl_4OS_8$
F. Wt.	1999.97	599.45
Space group	*Pbcn*	*Pnna*
Crystal system	Orthorhombic	Orthorhombic
a (Å)	10.3727 (4)	12.3724 (9)
b (Å)	19.6588 (8)	12.3738 (9)
c (Å)	36.2145 (13)	13.7726 (13)
$V/Å^3$	7384.7 (5)	2108.5 (3)
Z	4	4
T (K)	120	120
$\rho_{calc}/g \cdot cm^{-3}$	1.798	1.856
μ/mm^{-1}	1.199	1.923
F(000)	4052	1156
R(int)	0.1380	0.1089
θ range (deg)	2.910–25.053	2.958–25.044
Total reflections	59,518	14,081
Unique reflections	6534	1867
Data with $I > 2\sigma (I)$	6534	1867
N_{var}	462	114
R_1 [a] on $I > 2\sigma (I)$	0.0700	0.0509
wR_2 [b] (all)	0.1729	0.1006
GOF [c] on F^2	1.011	1.080
$\Delta\rho_{max}$ (eÅ^{-3})	0.561	1.130
$\Delta\rho_{min}$ (eÅ^{-3})	−0.839	−0.553

[a] $R_1 = \Sigma||F_o| - |F_c||/\Sigma|F_o|$. [b] $wR_2 = [\Sigma w(F_o^2 - F_c^2)^2/\Sigma w(F_o^2)^2]^{1/2}$. [c] GOF $= [\Sigma[w(F_o^2 - F_c^2)^2/(N_{obs} - N_{var})]^{1/2}$.

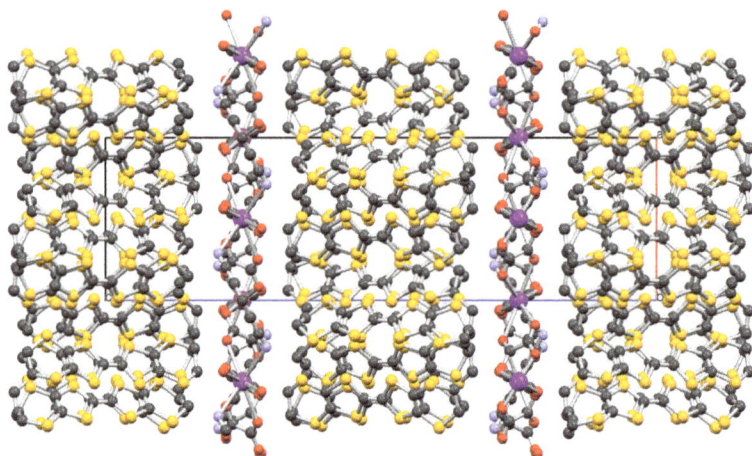

Figure 2. View of the cationic and anionic layers alternating along the *c*-direction in **1**.

The anionic layers form a honeycomb structure with hexagonal cavities that are occupied by the benzonitrile guest molecules (Figure 3a). The –CN group of the benzonitrile molecule presents a disorder over two positions related by the C_2 axis passing through the centre of the aromatic ring. The –CN group in both positions lies very close to a K^+ cation (K1-N100 = 3.055(15) Å) and, therefore, we can consider that the K^+ ions present a 6 + 2 coordination. This double orientation of the –CN groups and the close distance from the N atom to the monovalent cation is also observed in all the other reported orthorhombic structures with PhCN as solvent (Table 2) [7,32,47]. The Mn$\cdots$K distances (6.308, 6.239 and 6.308 Å) reflect a slight elongation of the hexagonal cavities parallel to the C_2 axis to accommodate the –CN group of the PhCN guest molecule. Similar elongations are also observed in all the reported orthorhombic $(ET)_4[AM(C_2O_4)_3]\cdot G$ phases except in the Al-NH$_4^+$ and Ru-H$_3$O$^+$/K$^+$ compounds.

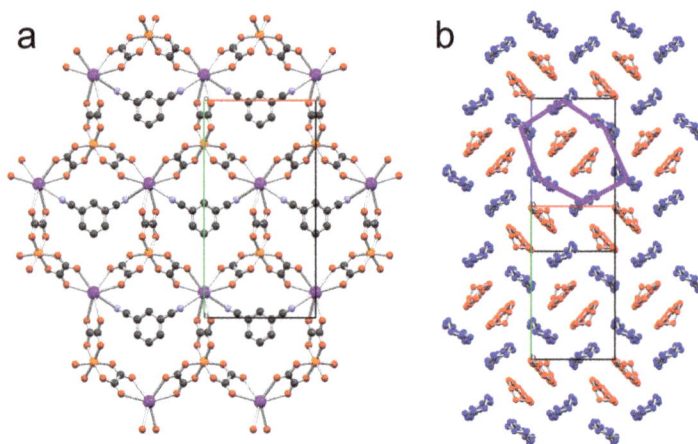

Figure 3. Structure of compound **1**: (**a**) view of the anionic layer showing the two possible positions of the –CN groups and the K–N bond. (**b**) view of the pseudo-κ packing of the bis(ethylenedithio)tetrathiafulvalene (ET) molecules in the cationic layer showing the A-type dimers (in red) surrounded by six B-type monomers (in blue). H atoms have been omitted for clarity.

The cationic layers are formed by ET dimers surrounded by six ET monomers in the so-called pseudo-κ phase (Figure 3b). The ET dimers are formed by A-type ET molecules, whereas the isolated ET molecules correspond to the B-type ones. As observed in other similar pseudo-κ phases, there are several short S⋯S contacts shorter that the sum of the Van der Waals radii (3.60 Å) (Table 6).

Table 6. Intermolecular S⋯S contacts shorter that the sum of the Van der Waals radii in **1**.

Atoms	Distance (Å)	Atoms	Distance (Å)
S6A^i⋯S6B^{ii}	3.563	S3A^{ii}⋯S7B^i	3.294
S8A^i⋯S8B^{ii}	3.571	S5A^{ii}⋯S5B^i	3.446
S2A^i⋯S8B^{iii}	3.497	S7A^{ii}⋯S7B^i	3.539
S8A^i⋯S2B^{iii}	3.564	-	-

Symmetry codes: i = 1.5 − x, 1.5 − y, −1/2 + z; ii = 1 − x, y, 1.5 − z; iii = x, 1 − y, −1/2 + z.

The estimation of the charge on the ET molecules in compound **1** using the formula proposed by Guionneau et al. [65] gives values of ca. +1 and ca. 0 for A- and B-type ET molecules (Table 7), respectively, as also found in all the reported orthorhombic (ET)$_4$[AM(C$_2$O$_4$)$_3$]·G phases [7,32,34,47,51].

Table 7. Bond distances (Å) and calculated charges of the ET molecules in **1** and **2**.

Compound	Molecule	a	b	c	d	δ	Q
1	A	1.392	1.723	1.7468	1.3415	0.7363	0.85
	B	1.353	1.7542	1.7565	1.337	0.8207	0.22
2	A	1.424	1.694	1.718	1.380	0.6080	1.81

δ = (b + c) − (a + d); Q = 6.347 − 7.463 × δ.

Structure of (ET)[Mn^{II}Cl$_4$]·H$_2$O (**2**). Compound **2** crystallizes in the orthorhombic space group *Pnna* (Table 5). The asymmetric unit contains a half ET molecule, half [MnCl$_4$]$^{2−}$ anion and half water molecule lying on special positions. Figure 4 shows the ellipsoid diagram of the molecules in **2** together with the atom-labelling scheme.

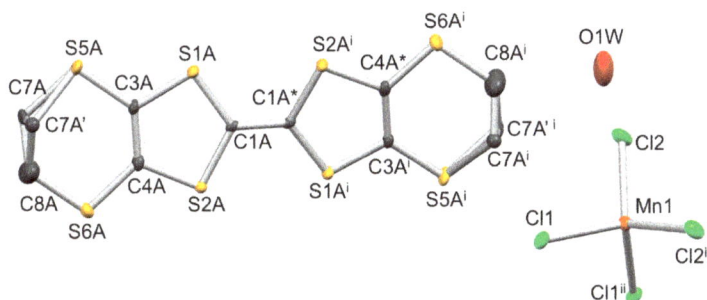

Figure 4. Thermal ellipsoid diagram (at 50% probability) of the molecules in compound **2**. Symmetry code: i = 1/2 − x, −y, z; ii = x, 1.5 − y, 1.5 − z.

The crystal structure of compound **2** consists of layers of ET molecules lying parallel to the plane alternating with layers of [MnCl$_4$]$^{2−}$ anions (Figure 5a,d). The anions adopt a square lattice with

Mn···Mn distances of 8.321 Å (Figure 5b) and with a shortest Cl···Cl intermolecular contact of 4.877 Å, well above the sum of the Van der Waals radii (3.50 Å). The cationic layer contains ET molecules lying parallel to the layer forming double layers. The ET molecules are parallel to each other inside the double layers but are orthogonal to the ET molecules of consecutive double layers (Figure 5c). This very unusual packing of the ET molecules parallel to the layer may be due to the +2 charge of the ET molecules (Table 7), precluding the usual packing in columns or dimers due to the coulombic repulsions. The short anion–cation contacts in **2** (Table 8) are also a consequence of this double charge on the ET molecules. Additionally, there is a Cl···O short contact (3.336 Å) that suggests the presence of hydrogen bonds of the type O–H···Cl connecting neighbouring $[MnCl_4]^{2-}$ anions. Unfortunately, the H atoms of the water molecules could not be located in the single crystal structural analysis.

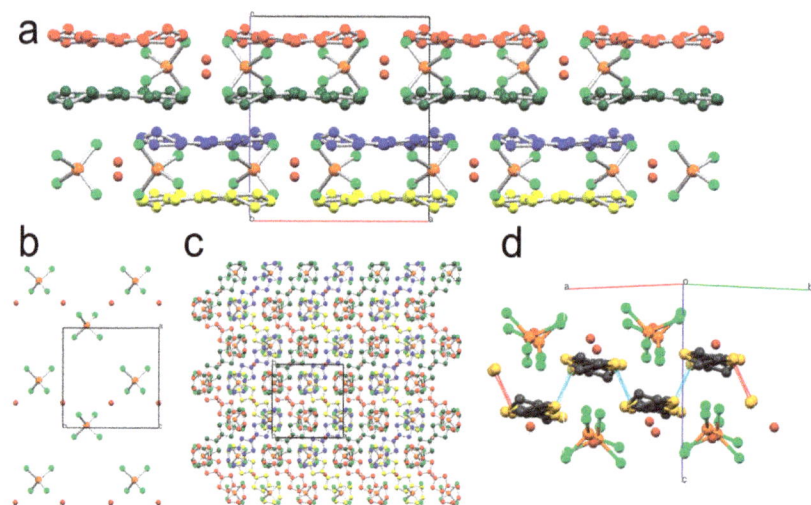

Figure 5. Structure of compound **2**: (**a**) view of the cationic and anionic layers alternating along the *c* direction in **2**. Green and blue (or red and yellow) ET molecules form one double layer; (**b**) view of the anionic layer; (**c**) view of two consecutive double ET and anionic layers down the *c* direction; (**d**) view of the zigzag chains in the *ab* plane showing the short S···S intermolecular contacts.

Table 8. Cation–anion contacts shorter than the sum of the Van der Waals radii in **2**.

Atoms	Distance (Å)	Atoms	Distance (Å)	Atoms	Distance (Å)
Cl1-S5A	3.250	Cl1^{ii}-S2A^{iii}	3.417	S2A^v-S2A^{vi}	3.412
Cl1-C7A′	3.437	Cl2^{ii}-S2A^{iii}	3.446	S2A^v-S6A^{vi}	3.597
Cl1-S1A^i	3.353	Cl2^{ii}-S6A^{iv}	3.295	C7A$'^v$-O1W^{vii}	3.147

Symmetry codes: i = $1/2 - x$, $1 - y$, z; ii = x, $1.5 - y$, z; iii = x, $1.5 - y$, $1.5 - z$; iv = $1.5 - x$, $1 - y$, z; v = $1/2 - x$, $-y$, z; vi = $-1/2 + x$, y, $1 - z$; vii = $-1 + x$, $-1 + y$, z.

The presence of ET^{2+} di-cations is very unusual. In fact, only six ET salts with ET^{2+} di-cations have been reported to date [66–69]. Its presence in **2** implies that the anion must be $[MnCl_4]^{2-}$, i.e., that the precursor Mn(III) salt has been reduced to Mn(II). The oxidation state of the Mn ion in this anion is confirmed by the magnetic measurements (see below) and by the Mn–Cl bond distances in the anion (Mn1–Cl1 = 2.3724 (15) Å and Mn1–Cl2 = 2.3638 (15) Å). These distances are very similar to those reported for the $[Mn^{II}Cl_4]^{2-}$ dianion in all the reported ET salts with this anion (2.348–2.363 Å, Table 9). Furthermore, the hypothetical $[Mn^{III}Cl_4]^-$ monoanion has never been reported and the Mn–Cl bond distances should be ca. 0.2 Å shorter, (i.e., around 2.15–2.17 Å).

Table 9. Average Mn–Cl bond distances (Å) in all the ET salts with the $[MnCl_4]^{2-}$ anion.

Compound	Formula	Mn–Cl (Å)	Ref
ECIQEM	β''-$(ET)_3MnCl_4 \cdot TCE$	2.360	[70]
FEWJAT	α-$(ET)_7[MnCl_4]_2 \cdot TCE$	2.348	[71]
GAMSOC	$(ET)_3[MnCl_4]_2$	2.363	[67]
2	$(ET)[MnCl_4] \cdot H_2O$	2.368	this work

TCE = 1,1,2-trichloroethane = CCl_2HCClH_2.

2.3. Magnetic Properties

The product of the magnetic susceptibility times the temperature $(\chi_m T)$ per Mn(III) ion for compound **1** shows a value of ca. 3.2 $cm^3 \cdot K \cdot mol^{-1}$, close to the expected one (3.0 $cm^3 \cdot K \cdot mol^{-1}$) for an $S = 2$ isolated Mn(III) ion with $g = 2$ (Figure 6). When the temperature is lowered, $\chi_m T$ remains constant down to ca. 50 K where a progressive decrease starts to reach a value of ca. 1.0 $cm^3 \cdot K \cdot mol^{-1}$ at 2 K. This behaviour indicates that compound **1** is essentially paramagnetic and presents the contribution expected for the anionic lattice, in agreement with the crystal structure that shows magnetically isolated $[Mn(C_2O_4)_3]^{3-}$ anions since the K^+ ions are diamagnetic. The decrease at low temperatures is simply due to the presence of a zero field splitting of the $S = 2$ spin ground state. The lack of magnetic contribution of the cationic lattice indicates that the spins on the ET molecules of the $(ET_2)^{2+}$ dimers are strongly antiferromagnetically coupled and the neutral isolated ET monomers are also diamagnetic.

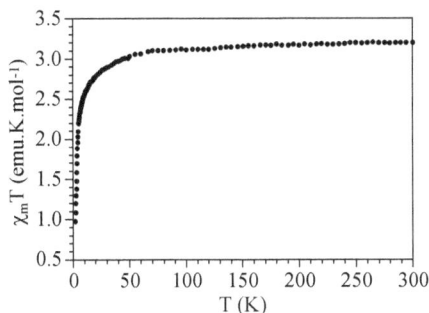

Figure 6. Thermal variation of the $\chi_m T$ product per Mn(III) ion for compound **1**.

For compound **2**, the $\chi_m T$ product per $[MnCl_4]^{2-}$ anion shows a value of ca. 4.5 $cm^3 \cdot K \cdot mol^{-1}$, close to the expected one (4.375 $cm^3 \cdot K \cdot mol^{-1}$) for an $S = 5/2$ isolated Mn(II) ion with $g = 2$ (Figure 7). When the sample is cooled, $\chi_m T$ shows a progressive decrease to reach a value of ca. 1.0 $cm^3 \cdot K \cdot mol^{-1}$ at 2 K. This behaviour indicates that compound **2** presents a weak antiferromagnetic coupling that might be attributed to a relatively short intermolecular Cl$\cdots$Cl contact (4.877 Å) or to the short O–H$\cdots$Cl H-bonds present in the anionic layer. Note that weak antiferromagnetic couplings through Cl$\cdots$H–N contacts with similar distances have already been observed and confirmed with theoretical calculations [72]. Accordingly, we have fit the magnetic properties to a simple Curie–Weiss law $[\chi = C/(T - \theta)]$ in order to estimate the weak magnetic coupling in **1**. Thus, the χ_m^{-1} vs. T plot can be fit in the 30–300 K range with a Curie constant, $C = 4.57$ $cm^3 \cdot K \cdot mol^{-1}$ and a Weiss temperature, $\theta = -9.4$ cm^{-1} (solid line in insert in Figure 7), confirming the presence of a weak antiferromagnetic coupling. As in **1**, we do not observe any magnetic contribution of the cationic lattice, suggesting that, as expected, the ET^{2+} cations are diamagnetic.

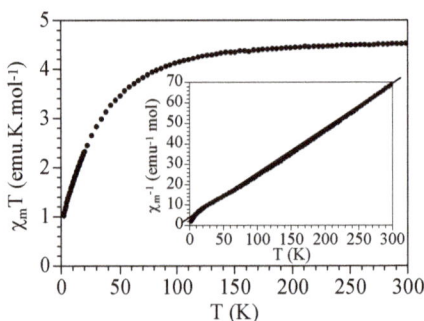

Figure 7. Thermal variation of the $\chi_m T$ product per $[MnCl_4]^{2-}$ ion for **2**. Inset shows the Curie–Weiss fit.

2.4. Electrical Properties

Compound **1** is a semiconductor with a room temperature conductivity value of ca. 2×10^{-5} S/cm and an activation energy of ca. 180 meV (Figure 8). This behaviour is very similar to that observed in all the similar orthorhombic salts of the type $(ET)_4[AM(C_2O_4)_3] \cdot G$ that are semiconductors with activation energies in the range 140–225 meV (Table 2). The semiconducting behaviour is attributed to the presence of completely ionized $(ET_2)^{2+}$ dimers surrounded by neutral ET monomers.

Compound **2** is also a semiconductor with a conductivity at room temperature of ca. 8×10^{-5} S/cm and an activation energy of ca. 210 meV (Figure 8). Note that this behaviour can be attributed to two possible reasons: (i) a charge transfer between the Cl ligands of the $[MnCl_4]^{2-}$ anion through the six short Cl$\cdots$S contacts (see Table 8); and (ii) the presence of a small degree of mixed valence in the ET molecules due to the presence of neutral or (most probably) mono-cationic ET molecules. Although most of the ET molecules are doubly oxidized, we cannot discard that, during the electro-crystallization process, some mono cationic ET$^+$ molecules enter in the structure. This is in agreement with the average charge of ca. 1.8 found for the ET molecules in **2** (see Table 7). The weak electron delocalization would take place through the two short S$\cdots$S intermolecular contacts present in compound **2** (Figure 5d and Table 8).

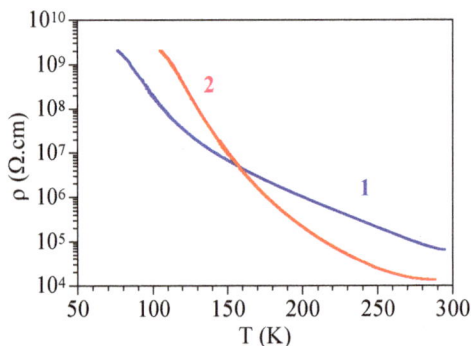

Figure 8. Thermal variation of the electrical resistivity of compounds **1** and **2**.

3. Experimental Section

3.1. Starting Materials

The organic donor bis(ethylenedithio)tetrathiafulvalene (ET), the 18-crown-6 ether, benzoic acid and all the solvents used in this work are commercially available and were used as received.

The potassium salt $K_3[Mn(C_2O_4)_3]$ was prepared as previously reported [73] and was recrystallized several times from water. The radical salts were prepared by electrochemical oxidation of ET on platinum wire electrodes (1 mm diameter) in U-shaped cells under low constant current (Table 10). The anodic and cathodic compartments are separated by a porous glass frit. The exact conditions for the synthesis of each particular radical salt are described in Table 10.

Table 10. Synthetic conditions used for salts **1** and **2**.

Compound	Anode	Cathode	Current	Time
(ET)₄[KMn(C₂O₄)₃]·PhCN (**1**)	ET (10 mg) PhCN (10 mL) MeOH (1 mL)	K₃[Mn(C₂O₄)₃] (0.1 mmol) 18-crown-6 (90 mg) PhCOOH (0.147 mmol) PhCN (10 mL) MeOH (1.5 mL)	3 µA	1 week
(ET)[MnCl₄] H₂O (**2**)	ET (10 mg) TCE (10 mL) MeOH (1 mL)	K₃[Mn(C₂O₄)₃] (0.1 mmol) 18-crown-6 (90 mg) TCE (10 mL) MeOH (1 mL)	2 µA 4 µA 5 µA	3 weeks 1 week 1 week

TCE = 1,1,2-trichloroethane = CCl_2HCClH_2.

3.2. Synthesis of (ET)₄[KMn(C₂O₄)₃]·PhCN (**1**)

A solution of racemic $K_3[Mn(C_2O_4)_3]$ (43.6 mg, 0.1 mmol), PhCOOH (18 mg, 0.15 mmol) and 18-crown-6 ether (90 mg, 0.35 mmol) in a mixture of 10 mL of PhCN and 1.5 mL of MeOH was placed in the cathode of a U-shaped electrochemical cell. A solution of ET (10 mg, 0.026 mmol) in a mixture of 10 mL of PhCN and 1.5 mL of MeOH was placed in the anode of the U-shaped cell and a constant current of 3 µA was applied. Black plate single crystals were collected from the anode after one week.

3.3. Synthesis of (ET)[MnCl₄]·H₂O (**2**)

A solution of racemic $K_3[Mn(C_2O_4)_3]$ (43.6 mg, 0.1 mmol) and 18-crown-6 ether (90 mg, 0.35 mmol) in a mixture of 10 mL of 1,1,2-trichloroethane and 1.5 mL of MeOH was placed in the cathode of a U-shaped electrochemical cell. A solution of ET (10 mg, 0.026 mmol) in a mixture of 10 mL of 1,1,2-trichloroethane and 1.5 mL of MeOH was placed in the anode of the U-shaped cell and a constant current of 2 µA was applied during three weeks. The intensity was increased to 4 µA for one week more and finally to 5 µA. Dark green prismatic crystals were collected from the anode after one week at 5 µA.

3.4. Physical Measurements

Magnetic susceptibility measurements were carried out in the temperature range 2–300 K with an applied magnetic field of 0.5 T on polycrystalline samples of compounds **1** and **2** with a MPMS-XL-5 SQUID susceptometer (Quantum Desing, San Diego, CA, USA). The susceptibility data were corrected for the sample holders previously measured using the same conditions and for the diamagnetic contributions of the salt as deduced by using Pascal's constant tables [74].

The temperature dependence of the DC electrical conductivity was measured with the four contact method on different single crystals of compounds **1** and **2** in cooling and warming scans with similar results within experimental errors. The contacts were made with Pt wires (25 µm diameter) using graphite paste. The samples were measured in a PPMS-9 equipment (Quantum Desing, San Diego, CA, USA) connected to an external voltage source model 2450 source-meter (Keithley, Cleveland, OH, USA) and amperometer model 6514 electrometer (Keithley, Cleveland, OH, USA). The conductivity quoted values have been measured in the voltage range where the crystals are Ohmic conductors. The cooling and warming rates were 1 and 2 $K \cdot min^{-1}$.

3.5. Crystallographic Data Collection and Refinement

Suitable single crystals of compounds **1** and **2** were mounted on a glass fibre using a viscous hydrocarbon oil to coat the crystal and then transferred directly to the cold nitrogen stream for data collection. X-ray data were collected at 120 K on a Supernova diffractometer (Agilent, Santa Clara, CA, USA) equipped with a graphite-monochromated Enhance (Mo) X-ray Source (λ = 0.71073 Å). The program CrysAlisPro v38.43, (Rigaku, Tokyo, Japan), was used for unit cell determinations and data reduction. Empirical absorption correction was performed using spherical harmonics, implemented in the SCALE3 ABSPACK scaling algorithm. Crystal structures were solved with direct methods with the SIR97 program [75], and refined against all F^2 values with the SHELXL-2014 program [76], using the WinGX graphical user interface [77]. All non-hydrogen atoms were refined anisotropically, and hydrogen atoms were placed in calculated positions and refined isotropically with a riding model. There is a disorder in the CH_3CN solvent molecules that appears with two possible orientations with a common N atom located on a C_2 axis. Data collection and refinement parameters are given in Table 5.

CCDC-1527866 and 1527859 contain the supplementary crystallographic data for compounds **1** and **2**, respectively. These data can be obtained free of charge from The Cambridge Crystallographic Data Centre at www.ccdc.cam.ac.uk/data_request/cif.

4. Conclusions

The combination of the magnetic anion $[Mn(C_2O_4)_3]^{3-}$ with the organic donor ET under different synthetic conditions has resulted in the synthesis of two very original magnetic and conducting radical salts: $(ET)_4[KMn(C_2O_4)_3]\cdot PhCN$ (**1**) and $(ET)[MnCl_4]\cdot H_2O$ (**2**). The radical salt with the anion $[Mn(C_2O_4)_3]^{3-}$ is the first reported member with Mn(III) of the Day's huge family of magnetic conductors and superconductors formulated as $(ET)_4[A^IM^{III}(C_2O_4)_3]\cdot G$ (A^I = H_3O^+, K^+, NH_4^+, Na^+, ...; M^{III} = Fe, Cr, Ga, Co, Al and Ru; G = PhCN, $PhNO_2$, PhCl, PhBr, py, ...). This compound crystallizes in an orthorhombic pseudo-κ phase where $(ET_2)^{2+}$ dimers are surrounded by isolated neutral ET monomers. The change of PhCN by 1,1,2-trichloroethane as a solvent gives rise to the radical salt $(ET)[MnCl_4]\cdot H_2O$, where the ET molecules have been oxidized to a very unusual oxidation state of +2 and the Mn(III) metal atom has been reduced to Mn(II). The degradation of the chlorinated solvent furnishes the Cl^- ligands for the in situ formation of the anion $[MnCl_4]^{2-}$. Both salts are semiconductors (with activation energies of ca. 180 and ca. 210 meV, respectively) and paramagnetic with magnetic moments corresponding to the anionic complexes, since the organic lattices do not contribute to the magnetic moment.

Acknowledgments: We thank the Generalitat Valenciana (projects PrometeoII/2014/076 and ISIC) for the financial support.

Author Contributions: S.B. designed the synthesis and performed the X-ray structural analysis. Y.S.-M. performed the synthesis of the precursor salts and of the radical salts. C.J.G.-G. performed the magnetic and conductivity measurements. All authors contributed to the writing of the manuscript.

Conflicts of Interest: The authors declare no conflict of interest.

References

1. Coronado, E.; Day, P. Magnetic Molecular Conductors. *Chem. Rev.* **2004**, *104*, 5419–5448. [CrossRef] [PubMed]
2. Enoki, T.; Miyazaki, A. Magnetic TTF-Based Charge-Transfer Complexes. *Chem. Rev.* **2004**, *104*, 5449–5478. [CrossRef] [PubMed]
3. Kobayashi, H.; Cui, H.; Kobayashi, A. Organic Metals and Superconductors Based on BETS (BETS = Bis(ethylenedithio)tetraselenafulvalene). *Chem. Rev.* **2004**, *104*, 5265–5288. [CrossRef] [PubMed]
4. Coronado, E.; Galán-Mascarós, J.R.; Gómez-García, C.J.; Laukhin, V. Coexistence of ferromagnetism and metallic conductivity in a molecule-based layered compound. *Nature* **2000**, *408*, 447–449. [CrossRef] [PubMed]

5. Kobayashi, H.; Kobayashi, A.; Cassoux, P. BETS as a source of molecular magnetic superconductors (BETS = bis(ethylenedithio)tetraselenafulvalene). *Chem. Soc. Rev.* **2000**, *29*, 325–333. [CrossRef]

6. Graham, A.W.; Kurmoo, M.; Day, P. β″-(bedt-ttf)₄[(H₂O)Fe(C₂O₄)₃]·PhCN: The first molecular superconductor containing paramagnetic metal ions. *J. Chem. Soc. Chem. Commun.* **1995**, 2061–2062. [CrossRef]

7. Kurmoo, M.; Graham, A.W.; Day, P.; Coles, S.J.; Hursthouse, M.B.; Caulfield, J.L.; Singleton, J.; Pratt, F.L.; Hayes, W. Superconducting and Semiconducting Magnetic Charge Transfer Salts: (BEDT-TTF)₄*A*Fe(C₂O₄)₃·C₆H₅CN (*A* = H₂O, K, NH₄). *J. Am. Chem. Soc.* **1995**, *117*, 12209–12217. [CrossRef]

8. Kobayashi, H.; Fujiwara, E.; Fujiwara, H.; Tanaka, H.; Tamura, I.; Bin, Z.; Gritsenko, V.; Otsuka, T.; Kobayashi, A.; Tokumoto, M.; et al. Magnetic organic superconductors based on BETS molecules—Interplay of conductivity and magnetism. *Mol. Cryst. Liq. Cryst.* **2002**, *379*, 9–18. [CrossRef]

9. Kobayashi, H.; Tomita, H.; Naito, T.; Kobayashi, A.; Sakai, F.; Watanabe, T.; Cassoux, P. New BETS conductors with magnetic anions (BETS = bis(ethylenedithio)-tetraselenafulvalene). *J. Am. Chem. Soc.* **1996**, *118*, 368–377. [CrossRef]

10. Kobayashi, H.; Fujiwara, E.; Fujiwara, H.; Tanaka, H.; Otsuka, T.; Kobayashi, A.; Tokumoto, M.; Cassoux, P. Antiferromagnetic organic superconductors, BETS₂Fe*X*₄ (*X* = Br, Cl). *Mol. Cryst. Liq. Cryst.* **2002**, *380*, 139–144. [CrossRef]

11. Fujiwara, H.; Fujiwara, E.; Nakazawa, Y.; Narymbetov, B.Z.; Kato, K.; Kobayashi, H.; Kobayashi, A.; Tokumoto, M.; Cassoux, P. A novel antiferromagnetic organic superconductor κ-(BETS)₂FeBr₄ [where BETS = bis(ethylenedithio)tetraselenafulvalene]. *J. Am. Chem. Soc.* **2001**, *123*, 306–314. [CrossRef] [PubMed]

12. Kobayashi, H.; Tanaka, H.; Ojima, E.; Fujiwara, H.; Nakazawa, Y.; Otsuka, T.; Kobayashi, A.; Tokumoto, M.; Cassoux, P. Antiferromagnetism and superconductivity of BETS conductors with Fe³⁺ ions. *Synth. Met.* **2001**, *120*, 663–666. [CrossRef]

13. Kobayashi, H.; Tanaka, H.; Ojima, E.; Fujiwara, H.; Otsuka, T.; Kobayashi, A.; Tokumoto, M.; Cassoux, P. Coexistence of antiferromagnetic order and superconductivity in organic conductors. *Polyhedron* **2001**, *20*, 1587–1592. [CrossRef]

14. Tanaka, H.; Kobayashi, H.; Kobayashi, A.; Cassoux, P. Superconductivity, antiferromagnetism, and phase diagram of a series of organic conductors: λ-(BETS)₂Fe*x*Ga₁₋*x*Br*y*Cl₄₋*y*. *Adv. Mater.* **2000**, *12*, 1685–1689. [CrossRef]

15. Ojima, E.; Fujiwara, H.; Kato, K.; Kobayashi, H.; Tanaka, H.; Kobayashi, A.; Tokumoto, M.; Cassoux, P. Antiferromagnetic organic metal exhibiting superconducting transition, κ-(BETS)₂FeBr₄ [BETS = bis(ethylenedithio)tetraselenafulvalene]. *J. Am. Chem. Soc.* **1999**, *121*, 5581–5582. [CrossRef]

16. Alberola, A.; Coronado, E.; Galán-Mascarós, J.R.; Giménez-Saiz, C.; Gómez-García, C.J. A molecular metal ferromagnet from the organic donor bis(ethylenedithio)-tetraselenafulvalene and bimetallic oxalate complexes. *J. Am. Chem. Soc.* **2003**, *125*, 10774–10775. [CrossRef] [PubMed]

17. Coronado, E.; Galán-Mascarós, J.R.; Gómez-García, C.J. Charge transfer salts of tetrathiafulvalene derivatives with magnetic iron(III) oxalate complexes: [TTF]₇[Fe(ox)₃]₂·4H₂O, [TTF]₅[Fe₂(ox)₅]·2PhMe·2H₂O and [TMTTF]₄[Fe₂(ox)₅]·PhCN·4H₂O (TMTTF = tetramethyltetrathiafulvalene). *J. Chem. Soc. Dalton Trans.* **2000**, 205–210. [CrossRef]

18. Coronado, E.; Galán-Mascarós, J.R.; Giménez-Saiz, C.; Gómez-García, C.J.; Martínez-Agudo, J.M.; Martínez-Ferrero, E. Magnetic properties of hybrid molecular materials based on oxalato complexes. *Polyhedron* **2003**, *22*, 2381–2386. [CrossRef]

19. Coronado, E.; Curreli, S.; Giménez-Saiz, C.; Gómez-García, C.J.; Alberola, A. Radical salts of bis(ethylenediseleno)tetrathiafulvalene with paramagnetic tris(oxalato)metalate anions. *Inorg. Chem.* **2006**, *45*, 10815–10824. [CrossRef] [PubMed]

20. Martin, L.; Turner, S.S.; Day, P.; Guionneau, P.; Howard, J.A.K.; Uruichi, M.; Yakushi, K. Synthesis, crystal structure and properties of the semiconducting molecular charge-transfer salt (bedt-ttf)₂Ge(C₂O₄)₃·PhCN [bedt-ttf = bis(ethylenedithio)tetrathiafulvalene]. *J. Mater. Chem.* **1999**, *9*, 2731–2736. [CrossRef]

21. Martin, L.; Day, P.; Clegg, W.; Harrington, R.W.; Horton, P.N.; Bingham, A.; Hursthouse, M.B.; McMillan, P.; Firth, S. Multi-layered molecular charge-transfer salts containing alkali metal ions. *J. Mater. Chem.* **2007**, *17*, 3324–3329. [CrossRef]

22. Martin, L.; Day, P.; Nakatsuji, S.; Yamada, J.; Akutsu, H.; Horton, P. A molecular charge transfer salt of BEDT-TTF containing a single enantiomer of tris(oxalato)chromate(III) crystallised from a chiral solvent. *CrystEngComm* **2010**, *12*, 1369–1372. [CrossRef]

23. Martin, L.; Day, P.; Barnett, S.A.; Tocher, D.A.; Horton, P.N.; Hursthouse, M.B. Magnetic molecular charge-transfer salts containing layers of water and tris(oxalato)ferrate(III) anions. *CrystEngComm* **2008**, *10*, 192–196. [CrossRef]

24. Zhang, B.; Zhang, Y.; Liu, F.; Guo, Y. Synthesis, crystal structure, and characterization of charge-transfer salt: (BEDT-TTF)$_5$[Fe(C$_2$O$_4$)$_3$]·(H$_2$O)$_2$·CH$_2$Cl$_2$ (BEDT-TTF = bis(ethylenedithio)tetrathiafulvalene). *CrystEngComm* **2009**, *11*, 2523–2528. [CrossRef]

25. Martin, L.; Day, P.; Nakatsuji, S.; Yamada, J.; Akutsu, H.; Horton, P.N. BEDT-TTF Tris(oxalato)germanate(IV) Salts with Novel Donor Packing Motifs. *Bull. Chem. Soc. Jpn.* **2010**, *83*, 419–423. [CrossRef]

26. Rashid, S.; Turner, S.S.; Day, P.; Light, M.E.; Hursthouse, M.B. Molecular charge-transfer salt of BEDT-TTF [bis(ethylenedithio)tetrathiafulvalene] with the oxalate-bridged dimeric anion [Fe$_2$(C$_2$O$_4$)$_5$]$^{4-}$. *Inorg. Chem.* **2000**, *39*, 2426–2428. [CrossRef]

27. Coronado, E.; Galán-Mascarós, J.R.; Giménez-Saiz, C.; Gómez-García, C.J.; Ruiz-Perez, C. Hybrid organic/inorganic molecular materials formed by tetrathiafulvalene radicals and magnetic trimeric clusters of dimetallic oxalate-bridged complexes: The series (TTF)$_4${M^{II}(H$_2$O)$_2$[M^{III}(ox)$_3$]$_2$}·nH$_2$O (M^{II} = Mn, Fe, Co, Ni, Cu and Zn; M^{III} = Cr and Fe; ox = C$_2$O$_4^{2-}$). *Eur. J. Inorg. Chem.* **2003**, 2290–2298. [CrossRef]

28. Coronado, E.; Galán-Mascarós, J.R.; Giménez-Saiz, C.; Gómez-García, C.J.; Ruiz-Pérez, C.; Triki, S. Hybrid molecular materials formed by alternating layers of bimetallic oxalate complexes and tetrathiafulvalene molecules: Synthesis, structure, and magnetic properties of TTF$_4${Mn(H$_2$O)$_2$[Cr(ox)$_3$]$_2$}·14H$_2$O. *Adv. Mater.* **1996**, *8*, 737–740. [CrossRef]

29. Alberola, A.; Coronado, E.; Galán-Mascarós, J.R.; Giménez-Saiz, C.; Gómez-García, C.J.; Romero, F.M. Multifunctionality in hybrid molecular materials: Design of ferromagnetic molecular metals and hybrid magnets. *Synth. Met.* **2003**, *133*, 509–513. [CrossRef]

30. Galán-Mascarós, J.R.; Coronado, E.; Goddard, P.A.; Singleton, J.; Coldea, A.I.; Wallis, J.D.; Coles, S.J.; Alberola, A. A Chiral Ferromagnetic Molecular Metal. *J. Am. Chem. Soc.* **2010**, *132*, 9271–9273. [CrossRef] [PubMed]

31. Coronado, E.; Galán-Mascarós, J.R. Hybrid molecular conductors. *J. Mater. Chem.* **2005**, *15*, 66–74. [CrossRef]

32. Martin, L.; Turner, S.S.; Day, P.; Guionneau, P.; Howard, J.A.K.; Hibbs, D.E.; Light, M.E.; Hursthouse, M.B.; Uruichi, M.; Yakushi, K. Crystal Chemistry and Physical Properties of Superconducting and Semiconducting Charge Transfer Salts of the Type (BEDT-TTF)$_4$[$A^I M^{III}$(C$_2$O$_4$)$_3$]·PhCN (A^I = H$_3$O, NH$_4$, K; M^{III} = Cr, Fe, Co, Al; BEDT-TTF = Bis(ethylenedithio)tetrathiafulvalene). *Inorg. Chem.* **2001**, *40*, 1363–1371. [CrossRef] [PubMed]

33. Akutsu-Sato, A.; Akutsu, H.; Yamada, J.; Nakatsuji, S.; Turner, S.S.; Day, P. Suppression of superconductivity in a molecular charge transfer salt by changing guest molecule: β′′-(BEDT-TTF)$_4$[(H$_3$O)Fe(C$_2$O$_4$)$_3$] (C$_6$H$_5$CN)$_x$(C$_5$H$_5$N)$_{1-x}$. *J. Mater. Chem.* **2007**, *17*, 2497–2499. [CrossRef]

34. Prokhorova, T.G.; Buravov, L.I.; Yagubskii, E.B.; Zorina, L.V.; Khasanov, S.S.; Simonov, S.V.; Shibaeva, R.P.; Korobenko, A.V.; Zverev, V.N. Effect of electrocrystallization medium on quality, structural features, and conducting properties of single crystals of the (BEDT-TTF)$_4 A^I$[Fe^{III}(C$_2$O$_4$)$_3$]·G family. *CrystEngComm* **2011**, *13*, 537–545. [CrossRef]

35. Turner, S.S.; Day, P.; Malik, K.M.A.; Hursthouse, M.B.; Teat, S.J.; MacLean, E.J.; Martin, L.; French, S.A. Effect of included solvent molecules on the physical properties of the paramagnetic charge transfer salts β′′-(bedt-ttf)$_4$[(H$_3$O)Fe(C$_2$O$_4$)$_3$]·solvent (bedt-ttf = bis(ethylenedithio)tetrathiafulvalene). *Inorg. Chem.* **1999**, *38*, 3543–3549. [CrossRef] [PubMed]

36. Rashid, S.; Turner, S.S.; Day, P.; Howard, J.A.K.; Guionneau, P.; McInnes, E.J.L.; Mabbs, F.E.; Clark, R.J.H.; Firth, S.; Biggs, T. New superconducting charge-transfer salts (BEDT-TTF)$_4$[A·M(C$_2$O$_4$)$_3$]·C$_6$H$_5$NO$_2$ (A = H$_3$O or NH$_4$, M = Cr or Fe, BEDT-TTF = bis(ethylenedithio)tetrathiafulvalene). *J. Mater. Chem.* **2001**, *11*, 2095–2101. [CrossRef]

37. Sun, S.Q.; Wu, P.J.; Zhang, Q.C.; Zhu, D.B. The New Semiconducting Magnetic Charge Transfer Salt (BEDT-TTF)$_4$·H$_2$O·Fe(C$_2$O$_4$)$_3$·C$_6$H$_5$NO$_2$: Crystal Structure and Physical Properties. *Mol. Cryst. Liq. Cryst.* **1998**, *319*, 259–269. [CrossRef]

38. Sun, S.; Wu, P.; Zhang, Q.; Zhu, D. The new semiconducting magnetic charge transfer salt (BEDT-TTF)$_4$·H$_2$O·Fe(C$_2$O$_4$)$_3$·C$_6$H$_5$NO$_2$: Crystal structure and physical properties. *Synth. Met.* **1998**, *94*, 161–166. [CrossRef]

39. Zorina, L.; Prokhorova, T.; Simonov, S.; Khasanov, S.; Shibaeva, R.; Manakov, A.; Zverev, V.; Buravov, L.; Yagubskii, E. Structure and magnetotransport properties of the new quasi-two-dimensional molecular metal β''-(BEDT-TTF)$_4$H$_3$O[Fe(C$_2$O$_4$)$_3$]·C$_6$H$_4$Cl$_2$. *J. Exp. Theor. Phys.* **2008**, *106*, 347–354. [CrossRef]

40. Coronado, E.; Curreli, S.; Giménez-Saiz, C.; Gómez-García, C.J. A novel paramagnetic molecular superconductor formed by bis(ethylenedithio)tetrathiafulvalene, tris(oxalato) ferrate(III) anions and bromobenzene as guest molecule: ET$_4$[(H$_3$O)Fe(C$_2$O$_4$)$_3$]·C$_6$H$_5$Br. *J. Mater. Chem.* **2005**, *15*, 1429–1436. [CrossRef]

41. Prokhorova, T.G.; Khasanov, S.S.; Zorina, L.V.; Buravov, L.I.; Tkacheva, V.A.; Baskakov, A.A.; Morgunov, R.B.; Gener, M.; Canadell, E.; Shibaeva, R.P.; et al. Molecular Metals Based on BEDT-TTF Radical Cation Salts with Magnetic Metal Oxalates as Counterions: β''-(BEDT-TTF)$_4$A[M(C$_2$O$_4$)$_3$]·DMF (A = NH$_4^+$, K$^+$; M = Cr^{III}, Fe^{III}). *Adv. Funct. Mater.* **2003**, *13*, 403–411. [CrossRef]

42. Zorina, L.V.; Khasanov, S.S.; Simonov, S.V.; Shibaeva, R.P.; Bulanchuk, P.O.; Zverev, V.N.; Canadell, E.; Prokhorova, T.G.; Yagubskii, E.B. Structural phase transition in the β''-(BEDT-TTF)$_4$H$_3$O[Fe(C$_2$O$_4$)$_3$]·G crystals (where G is a guest solvent molecule). *CrystEngComm* **2012**, *14*, 460–465. [CrossRef]

43. Coronado, E.; Curreli, S.; Giménez-Saiz, C.; Gómez-García, C.J. The series of molecular conductors and superconductors ET$_4$[AFe(C$_2$O$_4$)$_3$]·PhX (ET = bis(ethylenedithio)tetrathiafulvalene; (C$_2$O$_4$)$^{2-}$ = oxalate; A$^+$ = H$_3$O$^+$, K$^+$; X = F, Cl, Br, and I): Influence of the halobenzene guest molecules on the crystal structure and superconducting properties. *Inorg. Chem.* **2012**, *51*, 1111–1126. [PubMed]

44. Akutsu-Sato, A.; Kobayashi, A.; Mori, T.; Akutsu, H.; Yamada, J.; Nakatsuji, S.; Turner, S.S.; Day, P.; Tocher, D.A.; Light, M.E.; et al. Structures and Physical Properties of New β'-BEDT-TTF Tris-Oxalatometallate (III) Salts Containing Chlorobenzene and Halomethane Guest Molecules. *Synth. Met.* **2005**, *152*, 373–376. [CrossRef]

45. Coronado, E.; Curreli, S.; Gimenez-Saiz, C.; Gómez-García, C.J. New magnetic conductors and superconductors based on BEDT-TTF and BEDS-TTF. *Synth. Met.* **2005**, *154*, 245–248. [CrossRef]

46. Kanehama, R.; Yoshino, Y.; Ishii, T.; Manabe, T.; Hara, H.; Miyasaka, H.; Matsuzaka, H.; Yamashita, M.; Katada, M.; Nishikawa, H.; et al. Syntheses and physical properties of new charge-transfer salts consisting of a conducting BEDT-TTF column and magnetic 1D or 2D Fe(III) networks. *Synth. Met.* **2003**, *133*, 553–554. [CrossRef]

47. Martin, L.; Turner, S.S.; Day, P.; Malik, K.M.A.; Coles, S.J.; Hursthouse, M.B. Polymorphism based on molecular stereoisomerism in tris(oxalato) Cr(III) salts of bedt-ttf [bis(ethylenedithio)tetrathiafulvalene]. *Chem. Commun.* **1999**, 513–514. [CrossRef]

48. Rashid, S.; Turner, S.S.; Le Pevelen, D.; Day, P.; Light, M.E.; Hursthouse, M.B.; Firth, S.; Clark, R.J.H. β''-(BEDT-TTF)$_4$[(H$_3$O)Cr(C$_2$O$_4$)$_3$]CH$_2$Cl$_2$: Effect of included solvent on the structure and properties of a conducting molecular charge-transfer salt. *Inorg. Chem.* **2001**, *40*, 5304–5306. [CrossRef] [PubMed]

49. Akutsu, H.; Akutsu-Sato, A.; Turner, S.S.; Le Pevelen, D.; Day, P.; Laukhin, V.; Klehe, A.; Singleton, J.; Tocher, D.A.; Probert, M.R.; et al. Effect of included guest molecules on the normal state conductivity and superconductivity of β''-(ET)$_4$[(H$_3$O)Ga(C$_2$O$_4$)$_3$]·G (G = pyridine, nitrobenzene). *J. Am. Chem. Soc.* **2002**, *124*, 12430–12431. [CrossRef] [PubMed]

50. Prokhorova, T.G.; Buravov, L.I.; Yagubskii, E.B.; Zorina, L.V.; Simonov, S.V.; Shibaeva, R.P.; Zverev, V.N. Metallic Bi- and Monolayered Radical Cation Salts Based on Bis(ethylenedithio)-tetrathiafulvalene (BEDT-TTF) with the Tris(oxalato)gallate Anion. *Eur. J. Inorg. Chem.* **2014**, 3933–3940. [CrossRef]

51. Prokhorova, T.G.; Zorina, L.V.; Simonov, S.V.; Zverev, V.N.; Canadell, E.; Shibaeva, R.P.; Yagubskii, E.B. The first molecular superconductor based on BEDT-TTF radical cation salt with paramagnetic tris(oxalato)ruthenate anion. *CrystEngComm* **2013**, *15*, 7048–7055. [CrossRef]

52. Prokhorova, T.G.; Buravov, L.I.; Yagubskii, E.B.; Zorina, L.V.; Simonov, S.V.; Zverev, V.N.; Shibaeva, R.P.; Canadell, E. Effect of Halopyridine Guest Molecules on the Structure and Superconducting Properties of β''-[Bis(ethylenedithio)tetrathiafulvalene]$_4$(H$_3$O)[Fe(C$_2$O$_4$)$_3$]·Guest Crystals. *Eur. J. Inorg. Chem.* **2015**, *2015*, 5611–5620. [CrossRef]

53. Martin, L.; Engelkamp, H.; Akutsu, H.; Nakatsuji, S.; Yamada, J.; Horton, P.; Hursthouse, M.B. Radical-cation salts of BEDT-TTF with lithium tris(oxalato)metallate(III). *Dalton Trans.* **2015**, *44*, 6219–6223. [CrossRef] [PubMed]

54. Zorina, L.V.; Khasanov, S.S.; Simonov, S.V.; Shibaeva, R.P.; Zverev, V.N.; Canadell, E.; Prokhorova, T.G.; Yagubskii, E.B. Coexistence of two donor packing motifs in the stable molecular metal α-pseudo-κ-(BEDT-TTF)$_4$(H$_3$O)[Fe(C$_2$O$_4$)$_3$]·C$_6$H$_4$Br$_2$. *CrystEngComm* **2011**, *13*, 2430–2438. [CrossRef]
55. Akutsu, H.; Akutsu-Sato, A.; Turner, S.S.; Day, P.; Canadell, E.; Firth, S.; Clark, R.J.H.; Yamada, J.; Nakatsuji, S. Superstructures of donor packing arrangements in a series of molecular charge transfer salts. *Chem. Commun.* **2004**, 18–19. [CrossRef] [PubMed]
56. Martin, L.; Day, P.; Akutsu, H.; Yamada, J.; Nakatsuji, S.; Clegg, W.; Harrington, R.W.; Horton, P.N.; Hursthouse, M.B.; McMillan, P.; et al. Metallic molecular crystals containing chiral or racemic guest molecules. *CrystEngComm* **2007**, *9*, 865–867. [CrossRef]
57. Martin, L.; Akutsu, H.; Horton, P.N.; Hursthouse, M.B.; Harrington, R.W.; Clegg, W. Chiral Radical-Cation Salts of BEDT-TTF Containing a Single Enantiomer of Tris(oxalato)aluminate(III) and -chromate(III). *Eur. J. Inorg. Chem.* **2015**, 1865–1870. [CrossRef]
58. Martin, L.; Day, P.; Horton, P.; Nakatsuji, S.; Yamada, J.; Akutsu, H. Chiral conducting salts of BEDT-TTF containing a single enantiomer of tris(oxalato)chromate(III) crystallised from a chiral solvent. *J. Mater. Chem.* **2010**, *20*, 2738–2742. [CrossRef]
59. Martin, L.; Akutsu, H.; Horton, P.N.; Hursthouse, M.B. Chirality in charge-transfer salts of BEDT-TTF of tris(oxalato)chromate(III). *CrystEngComm* **2015**, *17*, 2783–2790. [CrossRef]
60. Coldea, A.I.; Bangura, A.F.; Singleton, J.; Ardavan, A.; Akutsu-Sato, A.; Akutsu, H.; Turner, S.S.; Day, P. Fermi-surface topology and the effects of intrinsic disorder in a class of charge-transfer salts containing magnetic ions: β″-(BEDT-TTF)$_4$[(H$_3$O)M(C$_2$O$_4$)$_3$]·Y (M = Ga, Cr, Fe; Y = C$_5$H$_5$N). *Phys. Rev. B* **2004**, *69*, 085112. [CrossRef]
61. Martin, L.; Turner, S.S.; Day, P.; Mabbs, F.E.; McInnes, E.J.L. New molecular superconductor containing paramagnetic chromium (III) ions. *Chem. Commun.* **1997**, 1367–1368. [CrossRef]
62. Rosseinsky, M.J.; Kurmoo, M.; Talham, D.R.; Day, P.; Chasseau, D.; Watkin, D. A Novel Conducting Charge-Transfer Salt-(bedt-ttf)$_3$Cl$_2$·2H$_2$O. *J. Chem. Soc. Chem. Commun.* **1988**, 88–90. [CrossRef]
63. Zhang, B.; Yao, Y.X.; Zhu, D.B. A new organic conductor (BEDT-TTF)$_5$Cl$_3$(H$_2$O)$_5$. *Synth. Met.* **2001**, *120*, 671–674. [CrossRef]
64. Mori, H.; Hirabayashi, I.; Tanaka, S.; Maruyama, Y. Preparation, Crystal and Electronic Structures, and Electrical Resistivity of (BEDT-TTF)$_3$Cl$_{2.5}$(H$_5$O$_2$). *Bull. Chem. Soc. Jpn.* **1993**, *66*, 2156–2159. [CrossRef]
65. Guionneau, P.; Kepert, C.J.; Bravic, G.; Chasseau, D.; Truter, M.R.; Kurmoo, M.; Day, P. Determining the charge distribution in BEDT-TTF salts. *Synth. Met.* **1997**, *86*, 1973–1974. [CrossRef]
66. Kanehama, R.; Umemiya, M.; Iwahori, F.; Miyasaka, H.; Sugiura, K.; Yamashita, M.; Yokochi, Y.; Ito, H.; Kuroda, S.; Kishida, H.; et al. Novel ET-Coordinated Copper(I) Complexes: Syntheses, Structures, and Physical Properties (ET = BEDT-TTF = Bis(ethylenedithio)tetrathiafulvalene). *Inorg. Chem.* **2003**, *42*, 7173–7181. [CrossRef] [PubMed]
67. Mori, T.; Inokuchi, H. A BEDT-TTF Complex Including a Magnetic Anion, (BEDT-TTF)$_3$(MnCl$_4$)$_2$. *Bull. Chem. Soc. Jpn.* **1988**, *61*, 591–593. [CrossRef]
68. Shibaeva, R.P.; Lobkovskaya, R.M.; Korotkov, V.E.; Kusch, N.D.; Yagubskii, E.B.; Makova, M.K. ET cation-radical salts with metal complex anions. *Synth. Met.* **1988**, *27*, A457–A463. [CrossRef]
69. Chou, L.; Quijada, M.A.; Clevenger, M.B.; de Oliveira, G.F.; Abboud, K.A.; Tanner, D.B.; Talham, D.R. Dication Salts of the Organic Donor Bis(ethylenedithio)tetrathiafulvalene. *Chem. Mater.* **1995**, *7*, 530–534. [CrossRef]
70. Naito, T.; Inabe, T.; Takeda, K.; Awaga, K.; Akutagawa, T.; Hasegawa, T.; Nakamura, T.; Kakiuchi, T.; Sawa, H.; Yamamoto, T.; et al. β″-(ET)$_3$(MnCl$_4$)(1,1,2-C$_2$H$_3$Cl$_3$) (ET = bis(ethylenedithio)tetrathiafulvalene); a pressure-sensitive new molecular conductor with localized spins. *J. Mater. Chem.* **2001**, *11*, 2221–2227. [CrossRef]
71. Naito, T.; Inabe, T. Structural, Electrical, and Magnetic Properties of alpha-(ET)$_7$[MnCl$_4$]$_2$·(1,1,2-C$_2$H$_3$Cl$_3$)$_2$ (ET = Bis(ethylenedithio)tetrathiafulvalene). *Bull. Chem. Soc. Jpn.* **2004**, *77*, 1987–1995. [CrossRef]
72. Willett, R.D.; Gómez-García, C.J.; Twamley, B.; Gómez-Coca, S.; Ruiz, E. Exchange coupling mediated by N–H···Cl hydrogen bonds: Experimental and theoretical study of the frustrated magnetic system in bis(o-phenylenediamine)nickel(II) chloride. *Inorg. Chem.* **2012**, *51*, 5487–5493. [CrossRef] [PubMed]
73. Palmer, W.G. *Experimental Inorganic Chemistry*; Cambridge University Press: Cambridge, UK, 1954.

74. Bain, G.A.; Berry, J.F. Diamagnetic corrections and Pascal's constants. *J. Chem. Educ.* **2008**, *85*, 532–536. [CrossRef]

75. Altomare, A.; Burla, M.C.; Camalli, M.; Cascarano, G.L.; Giacovazzo, C.; Guagliardi, A.; Moliterni, A.G.G.; Polidori, G.; Spagna, R. SIR97: A new tool for crystal structure determination and refinement. *J. Appl. Cryst.* **1999**, *32*, 115–119. [CrossRef]

76. Sheldrick, G.M. Crystal structure refinement with SHELXL. *Acta Cryst. C* **2015**, *71*, 3–8. [CrossRef] [PubMed]

77. Farrugia, L.J. WinGX and ORTEP for Windows: An update. *J. Appl. Cryst.* **2012**, *45*, 849–854. [CrossRef]

magnetochemistry

MDPI

Article

Charge Ordering Transitions of the New Organic Conductors δ_m- and δ_o-(BEDT-TTF)$_2$TaF$_6$

Tadashi Kawamoto [1,*], Kohei Kurata [2], Takehiko Mori [1,3] and Reiji Kumai [4]

[1] Department of Materials Science and Engineering, Tokyo Institute of Technology, Tokyo 152-8552, Japan; mori.t.ae@m.titech.ac.jp
[2] Department of Organic and Polymeric Materials, Tokyo Institute of Technology, Tokyo 152-8552, Japan; kohei.kurata@outlook.com
[3] ACT-C JST, Honcho, Kawaguchi, Saitama 332-0012, Japan
[4] Condensed Matter Research Center and Photon Factory, Institute of Materials Structure Science, High Energy Accelerator Research Organization (KEK), Tsukuba, Ibaraki 305-0801, Japan; reiji.kumai@kek.jp
* Correspondence: kawamoto@o.cc.titech.ac.jp; Tel.: +81-3-5734-3657

Academic Editor: Manuel Almeida
Received: 13 January 2017; Accepted: 22 February 2017; Published: 1 March 2017

Abstract: Structural, transport, and magnetic properties of new organic conductors composed of (BEDT-TTF)$_2$TaF$_6$, where BEDT-TTF is bis(ethylenedithio)tetrathiafulvalene, have been investigated. Two δ-type polymorphs, monoclinic and orthorhombic phases are obtained by the electrocrystallization. Both phases show a semiconductor-insulator phase transition at 276 K and 300 K for the monoclinic and orthorhombic phases, respectively; the ground state of both salts is a nonmagnetic insulating state. The low-temperature X-ray diffraction measurements show two-fold superlattice reflections in the intercolumnar direction. The low-temperature crystal structures show a clear charge ordered state, which is demonstrated by the molecular shape and intramolecular bond lengths. The observed checkerboard charge ordered state is in agreement with the charge ordering in a dimer Mott insulator. If we distinguish between the monoclinic and orthorhombic phases, the transition temperature of the δ-type (BEDT-TTF)$_2$MF$_6$ conductors (M = P, As, Sb, and Ta) increases continuously with increasing the anion volume.

Keywords: molecular conductor; electronic correlation; charge order

1. Introduction

Among molecular conductors, bis(ethylenedithio)tetrathiafulvalene (BEDT-TTF) salts with octahedral anions, so-called β-(BEDT-TTF)$_2$PF$_6$ type salts, have a unique arrangement of the donor molecules with a twisted stacking structure (Figure 1) [1–3]. This molecular arrangement is categorized into the δ type; the donor column consists of alternate stacking of two interaction modes: the twisted stacking mode ($a1$ in Figure 1c) and the parallel stacking mode ($a2$) slipped along the molecular short axis (ring-over-atom mode) [3]. The highly conducting direction is not the donor stacking direction but the intercolumnar direction [1]. After the discovery of the PF$_6$ salt, several δ-type BEDT-TTF salts with octahedral anions have been developed. The crystal system, however, differs between the PF$_6$ (orthorhombic) and SbF$_6$ (monoclinic) salts despite the same anion shape [4]. The AsF$_6$ salt has been known as a monoclinic compound, but later the orthorhombic crystal has been found [4–7].

The PF$_6$ salt has been investigated using X-ray diffraction and Raman spectra; the phase transition at around room temperature is not a $2k_F$ charge-density-wave but a charge order [8,9]. The charge ordering transition is due to the nearest neighbor Coulomb repulsion V [10–12]. The orthorhombic AsF$_6$ salt also shows a similar charge order to that of the PF$_6$ salt [8]. The charge order has been observed in both the monoclinic AsF$_6$ and SbF$_6$ salts using Raman spectra [13]. Although many investigations of the δ-type BEDT-TTF salts with octahedral anions have been carried out, the charge ordering transition temperature does not seem to change in the sequence of the anion volume.

In order to clarify the relationship between the transition temperature and the anion volume, the TaF_6 salt is prepared; TaF_6 is the largest anion among MF_6 anions ($M = $ P, As, Sb, and Ta) [14]. In the present paper, we report two δ-type polymorphs of the TaF_6 salts, monoclinic (δ_m) and orthorhombic (δ_o) phases. Although both salts show a charge ordering transition, the transition temperature of the orthorhombic phase is higher than that of the monoclinic phase. If we distinguish between the monoclinic and orthorhombic phases in the δ-type $(BEDT-TTF)_2 MF_6$ conductors, the transition temperature increases with increasing the anion volume; this indicates that the chemical pressure affects the charge ordering transition temperature.

2. Results

The crystallographic data of both δ_m- and δ_o-$(BEDT-TTF)_2 TaF_6$ at 298 K are shown in Table 1. For the monoclinic δ_m phase, we choose the space group $C2/c$. Although unusual space group $I2/c$ has been used in the SbF_6 salt [4], the unit cell transformation $a' = -a - c$ gives the monoclinic angle $\beta' = 93.99(4)°$ and the reported lattice. In the AsF_6 salt, another unit cell transformation gives the reported $A2/a$ lattice [5]. The orthorhombic δ_o-TaF_6 salt and the PF_6 salt are isostructural, and the space group is $Pnna$. The unit cell volume of the monoclinic phase is slightly larger than that of the orthorhombic phase.

Table 1. Crystallographic data of the $(BEDT-TTF)_2 TaF_6$ salts.

	δ_m **Phase**		δ_o **Phase**	
Formula		$C_{20}H_{16}S_{16}TaF_6$		
Formula weight		1064.24		
Temperature (K)	298	171	298	39
Crystal system	Monoclinic	Monoclinic	Orthorhombic	Monoclinic
Space group	$C2/c$	$P2_1/n$	$Pnna$	$P2/c$
a (Å)	35.825(12)	35.6108(6)	14.924(9)	13.2594(5)
b (Å)	6.679(4)	13.2887(2)	33.366(14)	14.5036(4)
c (Å)	14.953(8)	14.7148(3)	6.656(2)	33.8724(7)
β (deg)	110.62(3)	110.1387(8)	–	101.206(2)
V (Å^3)	3349(3)	6537.7(2)	3315(3)	6389.8(3)
Z	4	8	4	8
Total reflections	4877	11899	2921	27208
Reflections [$F^2 > 2\sigma(F^2)$]	3571	7889	1558	20909
$R1$ [$F^2 > 2\sigma(F^2)$]	0.0327	0.0818	0.0748	0.1387
$wR2$ (all reflections)	0.0944	0.2172	0.2498	0.3242
GOF	1.028	1.128	1.026	1.231

Figure 1a–d show the crystal structure of the TaF_6 salts. These two phases take similar δ-type structures. For the monoclinic phase, the BEDT-TTF molecular planes of the adjacent conducting layers along the a-axis are almost parallel to each other (Figure 1a). In the orthorhombic phase, however, donor planes are inclined alternately along the interlayer direction (Figure 1b). There is one crystallographically independent donor and a half anion, and a unit cell contains eight donors and four anions, affording the donor to anion ratio 2:1. The terminal donor ethylene-groups are disordered. The donors form a stack along the c-axis for the δ_m phase (a-axis for the δ_o phase) with the twisted mode $a1$ and the ring-over-atom mode $a2$. For the $a1$ mode, the interplanar distance is $d_\perp = 3.58$ Å and the twist angle is $\omega = 29.9°$ in the δ_m phase, and $d_\perp = 3.57$ Å and $\omega = 29.9°$ in the δ_o phase. For the $a2$ mode, the slip distance along the molecular short axis is $d_s = 2.00$ Å, and $d_\perp = 3.67$ Å in the δ_m phase. These values are $d_s = 2.02$ Å and $d_\perp = 3.67$ Å in the δ_o phase.

An anion is located on an inversion center for the δ_m phase. In the δ_o phase, an anion is on a two-fold rotation axis parallel to the a-axis, and fluorine atoms are on general positions.

The arrangement of anions is an isosceles triangle on an anion layer parallel to the *bc*-plane (*ac*-plane) in the δ_m (δ_o) phase. An interaction between the donor hydrogen and anion fluorine atoms shorter than the sum of the van der Waals radii (2.67 Å) [15], a so-called hydrogen bond, is found. Although both terminal ethylene groups of the donor have hydrogen bonds in the δ_o phase, one side terminal of the donor has a hydrogen bond in the δ_m phase. This is due to the difference of the unit cell volumes because the unit cell of the δ_o phase is smaller than that of the δ_m phase.

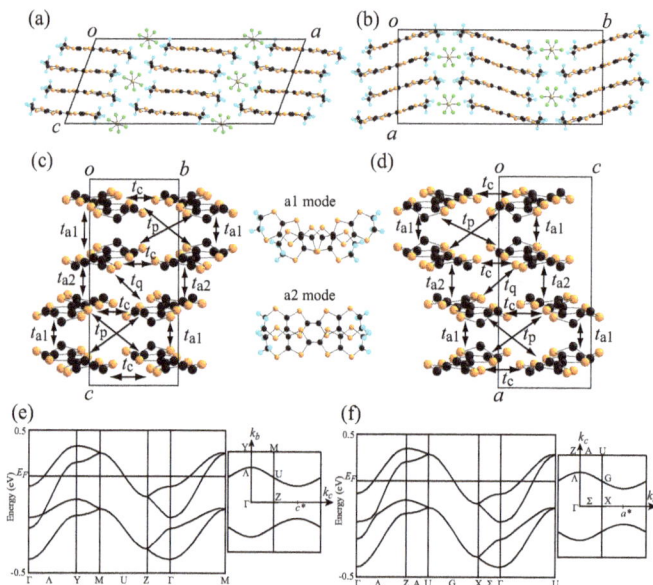

Figure 1. Crystal structure of δ_m-(BEDT-TTF)$_2$TaF$_6$ (**a**) and the δ_o phase (**b**) projected approximately along the molecular short axis. The donor layer of the δ_m phase (**c**) and the δ_o phase (**d**) projected approximately along the molecular long axis. The energy band structure and the Fermi surface of the δ_m phase (**e**) and the δ_o phase (**f**).

The calculated transfer integrals, the distance between the molecular centers *R*, and the twist angles ω are shown in Table 2. All interaction modes (*c*, *a1*, *a2*, *p*, and *q*) are the same as those of β-(BEDT-TTF)$_2$PF$_6$ [2,3]. The transfer integral of the twisted interaction mode t_{a1} is larger than that of the ring-over-atom mode t_{a2}; this twisted dimer structure is characteristic of this structure. The transverse interactions are large, and t_q is larger than t_{a1}. The energy band structure and Fermi surface are shown in Figure 1e,f. The quasi-one-dimensional Fermi surface is shown in an extended zone scheme because the energy bands degenerate at the zone boundary owing to the crystallographic symmetries.

Figure 2 shows the temperature dependence of the electrical resistivity. Both salts exhibit semiconducting behavior in the measurement temperature region, and show anomaly at approximately 300 K. The phase transition temperature T_c is determined from the peak of $d(\ln \rho)/d(1/T)$ at $T_c = 276$ K and $T_c = 300$ K for the δ_m and δ_o phases, respectively. As shown in the inset of Figure 2, the charge activation energy Δ_{charge} is estimated as 27 and 46 meV for the δ_m and δ_o phases, respectively.

Table 2. Transfer integrals and geometrical parameters of the δ-type (BEDT-TTF)$_2$TaF$_6$ salts.

Interaction Mode	t (meV)	R (Å)	ω (deg)
δ_m phase			
c	−75.6	6.68	
$a1$	−103.3	4.49	29.9
$a2$	38.4	4.18	
p	−23.8	8.05	
q	−147.8	5.72	
δ_o phase			
c	−79.4	6.66	
$a1$	−95.9	4.51	29.9
$a2$	37.1	4.19	
p	−24.5	8.04	
q	−141.2	5.72	

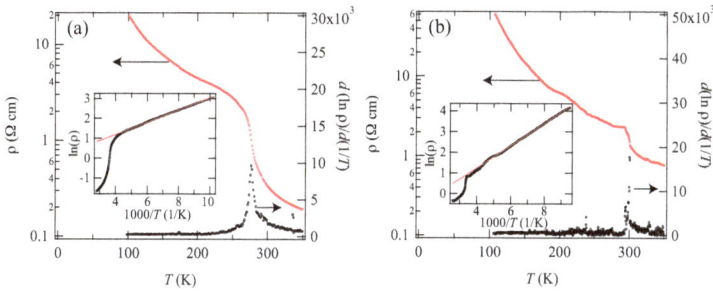

Figure 2. Temperature dependence of the resistivity of the δ_m phase (**a**) and the δ_o phase (**b**). The insets show the Arrhenius plots, and solid lines in the insets are fits to the data.

The Seebeck coefficient of the δ_m phase shows metallic behavior around 300 K, and has anomaly at approximately 280 K (Figure 3). For the δ_o phase, the thermopower shows semiconducting behavior even in the high temperature region and has no anomaly at 300 K where the resistivity has an anomaly. The value of the thermopowers around 300 K is approximately 52 μV/K. This large value is not in agreement with the energy band structure, indicating a strongly correlated electronic system [1,2,16,17].

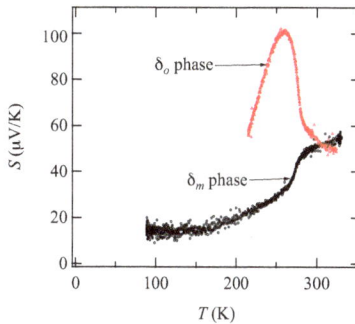

Figure 3. Temperature dependence of Seebeck coefficients of the δ_m and δ_o phases.

Figure 4 shows the temperature dependence of the electron spin resonance (ESR) g-values, peak-to-peak linewidths, and the normalized spin susceptibilities measured under the magnetic field perpendicular to the conducting layers. A single Lorentzian lineshape is observed for both phases.

The spin susceptibility of the δ_m phase clearly decreases below 280 K without a divergent increase of the linewidth, leading to a spin singlet state. The spin susceptibility of the δ_o phase gradually decreases with decreasing temperature, and the spin singlet state appears at low temperatures. However, the phase transition temperature is not clear. The g-values are almost independent of the temperature around the phase transition temperature. The spin activation energy Δ_{spin} is estimated from the simple singlet–triplet model given by $\chi_{spin} \propto (1/T) \exp(-\Delta_{spin}/T)$ in the 100–200 K range. The spin activation energy Δ_{spin} is about 30 meV. Δ_{spin} is in rough agreement with Δ_{charge} estimated from the electrical resistivity.

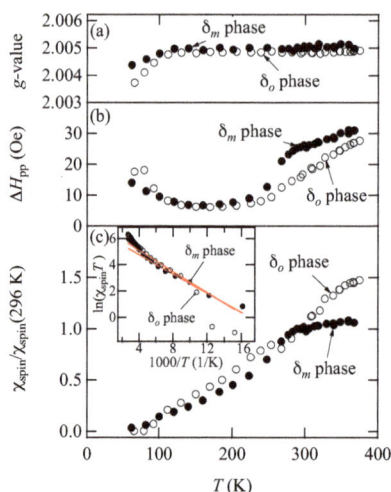

Figure 4. Temperature dependence of ESR g-values (**a**), linewidths (**b**), and the relative spin susceptibilities (**c**) of the δ_m and δ_o phases. The applied magnetic field is perpendicular to the conducting layer for both compounds. The inset shows the Arrhenius plots, and solid lines in the inset are fits to the data.

In order to clarify the origin of the phase transition, the low-temperature X-ray diffraction measurements were performed. Figure 5 shows the X-ray oscillation photographs of both δ_m and δ_o phases. At 171 K, superlattice reflections with the wave vector $q_m = (0, 1/2, 0)$ are observed for the δ_m phase (Figure 5a,b). Although the superlattice reflections are observed at 171 K for the δ_o phase, the wave vector is $q_o = (0, 1/2, 1/2)$ (Figure 5c,d). Figure 6 shows the temperature dependence of the integrated intensity of the superlattice reflections. The superlattice reflection disappears at 277 K and 302 K for the δ_m and δ_o phases, respectively. These results indicate that the temperatures at which the resistivity shows anomaly are the phase transition temperatures.

Figure 7a,b show the crystal structure of the δ_m phase at 171 K. The C-centered unit cell changes to a primitive cell with the space group $P2_1/n$ as shown in Table 1, which is a sub-group of $C2/c$; the superlattice unit cell is $a' = a$, $b' = 2b$, and $c' = c$ because superlattice reflections with $h + k \neq 2n$ appear for hkl, where n is an integer. Some reflections with $h + l \neq 2n$ are observed in $h0l$ reflections. However, in order to analyze the low-temperature average structure, we have chosen $P2_1/n$, because a centrosymmetric space group has been suggested by SHELXT; the program SHELXT proposes possible space groups on the basis of phases in the space group $P1$ without using systematic absences [18]. In the present space group, there are four crystallographic independent molecules along the stacking direction (Figure 7a). The molecules in the nearest neighbor columns in the conducting sheet is connected by a two-fold screw axis; this indicates that the charge ordering pattern is not related by the glide symmetry. Moreover, in the case of $P2_1$, the maximum shift/error for the parameter refinement does not converge, and the atomic coordinates of four additional independent molecules are connected by the inversion symmetry operation. Although the existence of the inversion center is important for electronic polarity, we do not treat this problem further in the present work.

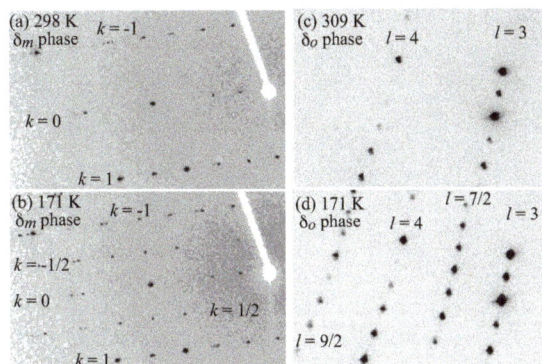

Figure 5. X-ray oscillation photographs of the δ_m phase at 298 K (**a**) and 171 K (**b**), and of the δ_o phase at 309 K (**c**) and 171 K (**d**).

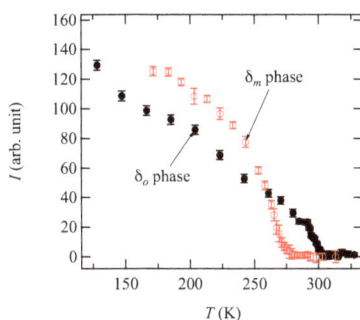

Figure 6. Temperature dependence of the integrated intensity of superlattice reflections. The background intensities are subtracted. Although integrated intensity is measured using a four circle diffractometer for the δ_o phase, this is measured using an imaging plate for the δ_m phase. The indices of the used reflections are $4, -1/2, -3$ for the δ_m phase, and $8, -3/2, -1/2$ for the δ_o phase.

Figure 7c,d show the crystal structure of the δ_o phase at 39 K. Some reflections with $l \neq 2n$ for $h0l$ are observed. However, we have chosen $P2/c$ as suggested by SHELXT owing to the same reason as the δ_m phase (Table 1). This space group is a sub-group of $Pnna$, where the superlattice is given by $a' = 2c$, $b' = a$, and $c' = -c - b$ on the basis of the superlattice wave vector $q_o = (0, 1/2, 1/2)$. The n-glide in the original cell corresponds to the c-glide in the new unit cell. In the δ_o phase, molecules in the nearest neighbor column in the conducting sheet are related by the inversion symmetry. If the space group has no inversion center ($P2$), the nearest neighbor column is composed of four other independent molecules. However, in the case of $P2$, the analysis does not converge similarly to $P2_1$ in the δ_m phase.

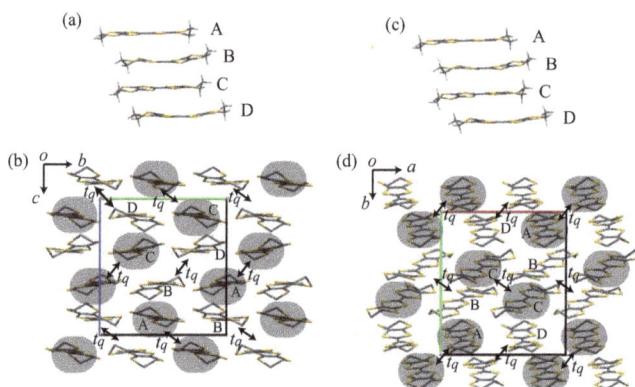

Figure 7. Crystal structure of the δ_m phase at 171 K (**a,b**) and of the δ_o phase at 39 K (**c,d**). Four crystallographically independent molecules (A, B, C, and D) exist for both phases (**a,c**). A and C (B and D) molecules are flat (bent). Black rounded rectangles denote hole-rich molecules, and t_q is the interaction mode q in the high temperature structure (**b,d**).

A terminal ethylene group at one side is disordered except for the D molecule in the δ_m phase. The A and C molecules are disordered in the δ_o phase. The completely ordered structure has been observed in the PF$_6$ and δ_o-AsF$_6$ salts. However, this conformational ordering depends on the cooling speed [8]. The cooling speed approximately 1–3 K/min is too fast to attain a completely ordered structure. Although A and C molecules are flat, B and D molecules are bent for both δ_m and δ_o phases (Figure 7a,c); the bent TTF skeleton is characteristic of the neutral BEDT-TTF molecule [19]. The BEDT-TTF molecules have a boat-like structure. The dihedral angles between the central and outer tetrathio-substituted ethylene moieties are approximately 2.4°–3.0° in the A and C molecules. These are approximately 8.5°–13.7° in the B and D molecules in the δ_m phase, and 7.0°–10.9° in the δ_o phase. This clearly shows that the low-temperature phase is a charge ordered state. The charges of the BEDT-TTF molecules are estimated empirically from the bond lengths [20]: $Q_A = +0.8(1)$, $Q_B = +0.1(1)$, $Q_C = +0.6(1)$, and $Q_D = -0.1(1)$ in the δ_m phase. In the δ_o phase, the charges are $Q_A = +0.8(2)$, $Q_B = +0.5(2)$, $Q_C = +0.7(2)$, and $Q_D = +0.2(2)$. The estimated Q values are in agreement with the molecular shape; the flat A and C molecules are hole rich, whereas the bent B and C molecules are nearly neutral. The calculated energy levels of the highest occupied molecular orbital (HOMO) of these molecules show that the averaged HOMO level of the bent molecules is lower than that of the flat molecules by ~0.1 eV. This is in agreement with the charge disproportionation between the flat and bent molecules. The hole-rich donor molecules are connected by the transfer integral t_q with the largest absolute value in the high temperature structure (Figure 7b,d); then, a spin-singlet state is realized. Although A and C molecules are connected in the δ_m phase ($R_q^{AC} = 5.71$ Å), A and A or C and C molecules are connected in the δ_o phase ($R_q^{AA} \approx R_q^{CC} = 5.70$ Å). The absolute value of the transfer integral between A and C molecules $|t_q^{AC}|$ increases (174.7 meV), but $|t_q^{BD}|$ decreases (114.1 meV) despite $R_q^{BD} < R_q^{AC}$ in the δ_m phase. This is due to the molecular shape; B and D molecules are bent. The side-by-side molecular center distance is slightly modulated; the averaged difference is approximately 0.10 Å. These tendencies are the same as those in the δ_o phase. The anion arrangement is also changed; the isosceles triangle pattern is deformed. There are many hydrogen bonds between the donors and anions in the low-temperature structure. The short hydrogen bonds ($\leq$2.50 Å) are observed in the B and D molecules for both phases.

3. Discussion

The difference between the δ_m and δ_o phases is the inclination of the donor plane in a conducting layer. In the δ_o phase, the plane inclination changes alternately along the normal direction to the conducting layer. The same difference is found in the AsF$_6$ salts, which indicates that the monoclinic phase of the AsF$_6$ salt differs from the orthorhombic phase discovered by Senadeera et al. [4–7]. Nogami et al. have confirmed the lattice constants of the orthorhombic phase of the AsF$_6$ salt and solved the structure at 110 K including the superlattice reflections; the superlattice unit cells of both δ_o-PF$_6$ and δ_o-AsF$_6$ salts are also orthorhombic [8]. Leung et al. have found two-fold superlattice reflections in the monoclinic AsF$_6$ salt above 125 K, where the wave vector $b^*/2$ is the same as that of the δ_m-TaF$_6$ salt. The Raman spectra of the monoclinic AsF$_6$ salt show the charge ordering at approximately 260 K [13]. Therefore, we have to distinguish between the monoclinic and orthorhombic phases in the AsF$_6$ salts; the charge ordering transition temperatures are $T_c = 264$ K and 298 K for the monoclinic and orthorhombic phases, respectively. The transition temperature T_c of the orthorhombic phases is generally higher than that of the monoclinic phases in the other δ-type (BEDT-TTF)$_2$MF$_6$ salts as shown in Table 3. If we distinguish the crystal system, the transition temperature increases independently with increasing the anion size (Figure 8). This is in agreement with the concept of chemical pressure; the smaller the anion volume, the higher the pressure region. Actually, the transition temperature decreases with increasing the external pressure in the PF$_6$ salt [6].

Table 3. Summary of the δ-type (BEDT-TTF)$_2$MF$_6$ salts.

	δ_m-PF$_6$	δ_o-PF$_6$	δ_m-AsF$_6$	δ_o-AsF$_6$	δ_m-SbF$_6$	δ_o-SbF$_6$	δ_m-TaF$_6$	δ_o-TaF$_6$
existence	[b]	yes	yes	yes	yes	[b]	yes	yes
spin singlet	–	yes	yes	yes	yes	–	yes	yes
superlattice [a]	–	$c^*/2$	$b^*/2$	$c^*/2$	[b]	–	$b^*/2$	$b^*/2 + c^*/2$
charge ordering	–	yes	yes	yes	yes	–	yes	yes
T_c (K)	–	297	264	~ 298 [c]	273	–	276	300
Ref.	–	[1,6,8,9]	[4,5,13]	[6,8]	[4,13]	–	[d]	[d]

[a] For the δ_m-AsF$_6$ salt, the unit cell is transformed to that of the TaF$_6$ salt, C2/c; [b] unknown; [c] T_c is estimated from the temperature dependence of the superlattice reflection in Ref. [6]; [d] present work.

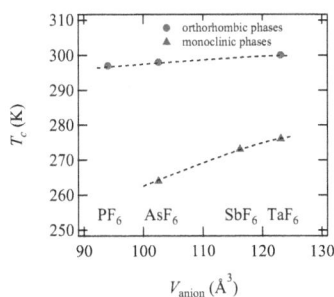

Figure 8. Anion volume dependence of the charge ordering temperature of δ-type (BEDT-TTF)$_2$MF$_6$ (M = P, As, Sb, and Ta). Anion volumes are estimated using Figure 2 in Ref. [14]. Dashed lines are guides to the eye.

The relationship between the charge order and anion has been investigated in α-(BEDT-TTF)$_2$I$_3$ [21]. The hydrogen bonds of the hole-rich donors are shorter than those of the hole-poor donors in α-(BEDT-TTF)$_2$I$_3$; the hydrogen bonds between the donors and anions affect the charge ordering transition. In the present compounds, however, the donors with short hydrogen bonds are hole poor molecules, i.e., bent neutral molecules. This indicates that the anion shift does not affect the present phase transition. Moreover, the shift of the donors is also small; this molecular displacement is known as a bond order wave [22,23]. The slight displacement is not dominant in the phase transition that leads to the charge ordered state. The above results show that the origin of the phase transition is due to the electronic correlation between the donor molecules.

The experimentally obtained charge ordered pattern is estimated from the high temperature crystal structure. In the θ-type BEDT-TTF compounds, theoretical investigation has revealed several charge ordering patterns depending on t/V, which have been verified by different kinds of experiments [10–12]. However, a simple method to expect the charge ordering pattern has been proposed, since the point charge approximation is appropriate; V is proportional to $1/R$ for $R \geq 4$ Å, where R is the distance between the molecular centers [24]. Table 4 shows the transfer integrals and intermolecular distances in the high temperature structure. This shows a trend for the molecular center distances ($R_{a2} < R_{a1} < R_q < R_c < R_p$), and $1/R_{a1} \sim 1/R_{a2}$. Therefore, we use V_a as the stacking direction V and others (V_q, V_c, and V_p) to calculate the potential energy for several charge ordering patterns by the point charge approximation (Figure 9).

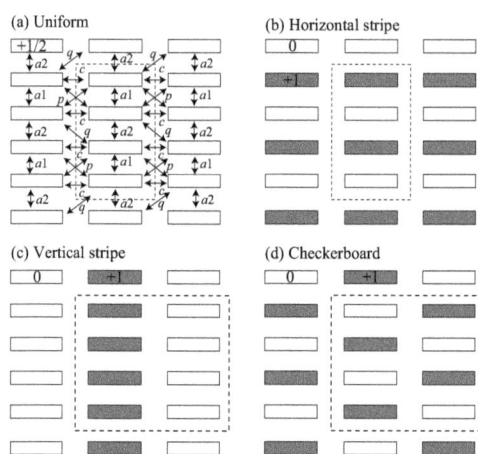

Figure 9. Charge order patterns of the δ-type salts. (**a**) uniform; (**b**) horizontal stripe; (**c**) vertical stripe; and (**d**) checkerboard. The dashed line shows a unit cell.

The extended Hubbard model is written as follows [10,12]:

$$H = - \sum_{\langle ij \rangle \sigma} (t_{ij} c_{i\sigma}^{\dagger} c_{j\sigma} + \text{h.c.}) + \sum_i U n_{i\uparrow} n_{i\downarrow} + \sum_{\langle ij \rangle} V_{ij} n_i n_j, \qquad (1)$$

where $\langle ij \rangle$ denotes pairs of the lattice sites i and j, $c_{i\sigma}^{\dagger}(c_{i\sigma})$ is the creation (annihilation) operator for a hole with spin $\sigma(=\uparrow,\downarrow)$ at the i site, and $n_{i\sigma} = c_{i\sigma}^{\dagger} c_{i\sigma}$ is the number operator with $n_i = n_{i\uparrow} + n_{i\downarrow}$. U is the onsite Coulomb repulsion, and t_{ij} and V_{ij} are the transfer integral and the nearest neighbor Coulomb repulsion between sites i and j. We neglect t_{ij} and assume the static limit in order to investigate complicated charge-order patterns [24–26]. The static-limit potential energies are listed in Table 5. The ratios $V_p/V_a = R_a/R_p \sim 0.54$, $V_c/V_a \sim 0.65$, and $V_q/V_a \sim 0.76$ give the energies on the right side (V_a energy) in Table 5. Since V_a is much larger than other Vs, the vertical pattern is unlikely. However, because of $V_c \leq V_q$, the horizontal pattern is more stable than the checkerboard in disagreement with the experiments. The checkerboard pattern is most stable when V_a and V_c make a square lattice [27]. Although V_a is obviously large, the relative importance of V_c and V_q is less clear. These stripe patterns have very close static energies, and the point charge approximation is insufficient to distinguish the stability of different stripe patterns. This is also in the case of β-(*meso*-DMBEDT-TTF)$_2$PF$_6$ with the checkerboard charge ordering, where DMBEDT-TTF stands for 2-(5,6-dihydro-1,3-dithiolo[4,5-b][1,4]dithiin-2-ylidene)-5,6-dihydro-5,6-dimethyl-1,3-dithiolo[4,5-b][1,4]dithiin [26,28].

Table 4. Transfer integrals and geometrical parameters of the δ-type (BEDT-TTF)$_2$AsF$_6$ and (BEDT-TTF)$_2$PF$_6$ salts.

Interaction Mode	t (meV)	R (Å)	ω (deg)
δ_m-AsF$_6$ [a]			
c	-77.7	6.67	
$a1$	-109.5	4.45	29.9
$a2$	33.2	4.14	
p	-23.4	8.01	
q	-156.4	5.71	
δ_o-AsF$_6$ [b]			
c	-74.8	6.69	
$a1$	-98.4	4.50	30.1
$a2$	32.3	4.20	
p	-22.4	8.07	
q	-139.5	5.76	
δ_o-PF$_6$ (From Ref. [2])			
c	-84	–	
$a1$	-104	–	–
$a2$	32	–	
p	-28	–	
q	-141	–	

[a] The atomic coordinates are in Ref. [5]; [b] The atomic coordinates are in Ref. [7].

Table 5. Potential energies of the charge-ordered patterns per unit cell in the δ-type.

Pattern	V Energy	V_a Energy
Uniform	$E_u = \frac{U}{4} + V_a + V_c + V_p + \frac{V_q}{2}$	$\frac{U}{4} + 2.57V_a$
Horizontal	$E_h = \frac{U}{2} + 2V_c$	$\frac{U}{2} + 1.3V_a$
Vertical	$E_v = \frac{U}{2} + 2V_a$	$\frac{U}{2} + 2V_a$
Checkerboard	$E_c = \frac{U}{2} + 2V_p + V_q$	$\frac{U}{2} + 1.84V_a$

We propose another explanation of the present results. The donor molecules have uniform side-by-side arrangement along the **b** direction in the δ_m-TaF$_6$ salt. This direction is the highest conducting direction, and the Fermi surface is perpendicular to the **b*** direction (Figure 1c,e). The value of $|t_{a1}|$ is smaller than $|t_q|$, and much smaller than the dimerization in the κ-type BEDT-TTF salts, where the dimer Mott insulating picture is realized ($t_{\text{dimer}} \sim 250$ meV). The molecular center distance $R_{a1} \sim 4.5$ Å is almost the same as that in the dimer of κ-type salts (~ 4.0 Å). Then, t/V of the $a1$ mode is half of the κ-type salt and comparatively small. However, t/V for q is 90% of that of the κ-type, and the dimerization due to q is important. The energy band in Figure 1 splits into two by taking the oblique dimerization due to q into account. The band splitting has been observed as the inter-band transition in the polarized reflectance spectra of the δ_o-PF$_6$ salt [29]. The present compounds show semiconducting behaviors even at 350 K sufficiently higher than the phase transition temperature. This indicates that the high temperature phase is a paramagnetic dimer Mott insulator. In the present compounds, a dimer is surrounded by eight neighboring dimers, i.e., a quasi-square lattice, where all dimers have one hole; the dimers are connected by the horizontal interaction composed of two c modes and one $a2$ mode, the vertical $a1$ mode, and the diagonal p mode. The interdimer Coulomb repulsion leads to the checkerboard type charge order $(2, 0, 2, 0)$ because of the relationship $V_p < V_c < V_a$, and the spin singlet state in the dimer is realized. The spin singlet charge ordering has been observed in the dimer Mott insulator κ-(BEDT-TTF)$_4$[Co(CN)$_6$][N(C$_2$H$_5$)$_4$]·2H$_2$O; the charged dimer composed of hole-rich flat donors and the neutral dimer of bent donors exist in the checkerboard charge ordered state [30].

The wave vector of the superlattice reflection of the δ_o-TaF$_6$ salt differs from those of other MF$_6$ salts, but this is only due to the alignment along the interlayer b-axis. The charge ordering transition temperature increases systematically as the transfer integral $|t_q|$ decreases, following a single line both for the monoclinic and orthorhombic phases (Figure 10). R_q is almost the same, and the ratio $|t_q|/V_q$ increases as $|t_q|$ increases. This suggests the importance of the dimerization q [3].

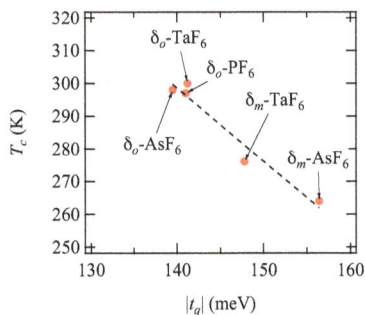

Figure 10. Charge ordering temperatures as a function of $|t_q|$. The dashed line shows linear dependence.

4. Materials and Methods

Single crystals of (BEDT-TTF)$_2$TaF$_6$ were grown by electrocrystallization. The crystal structures were determined by the X-ray single crystal structure analyses. The X-ray diffraction measurements were made on a Rigaku AFC7R four-circle diffractometer (Rigaku Corporation, Tokyo, Japan) with graphite monochromated Mo-$K\alpha$ radiation and a rotating anode generator ($\lambda = 0.71069$ Å). The X-ray oscillation photographs above 170 K were taken using a Rigaku R-AXIS RAPID II imaging plate with Cu-$K\alpha$ radiation from a rotating anode source with a confocal multilayer X-ray mirror (RIGAKU VM-Spider, $\lambda = 1.54187$ Å). For the low-temperature X-ray measurements, the samples were cooled by a nitrogen gas-stream cooling method. Low-temperature X-ray oscillation photographs at 39 K were taken using an imaging plate with Si monochromated synchrotron radiation ($\lambda = 0.99884(2)$ Å) at BL-8B of the Photon Factory, KEK, Tsukuba, Japan; the wavelength was calibrated using CeO$_2$. The sample was cooled by a helium gas-stream cooling method. The structures were solved by direct methods (SIR2008 and SHELXT) and refined by the full-matrix least-squares procedure (SHELXL) [18,31,32]. Crystallographic data have been deposited with Cambridge Crystallographic Data Center: deposition numbers CCDC 1526712 and 1526713. The energy band structures were calculated on the basis of the molecular orbital calculation and tight-binding approximation [33].

The electrical resistivities were measured by the four-probe method with low-frequency AC current (1.0 µA). Lock-in amplifiers were used for high-sensitivity detection. Thermoelectric power measurements were carried out by the two terminal method [34]. Electron spin resonance (ESR) spectra were measured using a conventional X-band spectrometer (JEOL JES-TE100, Tokyo, Japan). The sweep width of the magnetic field and the g-values were calibrated by the spectra of Mn^{2+}/MgO with a hyperfine structure constant of 86.77 Oe and g_0 of 2.00094. All measurements were performed after examining the lattice parameters using X-ray oscillation photographs.

5. Conclusions

Two polymorphs of (BEDT-TTF)$_2$TaF$_6$ have the same type of δ-type structure, and only the crystal system differs. The present compounds are categorized into a dimer Mott insulator because of the semiconducting behavior even at 350 K. The charge ordering transition is observed at 276 K and 300 K for the monoclinic (δ_m) and the orthorhombic (δ_o) phases, respectively. The bent and flat BEDT-TTF molecules exist in the low-temperature insulting phase, indicating the checkerboard charge ordering. The checkerboard pattern results in a spin singlet state due to the oblique dimerization q mode and the interdimer Coulomb repulsion. If we distinguish the crystal system, the charge ordering transition temperature increases systematically with increasing of the anion size.

Acknowledgments: This work was partially performed under the approval of the Photon Factory Program Advisory Committee (Proposal No. 2014G089) and supported by Japan Society for the Promotion of Science (JSPS) KAKENHI Grant Numbers 24540364 and 16K05436.

Author Contributions: T.K. and T.M. conceived and designed the experiments; K.K. and T.K. performed the experiments; K.K. and T.K. analyzed the data; R.K. contributed experiment and analysis tools for diffraction measurements using synchrotron radiation; and T.K. and T.M. wrote the paper.

Conflicts of Interest: The authors declare no conflict of interest.

References

1. Kobayashi, H.; Mori, T.; Kato, R.; Kobayashi, A.; Sasaki, Y.; Saito, G.; Inokuchi, H. Transverse conduction and metal-insulator transition in β-(BEDT-TTF)$_2$PF$_6$. *Chem. Lett.* **1983**, *12*, 581–584.
2. Mori, T.; Kobayashi, A.; Sasaki, Y.; Kato, R.; Kobayashi, H. Band structure of β-(BEDT-TTF)$_2$PF$_6$. One-dimensional metal along the side-by-side molecular array. *Solid State Commun.* **1985**, *53*, 627–631.
3. Mori, T. Structural genealogy of BEDT-TTF-based organic conductors III. twisted molecules: δ and α′ phases. *Bull. Chem. Soc. Jpn.* **1999**, *72*, 2011–2027.
4. Laversanne, R.; Amiell, J.; Delhaes, P.; Chasseau, D.; Hauw, C. A metal-insulator phase transition close to room temperature: (BEDT-TTF)$_2$SbF$_6$ and (BEDT-TTF)$_2$AsF$_6$. *Solid State Commun.* **1984**, *52*, 177–181.
5. Leung, P.C.; Beno, M.A.; Blackman, G.S.; Coughlin, B.R.; Miderski, C.A.; Joss, W.; Crabtree, G.W.; Williams, J.M. Structure of semiconducting 3,4;3′,4′-bis(ethylenedithio)-2,2′,5,5′-tetrathiafulvalene-hexafluoroarsenate (2:1), (BEDT-TTF)$_2$AsF$_6$, (C$_{10}$H$_8$S$_8$)$_2$AsF$_6$. *Acta Crystallogr. Sect. C* **1984**, *40*, 1331–1334.
6. Senadeera, G.K.R.; Kawamoto, T.; Mori, T.; Yamaura, J.; Enoki, T. 2k_F CDW transition in β-(BEDT-TTF)$_2$PF$_6$ family salts. *J. Phys. Soc. Jpn.* **1998**, *67*, 4193–4197.
7. Senadeera, G.K.R. *Report for "33rd International Course for Advanced Research Chemistry and Chemical Engineering"*; Tokyo Tech./United Nations Educational, Scientific and Cultural Organization (UNESCO): Tokyo, Japan, 1998.
8. Nogami, Y.; Mori, T. Unusual 2k_F CDW state with enhanced charge ordering in β-(BEDT-TTF)$_2$AsF$_6$ and PF$_6$. *J. Phys. IV Fr.* **2002**, *12*, 233–234.
9. Ding, Y.; Tajima, H. Optical study on the charge ordering in the organic conductor β-(BEDT-TTF)$_2$PF$_6$. *Phys. Rev. B* **2004**, *69*, 115121.
10. Seo, H. Charge ordering in organic ET compounds. *J. Phys. Soc. Jpn.* **2000**, *69*, 805–820.
11. Takahashi, T.; Nogami, Y.; Yakushi, K. Charge ordering in organic conductors. *J. Phys. Soc. Jpn.* **2006**, *75*, 051008.
12. Seo, H.; Merino, J.; Yoshioka, H.; Ogata, M. Theoretical aspects of charge ordering in molecular conductors. *J. Phys. Soc. Jpn.* **2006**, *75*, 051009.
13. Kowalska, A.; Wojciechowski, R.; Ulanski, J. Phase transitions in β-(BEDT-TTF)$_2$XF$_6$ (X = P, Sb or As) salts as seen by Raman spectroscopy. *Mater. Sci. Pol.* **2004**, *22*, 353–358.
14. Williams, J.M.; Beno, M.A.; Sullivan, J.C.; Banovetz, L.M.; Braam, J.M.; Blackman, G.S.; Carlson, C.D.; Greer, D.L.; Loesing, D.M.; Carneiro, K. Role of monovalent anions in organic superconductors. *Phys. Rev. B* **1983**, *28*, 2873–2876.
15. Bondi, A. van der Waals Volumes and Radii. *J. Phys. Chem.* **1964**, *68*, 441–451
16. Chaikin, P.M.; Beni, G. Thermopower in the correlated hopping regime. *Phys. Rev. B* **1976**, *13*, 647–651.
17. Kwak, J.F.; Beni, G. Thermoelectric power of a Hubbard chain with arbitrary electron density: Strong-coupling limit. *Phys. Rev. B* **1976**, *13*, 652–657.
18. Sheldrick, G.M. SHELXT—Integrated space-group and crystal structure determination. *Acta Crystallogr. Sect. A* **2015**, *71*, 3–8.
19. Kobayashi, H.; Kobayashi, A.; Sasaki, Y.; Saito, G.; Inokuchi, H. The crystal and molecular structures of bis(ethylenedithio)tetrathiafulvalene. *Bull. Chem. Soc. Jpn.* **1986**, *59*, 301–302.
20. Guionneau, P.; Kepert, C.J.; Bravic, G.; Chasseau, D.; Truter, M.R.; Kurmoo, M.; Day, P. Determining the charge distribution in BEDT-TTF salts. *Synth. Met.* **1997**, *86*, 1973–1974.

21. Alemany, P.; Pouget, J.P.; Canadell, E. Essential role of anions in the charge ordering transition of α-(BEDT-TTF)$_2$I$_3$. *Phys. Rev. B* **2012**, *85*, 195118.
22. Pouget, J.P. Bond and charge ordering in low-dimensional organic conductors. *Physica B* **2012**, *407*, 1762–1770.
23. Pouget, J.P. Interplay between electronic and structural degrees of freedom in quarter-filled low dimensional conductors. *Physica B* **2015**, *460*, 45–52.
24. Mori, T. Estimation of off-site Coulomb integrals and phase diagrams of charge ordered states in the θ-phase organic conductors. *Bull. Chem. Soc. Jpn.* **2000**, *73*, 2243–2253.
25. Mori, T. Non-stripe charge order in the θ-phase organic conductors. *J. Phys. Soc. Jpn.* **2003**, *72*, 1469–1475.
26. Mori, T. Non-stripe charge order in dimerized organic conductors. *Phys. Rev. B* **2016**, *93*, 245104.
27. Ohta, Y.; Tsutsui, K.; Koshibae, W.; Maekawa, S. Exact-diagonalization study of the Hubbard model with nearest-neighbor repulsion. *Phys. Rev. B* **1994**, *50*, 13594–13602.
28. Kimura, S.; Suzuki, H.; Maejima, T.; Mori, H.; Yamaura, J.; Kakiuchi, T.; Sawa, H.; Moriyama, H. Checkerboard-type charge-ordered state of a pressure-induced superconductor, β-(*meso*-DMBEDT-TTF)$_2$PF$_6$. *J. Am. Chem. Soc.* **2006**, *128*, 1456–1457.
29. Tajima, H.; Yakushi, K.; Kuroda, H.; Saito, G. Polarized reflectance spectrum of β-(BEDT-TTF)$_2$PF$_6$. *Solid State Commun.* **1985**, *56*, 251–254.
30. Ota, A.; Ouahab, L.; Golhen, S.; Yoshida, Y.; Maesato, M.; Saito, G.; Świetlik, R. Phase transition from Mott insulating phase into the charge ordering phase with molecular deformation in charge-transfer salts κ-(ET)$_4$[M(CN)$_6$][N(C$_2$H$_5$)$_4$]·2H$_2$O (M = Co^{III} and Fe^{III}). *Chem. Mater.* **2007**, *19*, 2455–2462.
31. Burla, M.C.; Caliandro, R.; Camalli, M.; Carrozzini, B.; Cascarano, G.L.; Caro, L.D.; Giacovazzo, C.; Polidori, G.; Siliqi, D.; Spagna, R. *IL MILIONE*: A suite of computer programs for crystal structure solution of proteins. *J. Appl. Crystallogr.* **2007**, *40*, 609–613.
32. Sheldrick, G.M. Crystal structure refinement with SHELXL. *Acta Crystallogr. Sect. C* **2015**, *71*, 3–8.
33. Mori, T.; Kobayashi, A.; Sasaki, Y.; Kobayashi, H.; Saito, G.; Inokuchi, H. The intermolecular interaction of tetrathiafulvalene and bis(ethylenedithio)tetrathiafulvalene in organic metals. Calculation of orbital overlaps and models of energy-band structures. *Bull. Chem. Soc. Jpn.* **1984**, *57*, 627–633.
34. Chaikin, P.M.; Kwak, J.F. Apparatus for thermopower measurements on organic conductors. *Rev. Sci. Instrum.* **1975**, *46*, 218–220.

magnetochemistry

MDPI

Article

The Highly Conducting Spin-Crossover Compound Combining Fe(III) Cation Complex with TCNQ in a Fractional Reduction State. Synthesis, Structure, Electric and Magnetic Properties

Yuri N. Shvachko [1,*], Denis V. Starichenko [1,*], Aleksander V. Korolyov [1], Alexander I. Kotov [2], Lev I. Buravov [2], Vladimir N. Zverev [3,4], Sergey V. Simonov [3], Leokadiya V. Zorina [3,*] and Eduard B. Yagubskii [2,*]

[1]   M.N. Miheev Institute of Metal Physics, Ural Branch of Russian Academy of Sciences, S. Kovalevskaya str., 18, Yekaterinburg 620137, Russia; korolyov@imp.uran.ru
[2]   Institute of Problems of Chemical Physics, Russian Academy of Sciences, Semenov ave., 1, Chernogolovka 142432, Moscow Region, Russia; kotov@icp.ac.ru (A.I.K.); buravov@icp.ac.ru (L.I.B.)
[3]   Institute of Solid State Physics, Russian Academy of Sciences, Ossipyan Street, 2, Chernogolovka 142432, Moscow Region, Russia; zverev@issp.ac.ru (V.N.Z.); simonovsv@rambler.ru (S.V.S.)
[4]   Department of General and Applied Physics, Moscow Institute of Physics and Technology, Dolgoprudnyi 141700, Moscow Region, Russia
*   Correspondence: yuri.shvachko@gmail.com (Y.N.S.); starichenko@imp.uran.ru (D.V.S.); zorina@issp.ac.ru (L.V.Z.); yagubski@icp.ac.ru (E.B.Y.); Tel.: +74965228386 (Y.N.S. & D.V.S. & L.V.Z.); +74965221185 (E.B.Y.)

Academic Editor: Manuel Almeida
Received: 23 January 2017; Accepted: 14 February 2017; Published: 22 February 2017

Abstract: Three systems $[Fe(III)(sal_2\text{-trien})](TCNQ)_n \cdot X$ (n = 1, 2, X = MeOH, CH_3CN, H_2O) showing spin-crossover transition, conductivity and ferromagnetic coupling were synthesized and studied by X-ray diffraction, Montgomery method for resistivity, SQUID magnetometry and X-band EPR. Spin-spin interactions between local magnetic moments of Fe(III) ions and electron spins of organic TCNQ network were discovered and discussed within the framework of intermolecular superexchange coupling.

Keywords: TCNQ radical anion salt; Fe(III) spin-crossover complex; crystal structure; conducting and magnetic properties

1. Introduction

The synthesis and investigation of multifunctional molecule materials combining conductivity and magnetism in the same crystal lattice attract considerable attention, because the interplay of these properties may lead to novel behavior [1–6]. Until recently, research in this direction has been focused on the family of the quasi-two-dimensional (super)conductors based on the radical cation salts of bis(ethylenedithio)tetrathiafulvalene (BEDT-TTF) and its derivatives with paramagnetic metal complex anions of different nature [2–6]. In such materials, conductivity is associated with mobile electrons in organic layers, whereas magnetism usually originates from localized spins of transition metal ions in insulating counterion layers. In particular, salts of BEDT-TTF and its selenium-substituted derivative bis(ethylenedithio)tetraselenafulvalene (BETS) have been shown to combine (super)conducting and paramagnetic [7–10] and even antiferromagnetic [11,12] and ferromagnetic properties [13]. Moreover, interaction between localized spins in insulating magnetic layers and itinerant spins in conducting organic layers was found to lead to new fascinating phenomena such as field-induced superconductivity observed on λ-$(BETS)_2FeCl_4$ [14] and κ-$(BETS)_2FeBr_4$ [15].

In the last decade, a new trend in choosing of magnetic subsystems for design of multifunctional materials combining conductivity and magnetism was outlined. The trend is associated with the use of the octahedral cation complexes of Fe(II), Fe(III) and Co(II), showing reversible spin-crossover (SCO) between high-spin (HS) and low-spin (LS) states of the metal ion, in combination with the radical anion conducting subsystems [16,17]. The latter could be represented by the systems based on $[M(dmit)_2]^{\delta-}$ complexes (M = Ni, Pd, Pt; dmit = 4,5-dithiolato-1,3-dithiole-2-thione; $0 < \delta < 1$) [18] and/or 7,7,8,8,-tetracyanoquinodimethane ($(TCNQ)^{\delta-}$, $0 < \delta < 1$) [19–23]. The availability of fractional oxidation ($[M(dmit)_2]^{\delta-}$) or reduction states ($(TCNQ)^{\delta-}$) is a necessary condition for the emergence of high conductivity in these systems. The SCO induced by temperature, pressure or light irradiation is accompanied by the changes in the coordination environment of the metal ion [24–26]. The electrical conductivity of the most molecular conductors is very sensitive to external and/or chemical pressure [27,28]. There is an every reason to believe that spin-crossover transition would affect the conductivity at least via a chemical compression or extension arising from structural perturbations in the process of SCO. Furthermore, magnetic interactions between the subsystems make possible a realization of a spin-dependent electronic transport. The interplay between spin-crossover and conductivity was already observed in some of such materials [17,29–33]. The most of conducting SCO compounds represent the Fe(III) cation complexes with $[M(dmit)_2]^{\delta-}$ anions [29,30,34–38]. There is only one compound with $TCNQ^{\delta-}$, described in publications: $[Fe(III)(acpa)_2](TCNQ)_2$, with σ_{rt} = 2.8 10^{-3} $\Omega^{-1} \cdot cm^{-1}$ [39]. The crystal structure of the latter has not been solved. Recently, four spin-crossover compounds, combining Fe(II) and Co(II) cation complexes with $TCNQ^{\delta-}$ anions, have been obtained [32,33,40] and showed the conductivity in the range of 10^{-2}–10^{-1} $\Omega^{-1} \cdot cm^{-1}$.

In this paper, we report synthesis, structure, and physical properties of the three new spin-crossover compounds based on the Fe(III) cationic complex, $[Fe(sal_2\text{-}trien)]^+$ ($H_2sal_2\text{-}trien$ = N,N'-bis[2-(salicylideneamino)ethyl]ethane-1,2-diamine), with TCNQ counterions in a fraction and fully reduced states, $[Fe(III)(sal_2\text{-}trien)](TCNQ)_2 \cdot CH_3OH$ (**1·MeOH**), $[Fe(III)(sal_2\text{-}trien)]$ $(TCNQ) \cdot CH_3CN$ (**2**) and $[Fe(III)(sal_2\text{-}trien)](TCNQ) \cdot H_2O$ (**3**). The compound **1·MeOH** possesses a record conductivity (5.0–6.0 $\Omega^{-1} \cdot cm^{-1}$) among known conducting SCO-compounds, whereas the magnetic response of the low conducting **2** and **3** demonstrate a hysteresis and a phase fractionation. A residual high-spin phase in **2** demonstrates a ferro(ferri)magnetic coupling at helium temperatures. Local magnetic moments in all three systems reveal spin-spin interactions in the SCO range.

2. Results and Discussion

2.1. Synthesis

The compound $[Fe(III)(sal_2\text{-}trien)](TCNQ)_2 \cdot CH_3OH$ (**1·MeOH**) were obtained by mixing CH_3CN solution of TCNQ with the solutions of $[Fe(III)(sal_2\text{-}trien)]I \cdot 1.5H_2O$ and LiTCNQ in a mixture of acetonitrile/methanol (see Materials and Methods). The resulting solution was left to stand in a refrigerator overnight. The black plate crystals of **1·MeOH** were formed (Figure S1). Figure S2 displays the thermogram of **1·MeOH**. With increasing temperature, a weight loss of 4.43% was observed in the temperature range 70–150 °C with endothermic peak at 126.7 °C, which is assigned to the loss of crystallization methanol molecule (calc. 3.77%). As this takes place, the ions with m/e 29 (CHO) and 31 (CH_3O) relating to the fragments of methanol molecule are observed in the mass spectrum. On heating above 200 °C, the complex begins to decompose (two *DSC*-peaks at 210.1 and 222.6 °C).

The compound $[Fe(III)(sal_2\text{-}trien)](TCNQ) \cdot H_2O$ (**3**) in the polycrystalline form was prepared by metathesis reaction between $[Fe(III)(sal_2\text{-}trien)]I \cdot 1.5H_2O$ and Li(TCNQ) in methanol. The recrystallization of **3** from acetonitrile yielded the crystals of $[Fe(III)(sal_2\text{-}trien)](TCNQ) \cdot CH_3CN$ (**2**) (Materials and Methods). The thermogram of **2** demonstrates a mass loss of 6.38% in the temperature range 70–100 °C with endothermic peak at 93 °C, which corresponds to the loss of lattice acetonitrile molecule (calc. 6.27%). (Figure S3). In the mass spectrum recorded in the gas phase, the peaks are observed at m/z 41, 15 and 26 relating to CH_3CN molecular ion and the fragments of acetonitrile molecule (CH_3: $m/z = 15$; CN: $m/z = 26$). The decomposition of **2** starts above 200 °C (two *DSC*-peaks at 227.1 and 232.0 °C) and accompanies by the release of CN-fragments ($m/z = 26$).

2.2. Crystal Structure

2.2.1. [Fe(III)(sal_2-trien)]$(TCNQ)_2$·CH_3OH (**1**·MeOH)

The structure of **1**·MeOH has been investigated at several different temperatures between 100 and 385 K. The asymmetric unit of the triclinic $P\bar{1}$ unit cell of initial **1**·MeOH crystal includes two independent $TCNQ^{\bullet\delta-}$ radical anions (denoted as **I** and **II**), one [Fe(sal_2-trien)]$^+$ cation and one methanol molecule in general positions (Figure 1). The crystal structure consists of layers of TCNQ radical anions parallel to the *ab*-plane which alternate with mixed Fe(sal_2-trien)-solvent layers along the *c*-axis (Figure 2).

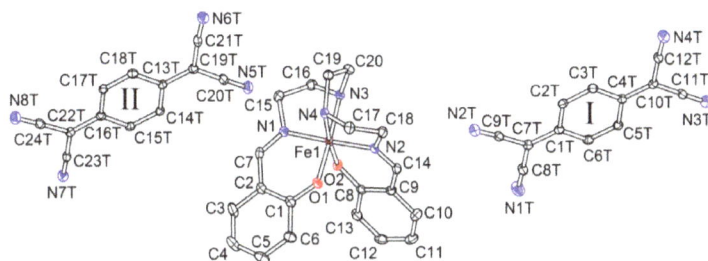

Figure 1. Asymmetric unit in **1**·MeOH (ORTEP drawing with 50% probability ellipsoids, solvent molecule is omitted).

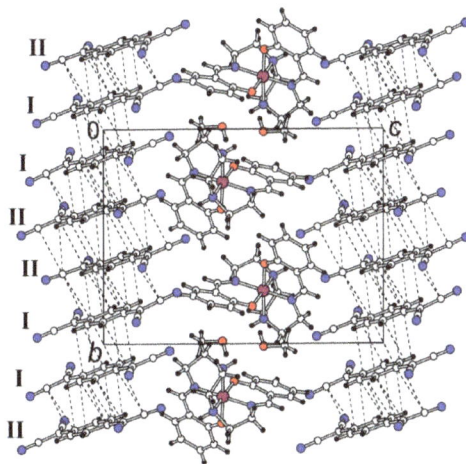

Figure 2. View of the structure **1**·MeOH at 100 K along *a*. Intrastack C...C contacts (<3.4 Å) are shown by dashed lines.

In the 100 K structure, charge state of both independent $TCNQ^{\bullet\delta-}$ radical anions **I** and **II** is slightly different and close to −0.4 and −0.6, respectively, according to calculations on the base of intramolecular bond lengths using Kistenmacher's empirical formula [41] (Table S1). π-stacking of TCNQ anions leads to a formation of infinite stacks along the [1$\bar{1}$0] direction. The stacks are composed of tetrads ...-(**II-I-I-II**)-(**II-I-I-II**)-... (Figure 2) with ring-over-bond overlap inside the tetrad due to longitudinal shift of adjacent anions and ring-over-ring one between the tetrads (longitudinal shift is absent, Figure S4). Interplane separations in the stack at 100 K are 3.274(4) Å (**I**...**I**), 3.19(4) Å (**I**...**II**) and 3.297(6) Å (**II**...**II**), the angle between mean planes of **I** and **II** is 2.37(5)°. There are a lot of intrastack C...C contacts shorter than 3.4 Å, while adjacent stacks interact only by weak hydrogen contacts of C–H...N type with H...N distances of 2.58–2.73 Å.

The [Fe(III)(sal$_2$-trien)]$^+$ cations are located between the TCNQ layers and connected by hydrogen contacts with surrounding TCNQ and methanol molecules. The Fe(III) ion has a slightly distorted octahedral coordination geometry formed by two oxygen and four nitrogen atoms of ligand. Key bond distances and angles in the Fe(III)(sal$_2$-trien)$^+$ cation at 100 K are listed in Table S2. Mean values of the Fe–O, Fe–N$_{imine}$ and Fe–N$_{amine}$ bond distances in the FeN$_4$O$_2$ chromophore (1.885(8), 1.936(8) and 2.007(2) Å, respectively) and shape of the cation with the dihedral angle α between two phenoxy groups of 53.87(6)° (Figure 3a) correspond to the low-spin state of the Fe(III) ion. The diagonals of the octahedron at Fe centre are close to linear (N–Fe–O and N–Fe–N angles are 174°–177°, Table S2) that is also characteristic for the LS state. The coordination FeN$_4$O$_2$ octahedrons are much more regular in LS state than in HS one [24–26,34]. There is no π-stacking between phenolate ligands of the nearest cations. Most of the cation-anion hydrogen contacts are formed with TCNQ **II** that is in agreement with higher charge of the latter in comparison with TCNQ **I**. There are two N–H$_{cat}$...N$_{II}$ contacts with H...N distances of 2.35 and 2.52 Å and three C–H$_{cat}$...N$_{II}$ contacts with H...N = 2.44, 2.53 and 2.65 Å and only two C–H...N$_I$ contacts to TCNQ **I** with longer H...N distances of 2.69 and 2.73 Å. The [Fe(III)(sal$_2$-trien)]$^+$ cations are also hydrogen bonded to the methanol solvent molecules (N–H$_{cat}$...O$_{solv}$, C–H$_{cat}$...O$_{solv}$ and O$_{cat}$...H–O$_{solv}$ contacts with H...O distances of 2.04, 2.46 and 1.80 Å, respectively).

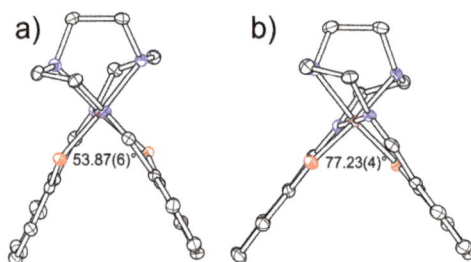

Figure 3. Molecular conformation of the [Fe(III)(sal$_2$-trien)]$^+$ cation at 100 K in **1**·MeOH (**a**) and **2** (**b**).

In order to follow the spin-crossover transition in the crystal of **1**·MeOH, additional X-ray diffraction experiments were performed at several temperatures between 100 and 385 K. The X-ray study showed single crystal to single crystal phase transition with solvent loss at annealing the crystal at 350 K in nitrogen stream. According to crystal structure refinement, the methanol site is fully occupied at 295 K [refining the occupancy gives value of 0.988(4)]. Process of solvent removal has started at the end of the 24-h X-ray experiment at 340 K [refined MeOH occupancy value is 0.956(6)]. It has been empirically found that annealing of the crystal at 350 K during one day transforms the crystal completely into a new phase with chemical formula [Fe(III)(sal$_2$-trien)](TCNQ)$_2$ (**1**), i.e., totally free of solvent. All further diffraction experiments were done for the new phase **1**, overall sequence of temperature changes being following: warming to 350 and 385 K, further cooling to 295 and 220 K and again warming to 260 and 325 K with full data collection at all these temperatures. The treatment to carry out X-ray study at 400 K was unsuccessful due to melting of the crystal. Thus, temperatures of solvent loss and crystal decay in the X-ray experiments appeared to be about 50–80 K lower than the temperatures of corresponding peaks in *DSC* spectrum that is associated with the distinctly different conditions of heating in X-ray and thermogravimetric experiments.

Symmetry of the crystal (triclinic $P\bar{1}$) persists upon the transition, but solvent loss causes significant distortion of the lattice (Materials and Methods). Relatively large temperature-dependent structural changes have been found in both cationic magnetic and anionic conducting layers. They are accumulated in Table 1 and Figure 4. All the structural parameters investigated show discontinuous changes between 340 and 350 K at the transition from solvate to solvent-free phase. In the magnetic subsystem, the Fe–N bond lengths and dihedral angle α between two phenoxy groups in the $[Fe(III)(sal_2\text{-trien})]^+$ cation, both of which are sensitive to spin state of the complex, show linear temperature dependences in the 100–340 K range and notably increase at 350 K (Figure 4a). An abrupt growth of these parameters reflects transition of Fe(III) from LS ($S = 1/2$) to HS state ($S = 5/2$). The mean Fe–N$_{imine}$ and Fe–N$_{amine}$ bond length values increase from 1.94 and 2.01 Å, typical for the LS Fe(III), to 2.00 and 2.08 Å, respectively, at 385 K. Taking into account corresponding average Fe–N distances in high-spin state, 2.12 and 2.20 Å [26,34], one can presume that about one third of iron centers are HS at this temperature. Stronger distortion of the FeN$_4$O$_2$ octahedron at 385 K (diagonal angles are 164°–177°, Table S2) is an additional evidence of partial transition to HS state. It should be emphasized that in the temperature range 100–340 K, in which the solvate phase 1·MeOH exists, the Fe–N bonds correspond to LS state and do not show any notable growth. In the solvent-free phase 1 a reversible character of SCO transition is observed at further temperature cycling. The Fe–N bond lengths return to initial, typical for LS Fe(III) values on cooling to 220 K and go again to HS values on a new cycle of warming (empty marks in Figure 4a) but in the phase 1 these changes are smooth, without jump. The angle α in 1 at low temperatures is larger than it was in 1·MeOH (Figure 4a) reflecting appearance of an additional space in the crystal lattice of the solvent-free phase.

Table 1. The average Fe–N$_{imine}$ and Fe–N$_{amine}$ bond length values and dihedral angle α between two phenoxy groups in $[Fe(III)(sal_2\text{-trien})]^+$ cation in 1·MeOH and 1, the charges (δ) of different TCNQ$^{\bullet\delta-}$ species I and II estimated from Kistenmacher's empirical formula (Table S1) [41] and then scaled to make the total charge on TCNQ pair exactly equal to −1 and intrastack I-I, I-II and II-II interplanar separations at different temperatures.

Parameter	1·MeOH 100 K	1·MeOH 220 K	1·MeOH 295 K	1·MeOH 340 K	1 350 K	1 385 K	1 325 K	1 295 K	1 260 K	1 220 K
av. (Fe–N$_{im}$), Å	1.936(8)	1.935(7)	1.936(8)	1.942(9)	1.97(2)	2.00(2)	1.96(2)	1.945(15)	1.931(13)	1.934(9)
av. (Fe–N$_{am}$), Å	2.007(2)	2.006(4)	2.010(4)	2.014(2)	2.05(2)	2.08(2)	2.035(17)	2.021(10)	2.005(8)	2.009(6)
α, °	53.87(6)	55.57(5)	56.71(4)	57.65(8)	64.00(11)	66.12(12)	62.93(9)	62.01(7)	61.39(6)	60.81(5)
δ (I)	−0.39	−0.46	−0.47	−0.44	−0.49	−0.51	−0.47	−0.44	−0.44	−0.39
δ (II)	−0.61	−0.54	−0.53	−0.56	−0.51	−0.49	−0.53	−0.56	−0.56	−0.61
I-I, Å	3.274(4)	3.254(5)	3.272(8)	3.293(9)	3.313(5)	3.322(6)	3.317(7)	3.316(7)	3.301(7)	3.310(9)
I-II, Å	3.19(4)	3.27(4)	3.31(4)	3.33(4)	3.31(4)	3.34(4)	3.31(5)	3.29(5)	3.26(5)	3.24(5)
II-II, Å	3.297(6)	3.344(6)	3.377(6)	3.404(5)	3.457(8)	3.480(7)	3.450(8)	3.432(10)	3.394(9)	3.371(9)

In contrast to the magnetic subsystem, changes in the conducting TCNQ layer occur in the entire temperature range investigated. Two independent TCNQ$^{\bullet\delta-}$ radical anions show tendency to a leveling of their charge state on warming (Figure 4b). Charges on TCNQ$^{\bullet\delta-}$ I and II radicals were calculated at different temperatures using the Kistenmacher formula (Table S1) and then scaled to make the total charge on the TCNQ pair exactly equal to −1 according to stoichiometry of the compound. The resulted scaled values are used in Table 1 and Figure 4b. Difference in charge on TCNQ$^{\bullet\delta-}$ I and II observed at 100 K fully disappears at 385 K and both of the anions become half-charged. Evolution of the charge is not uniform in 100–340 K range showing minimum for I and maximum for II at about room temperature. There is a jump at 350 K equalizing the charges. Further cooling leads to a linear divergence of charges to the starting values. Temperature evolution of the interplane separations along the TCNQ stack is not similar for three types of interactions: I-II and II-II distances grow on warming whereas I-I has a minimum at 220 K (Figure 4c). Again there are clear jumps on the heating curves in Figure 4c in the 340–350 K region and near linear reverse way on cooling. In absolute scale growth of II-II separation between the TCNQ tetrads is maximal in comparison with I-I and I-II inside the tetrad. No C…C contact shorter than 3.4 Å is observed in the **II**…**II** interaction at 385 K. It is believed that the tetrads in HS phase are more isolated from one another along the stacks. Overlap modes in the stack change insignificantly.

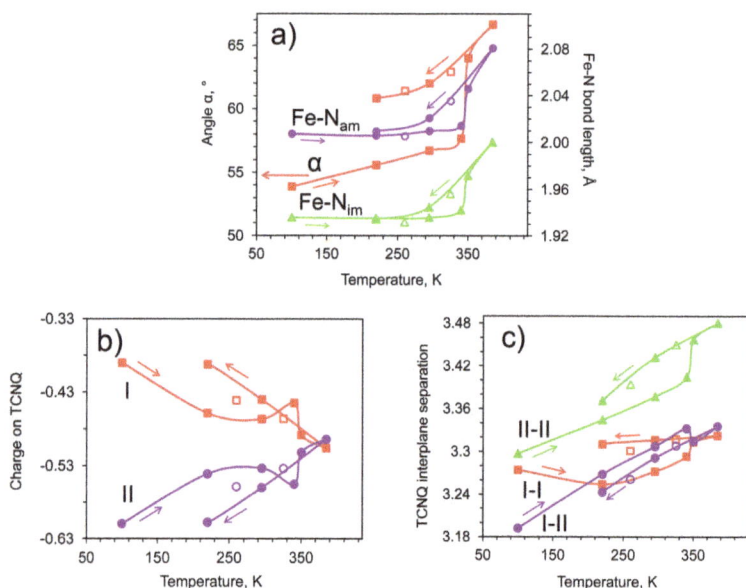

Figure 4. Temperature dependences of mean Fe–N$_{amine}$ and Fe–N$_{imine}$ bond length values and dihedral angle α between two phenoxy arms in the cation (**a**); TCNQ$^{•δ-}$ **I** and **II** charges (**b**) and interplane separations **I-I**, **I-II** and **II-II** in TCNQ stack (**c**). Starting point was at 100 K, warming from 100 to 385 K was followed by cooling from 385 to 220 K (see arrows) and new warming from 220 to 325 K (points corresponding to second warming are shown by empty marks). The data points obtained while warming in the range 100 K to 340 K belong to the solvate phase **1**·MeOH. The parameters obtained while warming above 350 K and further cooling correspond to the solvent-free phase **1**.

2.2.2. [Fe(III)(sal$_2$-trien)](TCNQ)·CH$_3$CN (2)

The asymmetric unit of the triclinic $P\bar{1}$ unit cell at 100 K includes one TCNQ$^-$ anion, one [Fe(sal$_2$-trien)]$^+$ cation and one acetonitrile molecule in general positions (Figure 5).

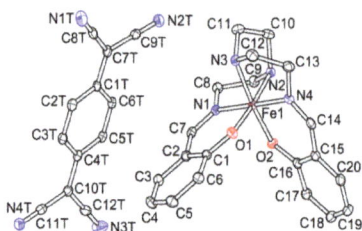

Figure 5. Asymmetric unit in **2** (ORTEP drawing with 50% probability ellipsoids, solvent molecule is omitted).

The crystal structure is built of the dianionic (TCNQ$_2$)$^{2-}$ dimers surrounded by [Fe(sal$_2$-trien)]$^+$ cations (Figure S5). According to the 1:1 stoichiometry of the complex, the TCNQ charge is −1 that agrees with charge calculations on the base of bond lengths values (Table S1). The adjacent dimers interact along *a* by hydrogen C–H...N≡C contacts (H...N distances of 2.59, 2.62 Å). The Fe(III) cation in [Fe(sal$_2$-trien)]$^+$ is octahedrally surrounded by four N and two O atoms of the ligand. Short Fe–N$_{imine}$, Fe–N$_{amine}$ and Fe–O bonds in the octahedron [average values are 1.931(3), 2.006(1) and 1.880(1) Å, respectively, Table S3], minor distortion of the octahedron (the diagonal N–Fe–N, N–Fe–O angles are close to linear, Table S3) and small dihedral angle between the two phenoxy mean planes (Figure 3b) are sings of the LS state of Fe(III) at 100 K. The nearest cations are coupled by π...π stacking of the aromatic rings with seven C...C contacts in the range 3.369(3)–3.583(4) Å (Figure S6). There is a hydrogen bonding between cation and anion (N–H...N, H...N = 2.17 Å; 7 C–H...N with H...N = 2.53–2.74 Å, Figure S6) as well as between cation and solvent (N–H...N, H...N = 2.18 Å; C–H...N, H...N = 2.67 Å).

In order to investigate temperature behavior of the structure, additional X-ray diffraction experiments were carried out on the crystal **2** at 220, 295 and 325 K. There was no indication of spin transition in the 100–325 K range because the mean Fe–N$_{imine}$ and Fe–N$_{amine}$ bond length values as well as dihedral angle α between two phenoxy groups in the [Fe(III)(sal$_2$-trien)]$^+$ cation remain almost unchanged (Table 2). At 350 K the diffraction quality sharply decreases, apparently, due to solvent loss. Disappearance of the long range order in the crystal makes impossible further structural investigation of the spin-crossover transition.

Table 2. The average Fe–N$_{imine}$ and Fe–N$_{amine}$ bond length values and dihedral angle α between two phenoxy groups in [Fe(III)(sal$_2$-trien)]$^+$ cation in **2** and interplanar separations *d* in the TCNQ dimer at different temperatures.

Parameter	2 100 K	2 220 K	2 295 K	2 325 K
av. (Fe–N$_{im}$), Å	1.931(3)	1.927(1)	1.927(2)	1.931(7)
av. (Fe–N$_{am}$), Å	2.006(1)	2.003(2)	2.007(2)	2.013(3)
α, °	77.23(4)	77.85(5)	78.02(5)	77.96(6)
d, Å	3.044(4)	3.071(9)	3.097(12)	3.112(13)

2.3. Conductivity and Magnetic Properties

The normalized dc resistance, *R*(*T*)/*R*(300 K), was measured by a standard four-probe method for the single crystal [Fe(III)(sal$_2$-trien)](TCNQ)$_2$·CH$_3$OH (**1**·MeOH) along the TCNQ stacks. It demonstrated a semiconducting type behavior (Figure 6). The placement of the electrodes on the crystal is shown in Figure S1. The value of the conductivity at room temperature is 1.5 Ω^{-1}·cm^{-1}. Below 110 K the data points were well described by the exponential law $R_{theor}(T) = \exp(\Delta E/kT)$ with the energy gap $\Delta E = 0.05$ eV (inset, Figure 6).

The results of the Montgomery method measurements (see Materials and Methods) at 300 K for the in-plane ($\sigma_{||TCNQstacks}$) and out-of-plane ($\sigma_{\perp}$) conductivity tensor components accounted for 5.4 Ω^{-1}·cm^{-1} and 3 × 10^{-3}·Ω^{-1}·cm^{-1}, respectively. Thus, the conductivity anisotropy ($\sigma_{||TCNQstacks}/\sigma_{\perp}$) was equal to 1.8 × 10^3, while the anisotropy in the plane was noticeably less ($\sigma_{||TCNQstacks}/\sigma_{\perp TCNQstacks} = 30$).

The conductivity at room temperature for crystals **1**·MeOH was 1–2 order of magnitude higher, and the activation energy was an order less, than those reported for conducting SCO complexes of Fe(II) and Co(II) with TCNQ in a fractional reduction state [32,33,40]. This difference was associated with the structure of the TCNQ stacks. In **1**·MeOH, the stacks are rather regular by charge distribution and intermolecular separation, while in the other structures the stacks are subdivided into the pronounced TCNQ triads. It should be noted, that the complex **1**·MeOH along with [Fe(III)(qsal)$_2$][Ni(dmit)$_2$]$_3$·CH$_3$CN·H$_2$O (σ = 2.0 Ω^{-1}·cm^{-1}) [29] possesses a record conductivity among known conducting SCO complexes. In contrast to **1**·MeOH, complex **2** was practically an insulator ($\sigma_{300K} = 10^{-8}$ Ω^{-1}·cm^{-1}), that is characteristic of many TCNQ salts with a fully reduced

TCNQ. The low conductivity is associated primarily with a strong dimerization of TCNQ radicals and absence of a long-range stacking.

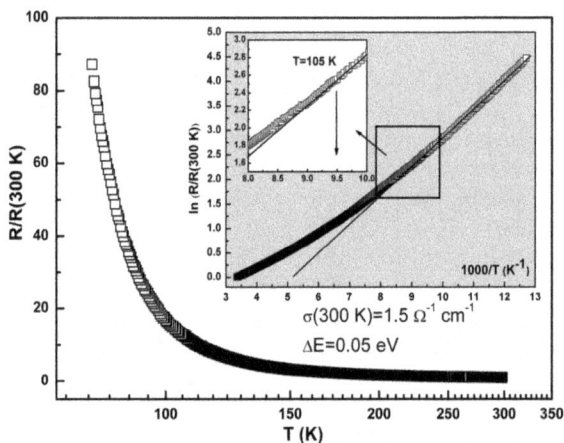

Figure 6. Temperature dependence of the normalized resistance, $R(T)/R(300 \text{ K})$ for **1**·MeOH, in the logarithmic scale. Inset—the plot of $R(T)/R(300 \text{ K})$ vs. scaled reciprocal temperature, $1000/T$. Solid line is the best fit curve in the range $78 \text{ K} < T < 110 \text{ K}$, $R_{\text{theor}}(T) = \exp(\Delta E/kT)$, with the energy gap $\Delta E = 0.05 \text{ eV}$. Zoomed window—a deviation of the experimental data from the theoretical curve starting above 105 K.

The bulk static magnetic susceptibility χ of **1**·MeOH was measured on a polycrystalline sample in the temperature range of 2–400 K (Figure 7a). The entire range was scanned twice at rates 5 K/min in the interval 20–400 K and 0.5 K/min at the interval 2–20 K. The exposure time at the highest temperature 400 K was 10 min. Two cycles of measurements at external field values 0.1 T and 4.0 T provided coinciding data points above 20 K. The data of the latter cycle were omitted in Figure 7. The value of spin response below 100 K indicated that ~98% of Fe(III) ions exist in LS state. Spin response of conducting TCNQ sublattice appeared to be negligible within less than 10% ($\chi \leq 2 \times 10^{-2} \text{ cm}^3 \cdot \text{K} \cdot \text{mol}^{-1}$) [42,43]. This is in agreement with earlier measured spin response of TCNQ layers in conducting Fe(II) SCO system [40]. A temperature independent contribution (TIP) to the spin susceptibility was also found insignificant for the description of the low temperature evolution of $\chi(T)$ in Figure 7a. A total magnetization, $M(B)$, measured at 2 K (Figure 7b), was described by a Brillouin curve for $S = 1/2$ (solid line). Thus, a paramagnetic response of **1**·MeOH below 100 K is related to the local moments $S = 1/2$ of LS Fe(III).

Sharp growth of χT above ~150 K is consistent with the structural changes in TCNQ sublattices (Figure 4b,c) and deviation of the resistivity from the exponential law (Figure 6). It arises due to SCO transition and can be modeled by replacing $S = 1/2$ to $S = 5/2$ magnetic moments via Boltzmann activation mechanism [40]. A respective curve is shown in Figure 7 (inset). The SCO parameters were $T^* = 410 \text{ K}$ and $\Delta T_{1/2} = 108 \text{ K}$ for the mid point and the width of the transition at the levels $\pm 1/2$ of the midpoint. In contrast to the diffraction data, the behavior of the magnetic susceptibility was found completely reversible. This can be understood in terms of different experimental conditions (measurement time window: minutes vs. hours, atmosphere: helium vs. nitrogen, etc.). A fully realized HS state was not achieved, because it rests above the stability threshold for this structure as it follows from the thermogravimetric analysis (Figure S2). We concluded that solvate methanol did not leave the structure while heating-cooling cycles in the magnetic measurements. A calculated value, $\chi T = 4.4 \text{ cm}^3 \cdot \text{K} \cdot \text{mol}^{-1}$, seemed a reliable estimate for total the spin response in HS state even though a complete was not achieved experimentally (Figure 7a, inset). A continuous transition curve allows determining mutual concentrations of S = 1/2 and S = 5/2 moments at any temperature within the SCO range.

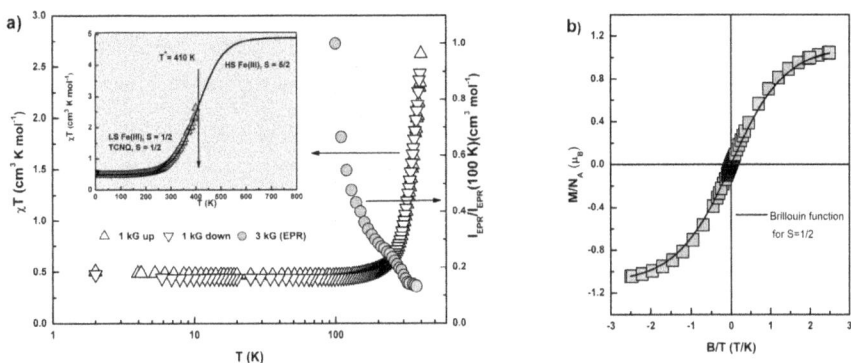

Figure 7. Temperature dependences (logarithmic scale) of the product χT measured for **1**·MeOH by SQUID magnetometer (B = 1 kG, heating ($\triangle$) and cooling ($\triangledown$) regimes) and the relative spin concentration $I_{\mathrm{EPR}}(T)/I_{\mathrm{EPR}}(100\ \mathrm{K})$ measured by EPR (B = 3 kG, heating ($\bullet$) regime) (**a**). Inset—detailed evolution of the χT in the range of spin crossover transition between LS Fe(III) S = 1/2 and HS Fe(III) S = 5/2. Solid line is a simulation by a Boltzmann distribution. Field dependence of the magnetization, $M(B)$, measured at $T = 2.0$ K (**b**). Solid line denotes the Brillouin function for S = 1/2.

The EPR spectrum of **1**·MeOH at 100 K was an intensive signal with anisotropic g-factor: $g_1 = 2.219(9)$, $g_2 = 2.171(0)$, $g_3 = 1.961(5)$. The evolution of the lineshape while heating up to 370 K is shown in Figure 8a. Temperature dependence of the g-components is presented in Figure 8b. The total spectrum was simulated at the temperatures of the measurement by using a standard lineshape model with anisotropic g-factor (WINEPR). The g-parameters were taken from the respective trend lines in Figure 8. An individual single line corresponding to the central g component has been extracted and further analyzed in terms of a relaxation rate. Qualitatively the linewidth behavior, $\Delta B(T)$, can be traced in Figure 8. Here we are showing the peak positions $B_{\mathrm{p+}}$ and $B_{\mathrm{p-}}$ for low- and high-field halves of the first derivative signal ($\Delta B = B_{\mathrm{p-}} - B_{\mathrm{p+}}$). Therefore, we were able to avoid a g-strain effect and only analyze relaxation.

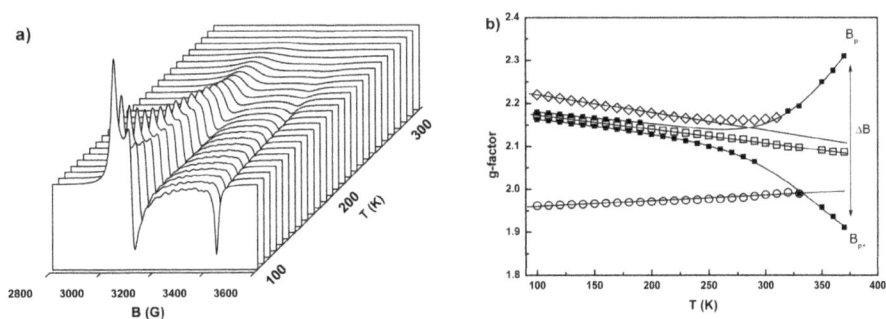

Figure 8. Temperature evolution of the EPR spectrum of **1**·MeOH in the range 100 K to 370 K (**a**). Temperature dependence of the g-values (g_1 ($\lozenge$), g_2 ($\square$), g_3 ($\circ$)) measured while heating in the range of SCO in **1**·MeOH: (**b**). Solid straight lines are linear fits. Solid bold lines connecting squares ($\blacksquare$) designate positions of the peaks $B_{\mathrm{p+}}$ and $B_{\mathrm{p-}}$ for a simulated central line of the spectrum corresponding to g_2. The peak-to-peak linewidth of the central line is $\Delta B = B_{\mathrm{p+}} - B_{\mathrm{p-}}$.

The observed signal is a typical X-band spectrum of LS Fe(III) with S = 1/2 in six-coordinated position. When a geometry of the ligand changes (for example a dihedral angle α in Figure 4a), the symmetry of hexa-coordination would also change, from rhombic distortions to axial symmetry. Axial symmetry arises in HS state due to Jahn-Teller effect. In various 4N-coordinated chelate sites the g_1 value reaches 3.0 following the ascending linear trend when coordination geometry gradually changes to axial symmetry [44]. In other words, g_1 is a sensitive probe of ligand geometry as well as a solvent presence. On contrary to expected, g_1 and g_2 components in our experiment demonstrated weakly descending trend with the temperature. This is in an indicative agreement with unchanged F–N_{am} and Fe–N_{im} distances in Figure 4a. This also implies that the coordination geometry in LS complexes [Fe(III)(sal_2-trien)] remains sustainable up to 370 K (63% LS, 37% HS). Variation of the dihedral angle between two phenoxy arms within ~5° does not cause a systematic effect on electronic structure of the metal cation. The HS configuration arises due to spontaneous thermal activation, without a precursor. The X-band spectrum for HS Fe(III), S = 5/2, usually shows a strong line at g = 6 (900 G < B < 1400 G) and a complimentary weak absorption at g ≈ 2.0 [45–47]. This is because the S = 5/2 multiplet ($^6A_1$) forms three Kramers' doublets that are separated by energies significantly larger than the microwave quantum, ~0.3 cm^{-1} for X-band. For higher rhombicities (E/D) an additional line arises at g = 4.3 due to quantum-mechanically mixed states. We detected none of the three lines in the range 100–370 K. Since the contribution of HS Fe(III) magnetic moments in 1·MeOH was determined by SQUID measurements, this is an indication of fast spin relaxation. As we show further, this signal was observed in the insulating compounds **2** and **3**. Hence, fast relaxation of S = 5/2 moments is caused by spin-spin interactions with mobile spins in TCNQ sublattice.

The concentration of LS [Fe(III)(sal_2-trien)] complexes was independently verified by EPR. According to Schumacher-Slichter method a double integrated intensity of the total EPR spectrum, I_{EPR}, is proportional to the respective spin concentration [48]. This method works best for spin S = 1/2 and detection of relative changes. The absolute values determined at high temperatures and/or for broad spectra are less reliable. The product $I_{EPR}T$, corresponding to the intensity $I_{EPR}(T)/I_{EPR}(100\,K)$ in Figure 7, has decreased 1.48 times by 370 K giving the estimate 67%. This is in good agreement with 63% portion of S = 1/2 moments obtained from SQUID data at the same temperature.

Local magnetic moments of LS Fe(III) can be considered as a useful probe for studying spin-spin interactions in the process of spin crossover. Indeed, the local environment remains stable as it follows from the behavior of g-tensor, whereas the linewidth experiences an abrupt broadening. The value of partial linewidth, ΔB, at 100 K was 18.8 G, that corresponds to a spin-lattice relaxation in an individual isolated complex LS [Fe(III)(sal_2-trien)] at finite temperature. Figure 9 demonstrates a sharp growth of ΔB at increasing relative concentration of the moments S = 5/2, n/N(%). The value of n/N was extracted from the magnetic susceptibility data in Figure 7a (inset). This growth was fitted well by a simple expression $\Delta B = \Delta B_0 + k(n/N)$, where ΔB_0 = −46 G is a base parameter relating to a threshold concentration (4.8%) and k = 21 G/% is a broadening factor. Weak and gradual broadening at the concentrations below 5% was reasonably affiliated with a trivial spin-phonon mechanism (spin-lattice relaxation). Above 5% the HS Fe(III) magnetic moments become a dominant source of relaxation. This is evidently a spin-spin relaxation. It looks credible that the relaxation rate in a reservoir A (S = 1/2) is proportional to the spin concentration in a reservoir B (S = 5/2). The question arises if this relaxation is of dipole-dipole nature, or a result of a weak superexchange coupling? The dipole-dipole could be a favorable mechanism once the dramatic broadening had been confirmed for other [Fe(III)(sal_2-trien)] complexes. So far, we do not have the proofs. Exchange coupling might not seem a realistic scenario, unless the counterion system of TCNQ in a fractional reduction state, possessing highly mobile delocalized electrons. The abrupt broadening of the conduction electron EPR signal was observed in the SCO system [Fe(II){HC(pz)$_3$}$_2$](TCNQ)$_3$ [40]. Note, that LS Fe(II) cation does not possess a magnetic moment. The broadening in the SCO range was related to the spin concentration of HS Fe(II) ions with S = 2 [40]. Owing to the fractional reduction state TCNQ triads, tetrads or even dimers might serve as efficient mediator of spin-spin interactions, when a direct overlapping of d-orbitals does not exist.

Absence of a direct exchange or superexchange does not eliminate a dynamic local spin density on ligands. In turn, ligand shell closely interacts with CN groups in TCNQ via a network of short contacts. This is a "bottleneck" junction, which facilitates cross relaxation and may affect the broadening factor k [49]. One more argument that is favorable is the absence of EPR signal from TCNQ sublattice in 1·MeOH. A typical signal in TCNQ semiconductors has width in the range 1–10 G that makes it very detectable even at low concentrations. It was shown for the system [Fe(II){HC(pz)$_3$}$_2$](TCNQ)$_3$ that at spin S = 2 concentrations above 5% in the SCO subsystem this signal rapidly spreads [40]. If so, there is no question why we did not observe it in 1·MeOH where 100% of S = 1/2 moments were presented at all temperatures below the SCO.

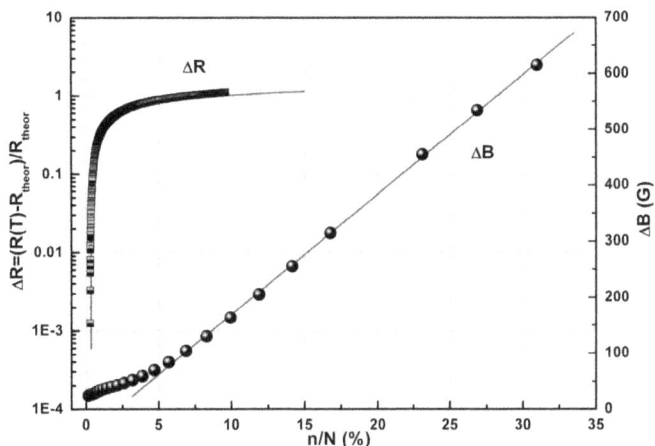

Figure 9. Partial EPR linewidth in 1·MeOH corresponding to the central g_2 component of g-factor (●), ΔB, vs relative concentration of [Fe(III)(sal$_2$-trien)] complexes in HS state, n/N(%). Solid line is the best fit curve $\Delta B = \Delta B_0 + k(n/N)$, where $|\Delta B_0| = 46$ G, k = 21 G/%. Relative deviation of the resistance (■), $\Delta R = (R(T) - R_{theor}(T))/R_{theor}(T)$, measured in the range 107 K < T < 300 K vs. the concentration n/N(%). Solid line is the best fit logarithmic curve $\Delta R(n/N) = A \cdot \ln(3n/N)$, where A = 0.3. The values n/N(%) were extracted from the modeling transition curve in Figure 7a (inset).

As soon as delocalized spins in anion sublattice of 1·MeOH were not detected by EPR, due to fast cross relaxation, and by SQUID, due to its negligible contribution to the total susceptibility, the only source of information about their reaction to SCO was an electron scattering. Below 110 K the resistance, $R(T)/R(300$ K), perfectly obeys thermal activation mechanism (Figure 6). One would expect that this mechanism also works at the higher temperatures, unless the structure of TCNQ stacks change substantially. Indeed, the gap of $\Delta = 0.05$ eV corresponds to 600 K (kT), which is higher than $T^* = 410$ K. Hence, theoretically the resistance was expected to follow the same trend in absence of SCO, $R_{theor}(T)$. However, the electron scattering can be affected by spin crossover due to local rearrangements inside the TCNQ tetrads **I-II** and **II-II** in the vicinity of HS complexes [Fe(III)(sal$_2$-trien)] (see Figure 4). For quantitative analysis we plotted a relative deviation of the actual resistance from its theoretical estimate, ΔR, against the concentration of the HS complexes, $\Delta R = (R(n/N) - R_{theor})/R_{theor}$, that was shown in Figure 9. The data were described by a logarithmic dependence $\Delta R = A \cdot \ln(B \cdot n/N)$, where A = 0.3 and B = 3.0. Note, that the argument $3n/N = n/(N/3)$ indicates that changes of resistivity occur in a reduced scale, $\sim(N/3)$. It is also worth to note, that the same type dependence was found for the in-plane components of the resistivity, $\rho_{||a}$ and $\rho_{||b}$, in the conducting SCO complex [Fe{HC(pz)$_3$}$_2$](TCNQ)$_3$ [40]. Moreover, it was shown in [40] that the transverse transport, $\rho_\perp(T)$, did not react to the changes in the cation layers. Therefore, we suggested that in 1·MeOH all the observed changes of the resistance took place for the in-plane transport. Comparing data on Figure 9 one can conclude, that the major effect of SCO on transport properties occurs at lower concentrations $n/N < 5\%$, whereas spin-spin depolarization is effective at $n/N \gg 5\%$. Similar effect was observed in [40]. This is comprehensible and has a qualitative explanation. At the very low concentrations of HS [Fe(III)(sal$_2$-trien)] complexes, the local displacements inside and/or between the TCNQ tetrads behave as randomly distributed defects. As the temperature approaches the midpoint of the transition, the anion sublattice turns into a modulated structure or to a certain extent into an amorphous one, which depends on heating/cooling rates and cooperative effects. Hence, a scattering rate gets less sensitive. On contrary, the efficiency of a cross relaxation $1/2 \rightarrow 5/2$ depends on a capacity of the spin reservoir of $S = 5/2$, i.e., concentration of the HS complexes within the interaction range. Thus, a momentum scattering and spin depolarization of conduction electrons in **1**·MeOH are decoupled to a large extent. In other words, this allows spin manipulation by an external source. It was found in [40], that for delocalized spin moments in the conducting layer depolarization is even more efficient, $\Delta B \sim (n/N)^2$. Once we obtain magnetically ordered local moments, we would get polarized conduction electrons.

The structures [Fe(III)(sal$_2$-trien)](TCNQ)·CH$_3$CN (**2**) and [Fe(III)(sal$_2$-trien)](TCNQ)·H$_2$O (**3**) could be viewed as the reference systems for distinguishing the effects of conduction electrons and solvent molecules in the magnetic and resonance properties of **1**·MeOH. Solvent effects are responsible for the difference in their SCO transitions. The static magnetic susceptibility of **2** and **3** was measured in the range of 2–400 K four times, in two sequential cycles. Heating and cooling rates were taken the same as for **1**·MeOH. The measurements begun at the room temperature in the external field $B = 4.0$ T. The virgin curves for the product χT are shown in Figures 10 and 11 (($\Box$) data points) for **2** and **3**, respectively. Upon cooling down to 2.0 K we measured the field dependences of the total magnetization by passing a complete field loop from $+4.0$ T $\rightarrow -5.0$ T $\rightarrow +5.0$ T $\rightarrow +0.1$ T. The exposure at liquid helium temperature lasted several hours. Then the magnetic susceptibility was measured second time at the field value 0.1 T (Δ). The measurements at heating were finished at 360 K, where the sample was exposed for about 10 min. A second cycle started at 360 K by measurements at cooling down to 2.0 K (∇) and further heating up to 360 K ($\bigcirc$). Repeating field measurements were not performed. The χT data for the heating curve of the cycle 2 in **3** are not shown in Figure 11, because we had to interrupt experiment due to technical reasons. We believe this was not critical for the discussion and conclusions.

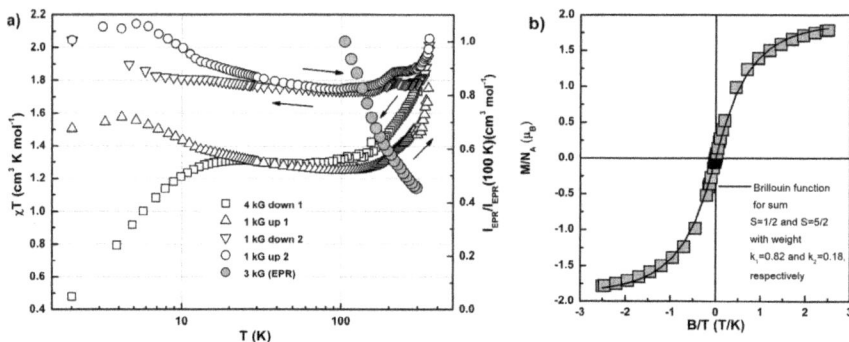

Figure 10. Temperature dependences (logarithmic OX scale) of the product χT for **2** obtained in two heating-cooling cycles (B = 1 kG, cycle 1: cooling ($\Box$), heating (Δ); cycle 2: cooling (∇) and heating ($\bigcirc$)) and the relative spin concentration of $S = 1/2$, $I_{EPR}(T)/I_{EPR}(100$ K), measured by EPR (B = 3 kG, heating ($\bullet$) regime) (**a**). Field dependence of the magnetization, $M(B)$, for **2** measured at $T = 2.0$ K (**b**). Solid line is the best fit curve corresponding to a weighted superposition of Brillouin functions for $S = 1/2$ and $S = 5/2$ (see main text for details).

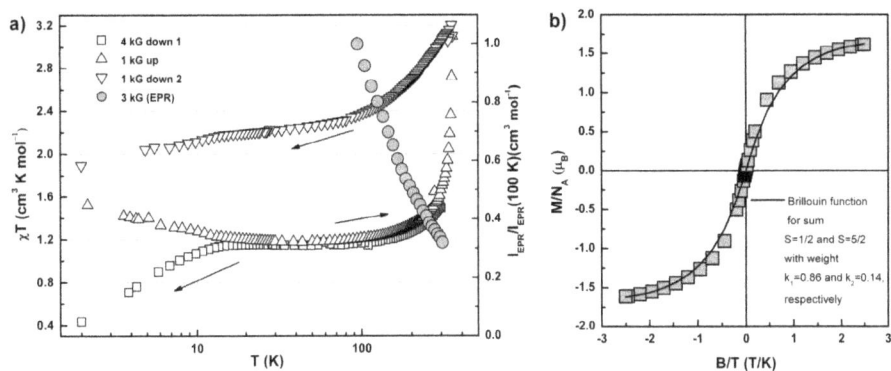

Figure 11. Temperature dependences (logarithmic OX scale) of the product χT for **3** (B = 4.0 T: cooling (□) regime; B = 0.1 T: heating (△) and cooling (▽) regimes) and the relative spin concentration of S = 1/2, $I_{EPR}(T)/I_{EPR}(100\ K)$, measured by EPR (B = 3 kG, heating (●) regime) (**a**). Field dependences of the magnetization, $M(B)$, for **3** measured at T = 2.0 K. Solid line is the best fit curve corresponding to a weighted superposition of Brillouin functions for S = 1/2 and S = 5/2 (see main text for details) (**b**).

It is worth to explain, how repeated measurements at the high and low fields, such as 4.0 T and 0.1 T, allow accounting magnetic impurities. If a sample contains impurities (for example iron oxide nanoparticles from the solvent) with a contribution χ_{IMP}, it is practically difficult to subtract it from the total magnetic response in experiment, $\chi_{exp} = \chi_{IMP} + \chi_{sample}(T)$. In an ideal case, when χ_{exp} coincides with the expected theoretical value, χ_{sample}^{theor}, the experimental data are entirely related to the spin system of the studied compound. This took place for **1**·MeOH, so we did not present the high field data. In other cases, proofs would be required. The absolute values of $\chi_{exp}T$, 1.25 and 1.75 cm^3·K·mol^{-1}, measured respectively for **2** and **3** at 100 K, were found considerably higher than their theoretical estimates for the LS state. The question is, whether this was a total spin response of SCO system, or it included the extrinsic component χ_{IMP}. Magnetic response of the solid particles usually does not depend on the external field of a measurement, whereas the actual signal χ_{sample} is proportional to its strength. By performing measurement at a higher field, we reduced contribution of the extrinsic part, ~χ_{IMP}/B. Thus, a divergence of $\chi_{exp}T$ data obtained at the various fields let us distinguish and account for external impurities, should they present in the sample. The negative side of the high field measurements is a descending trend, that arises due to insufficient population of upper spin state at higher Zeeman splitting (see the data points (□) below 15 K in Figures 10 and 11). Therefore, a temperature behavior below 20 K in Figures 10 and 12 was not discussed for the data points obtained at 4.0 T. In the range near 100 K the values $\chi_{exp}T$ measured at a broad span 0.1–4.0 G pretty much coincide for both **2** and **3**. That means the absence of extrinsic magnetic contributions. We also asserted the absence of single radicals TCNQ$^{\bullet-}$ as paramagnetic defects. Corresponding EPR signals [50] had negligible intensities, both in **2** and in **3** (Figure S8). These arguments as well as the measurement results allowed us to conclude, that the spin susceptibilities of **2** and **3** below 100 K were associated with the cation sublattice, namely with the magnetic moments of [Fe(III)(sal$_2$-trien)]$^+$ complexes.

The abrupt growth of χT above 100 K confirms SCO transition for both systems, which is in agreement with that for **1**·MeOH. Meanwhile, in both systems, **2** and **3**, the transition showed a thermal hysteresis. Intricate hysteresis of spin susceptibility is typical for mononuclear iron (III) SCO complexes [51]. The absolute values of χT contained a significant quantity of residual HS complexes at temperatures below 100 K. The measurements in **2** yielded 80% of LS and 20% of HS complexes at 100 K in the cycle 1, and 69% of LS and 31% of HS complexes in the cycle 2 (Figure 10). For **3** the χT data were fitted with 76% of LS and 24% of HS complexes in the first cycle, and 68% to 32% in the second cycle, respectively (Figure 11). This is close to the balance ratio 2:1, which is evidently determined by a complete loss of solvent (2/3 complexes in LS state and 1/3—in HS state). Field dependences shown in Figures 10 and 11 also confirm, that considerable ferric entity remains in the HS state at T = 2 K. Magnetization curves were successfully fitted with 82% and 18% for **2** (solid line in Figure 10b, and 86% and 14% for **3** (solid line in Figure 11b). Note, that these numbers were obtained from original samples (before heating). Different volumes of residual HS fractions and invariable gap between the temperature curves χT in Figure 10 ($\Delta\chi T \approx 0.5$ cm^3·K·mol^{-1} between the datasets (Δ) and ($\bigcirc$)) speak in favor of phase fractionation due to migration of solvent molecules. Similar fractionation was observed in [Fe(sal$_2$-trien)][Ni(dmit)$_2$] structures [34–38]. Though the dihedral angle between the two phenoxy groups in [Fe(III)(sal$_2$-trien)]$^+$ at 100 K corresponds to the LS state (73.24(4)$^\circ$, Figure 3b), the diffraction data might not resolve ~20% HS fraction. The *TG-DSC* curves in Figures S2 and S3 demonstrated more pronounced mass change at 93 °C, indicating a solvent loss in **2** in comparison with **1**·MeOH (126.7 °C).

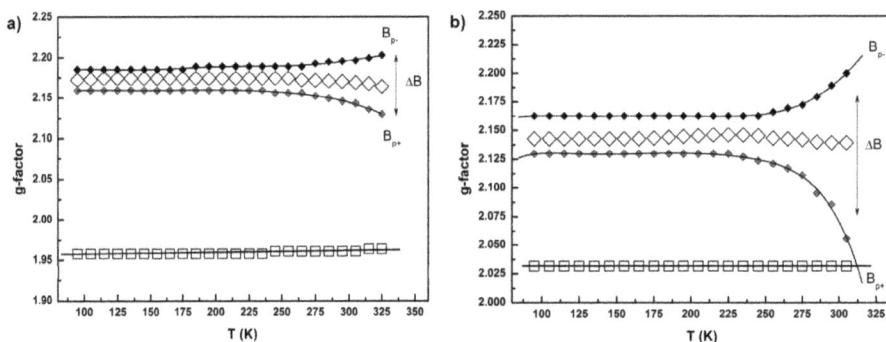

Figure 12. Temperature dependences of the g-parameters for **2** (**a**) and **3** (**b**) extracted from LS Fe(III) EPR signal while measurements in heating regime: $g_\perp$ ($\Diamond$), $g_{||}$ ($\Box$). Solid lines connecting squares (■) fit the positions of the peaks B_{p+} and B_{p-} for simulation the partial spectral line corresponding to $g_\perp$, $\Delta B = B_{p+} - B_{p-}$.

The concentration of LS Fe(III)(sal$_2$-trien) complexes in **2** and **3** was also verified by EPR. The measurements were carried out on the same samples that had passed through SQUID experiments. The absolute values, I_{EPR}, at 100 K were found close to 2/3 of the theoretical estimate to within ~20% accuracy. The relative intensities $I_{EPR}(T)/I_{EPR}(100$ K) plotted in Figures 10 and 11 (right ordinate axes) were obtained in the heating regime. In the range 95–325 K respective spin concentration of LS Fe(III) decreased with the temperature to 60% of its original value for **2**, and to 70% within 95–305 K range for **3**. Temperature evolution of the lineshapes is shown in the Figures S7 and S8. The extracted g-parameters are presented in Figure 12 ((**a**) for **2** and (**b**) for **3**). The character of anisotropy, as well as the weak temperature dependence, were discussed earlier for **1**·MeOH. A common peculiar feature, axial anisotropy, was described by two parameters $g_1 = g_2 = g_\perp$ and $g_3 = g_{||}$. At 95 K they were 2.172(3) and 1.957(8) for **2**, and 2.142(6) and 2.031(6) for **3**. As the complexes underwent a thermal SCO, the EPR signal broadened, while the g-parameters remained unchanged. A partial peak-to-peak linewidth, ΔB, corresponding to $g_\perp$ component was extracted and analyzed as a function of temperature like it had been done for **1**·MeOH. The guiding lines for B_{p+} and B_{p-} are shown in Figure 12.

The ΔB values at 100 K for **2** and **3** equaled 43.0 G and 45.3 G, which was broader than 18.8 G for **1**·MeOH. Meanwhile, in **1**·MeOH it quickly reached the value 46 G at HS Fe(III) concentration of 5%. It was somewhat unclear, why at presence of 1/3 (33%) complexes in HS state in **2** and **3** the EPR signal of S = 1/2 moments has the linewidth similar to that at 5% of HS complexes in **1**·MeOH. Consensus becomes apparent if we assume spatial inhomogeneity. Spin-spin relaxation becomes ineffective if the fraction with S = 1/2 moments of LS Fe(III) and the HS fraction are separated in the cation layers.

Besides that, insulating TCNQ dimers in **2** and **3** are unable to facilitate expanded interactions by mediating cross-relaxation via conduction electrons. At the same time, the LS fraction undergoes SCO transition, while the HS fraction remains unchanged. Indistinct n/N estimates together with less effective line broadening in **2**, did not result in finding an appropriate correlation between EPR linewidth and concentration of Fe(III) ions in HS state. However, a credible linkage between line broadening and χT growth was found in the system **3**. The transition was extrapolated by Boltzmann model. The exponential growth was superimposed by an ascending foothill segment (Figure S9). This feature arises because the thermal hysteresis is a kinetic effect, and spin states appear in result of cooperative interactions [52]. Spin concentration of S = 5/2, x, was obtained by solving the equation at every temperature point of the measurement, $\chi_{exp}(T) = (1 - x) \cdot \chi_{1/2}^{theor} + x \cdot \chi_{5/2}^{theor}$, where $\chi_{1/2}^{theor}$ and $\chi_{5/2}^{theor}$—theoretical values of the magnetic susceptibilities at 100% concentrations of LS or HS complexes, respectively. This procedure might seem arguable, because each data set was obtained in the different conditions, due to thermal irreversibility (spin fractionation). However, we believe that the linkage between EPR linewidth and spin concentration of S = 5/2 moments remains valid, if the latter was extracted from the χT data at heating in the first measurement cycle (data points ($\triangle$) in Figure 11). Indeed, it follows from Figure 13 that the value ΔB_{LS} increases proportionally to n/N above the threshold of 23%. Similar dependence took place in **1**·MeOH right at the beginning of the reversible SCO transition. Hence, a broadening part of the dependence in **3**, $\Delta B \sim k \cdot (n/N)$ (solid line in Figure 13a), where k = 30.1 G/%, occurs due to spin-spin relaxation. This contribution above the threshold is induced by thermally activated HS complexes with S = 5/2, homogeneously appearing inside the LS fraction. For reference, in Figure S10 we present the temperature dependences $\Delta B_{LS}(T)$ in **2** and **3** with respective fitting parameters in Table S4. The threshold is consistent with spatial in homogeneity of a spin ensemble, when spin probes of S = 1/2 do not exist inside the intact fraction of S = 5/2 moments.

The residual (intact) HS fraction remains spatially separated. EPR spectra from Fe(III) ions in the HS state in **2** and **3** are shown in Figure S8. Very weak temperature dependences of g-factor with $g = 5.6(0)$ at 300 K were observed for **2**, and with $g = 4.06$—for **3**, respectively. Line broadening in the SCO range, $\Delta B_{HS}(T)$, is shown in Figure 13b. The linewidth was found to follow a linear dependence $\Delta B_{HS} = \Delta B^{HS}_0 + k_{HS} \cdot T$, where $\Delta B^{HS}_0 = 5.66 \cdot 10^2$ G, and $k_{HS} = 1.7$ G/K for **2**, and $\Delta B^{HS}_0 = 5.09 \cdot 10^2$ G, and $k_{HS} = 0.91$ G/K for **3**. A proportional broadening $\Delta B_{HS}(T) \sim T$ was observed in the entire range of SCO on **3**, whereas in **2** it begun at higher temperatures, which correlates with the data in Figure 12a and in further discussed Figure 14a.

An assumption of a phase fractionation, promoted by a solvent loss, implies a partition of the total spin susceptibility to the fractional contributions and further analysis of their individual temperature behavior. The magnetic susceptibility of the LS phase in **2** was reconstructed by using the expression, $\chi_{LS}T = (\chi_{exp}(T) - y \cdot \chi_{5/2}^{theor}) \cdot T/(1 - y)$, where y is the concentration of the moments S = 5/2 in the HS phase. Similarly, the expression $\chi_{HS}T = (\chi_{exp}(T) - (1 - y) \cdot \chi_{1/2}^{theor}) \cdot T/y$ described the contribution of the HS phase. Here we took g = 2.00 and 2.17 for calculated values of $\chi_{1/2}^{theor}$ and $\chi_{5/2}^{theor}$, respectively. A comparative plot "$\chi_{LS}T$ vs. T" in Figure 14 demonstrates, that the thermal cycling affects a foothill domain of the SCO transition curve, leaving its steep slope unchanged. EPR line broadening correlates with the steep part, but remains insensitive to a foothill region (see also Figure S9). This is consistent with the threshold in **3** (Figure 13), that was attributed to the cooperative interactions [52]. The broadening is caused by spin-spin interactions between S = 1/2 spin probes with magnetic moments S = 5/2 in the neighboring HS [Fe(III)(sal$_2$-trien)]$^+$ complexes, which do not belong to the residual HS fraction (HS phase). After several cycles of measurements, the amount of [Fe(III)(sal$_2$-trien)]$^+$ complexes, taking part in the SCO transition, approached 2/3 of total.

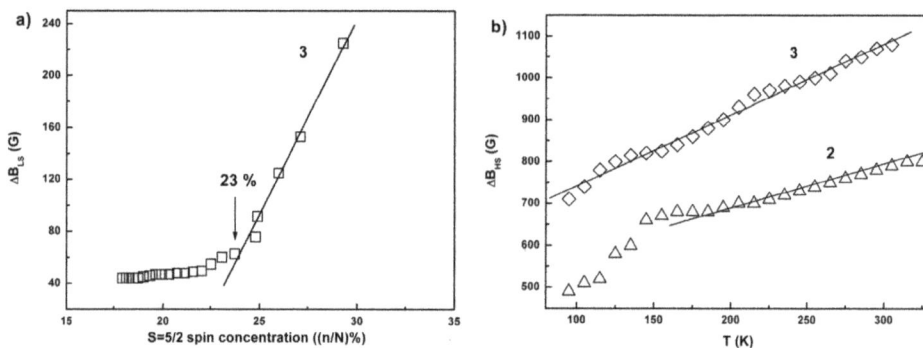

Figure 13. EPR linewidth, ΔB, associated with $g_\perp$ component of the total spectrum in **3** vs. concentration of the local moments S = 5/2, $n/N(\%)$ (**a**). Solid line is the best fit function for $n/N > 23\%$, $\Delta B_{LS} = \Delta B^{LS}_0 + k(n/N)$, where $|\Delta B^{LS}_0| = 6.60 \cdot 10^2$ G and k = 30.1 G/%. Temperature dependences of the EPR linewidth for the high-spin moments, S = 5/2, $\Delta B_{HS}(T)$ (**b**). Best-fit lines are given by the expression $\Delta B_{HS} = \Delta B^{HS}_0 + k_{HS} \cdot T$, where $\Delta B^{HS}_0 = 5.66 \cdot 10^2$ G, and $k_{HS} = 1.7$ G/K for **2**, and $\Delta B^{HS}_0 = 5.09 \cdot 10^2$ G, and $k_{HS} = 0.91$ G/K for **3**.

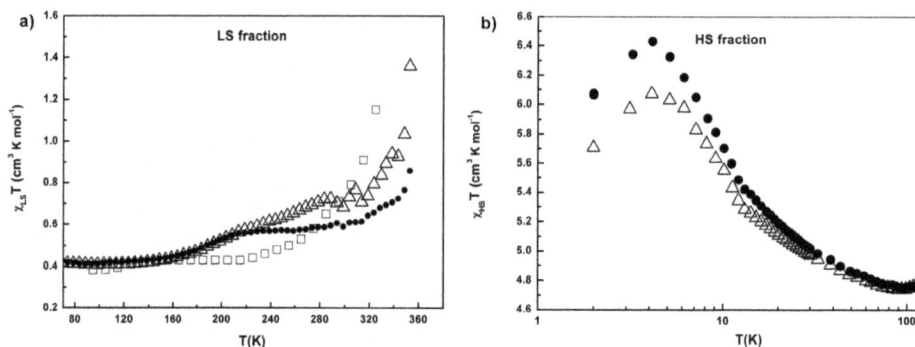

Figure 14. Temperature dependences of the product $\chi_{LS}T$ for the LS (SCO) fraction in **2** (measurements in heating regime, cycle 1 ($\triangle$), cycle 2 ($\bullet$)) (**a**). Temperature dependence of the EPR linewidth for $g_\perp$ component measured at heating in cycle 1 ($\square$). Temperature dependences of the product $\chi_{HS}T$ for the HS (residual) fraction (**b**).

The total spin response χT of **2** demonstrated unusual peaks at $T = 4.11$ K shown in Figure 10. System **3** also showed a significant growth of χT, though with no maximum, at temperatures $T < 10$ K (Figure 11). In this study, we attributed the observed peaks to the HS phase, $\chi_{HS}T$, as shown in Figure 14b. We incline to discuss their ferromagnetic nature rather than metastable high spin trapping. Our opinion is based on the reasoning below.

Enhancement of a spin response at the low temperatures is often a sign of a thermal-induced metastable spin-state trapping (TIESST) [53]. Indeed, some mononuclear Fe(III) complexes, especially those with two-step spin-crossover, exhibit a metastable spin-state HS* emerging after thermal quench from the true HS state of a complete SCO transition. A key feature is the rate of cooling, which determines the amount of quenched fraction. A drop of $\chi_{HS^*}T$ to the original LS level is, in fact, a relaxation of a metastable phase, while heating above the temperature, that corresponds to a respective energy barrier. A width of the transition depends on a heating rate, and usually consists tens of degrees in the transition range from 50 to 100 K [48]. A distinct feature of TIESST phase is the butte-like shape of the χT curve, where the flat top may reach ~50% of the total HS response, depending on a freezing rate. Here are the details, that contradict a mechanism of quenched HS* spin-state for **2**: (1) gradual refrigeration down to 2.0 K took 5 h (including measurement time), which is ~100 times slower than the flash freezing in SQUID chamber; (2) the difference $(\chi_{HS}T\text{-}\chi_{LS}T)$ at maximum exceeded 100% of the $\chi_{LS}T$ gain at 353 K; (3) $\chi_{HS}T$ curves had no plateau and further relaxation drop, associated with decaying HS*spin-state; (4) gradual 50% decrease of $\chi_{HS}T$ between 4 and 10 K was unlikely caused by released dynamics (thermal relaxation) of sal$_2$-trien ligand; (5) maximum value of χT in Figure 14b did grow at cycle 2, whereas LS phase has diminished. However, since the complete SCO transition has not been reached, we cannot reject this scenario completely. Importantly, the TIESST mechanism does not imply magnetic coupling in the HS* system.

The alternative scenario suggests the magnetic exchange interactions in the HS phase. A maximum in χT curve at helium temperatures is a characteristic feature of ferromagnetic or ferrimagnetic coupling in many molecular magneto-active systems, including HS Fe(III) complexes [52,54]. Ferromagnetic interactions were also found in numerous TCNQ based compounds with metallo-complex couenrions [55]. Ferro- and ferrimagnetic coupling was reported for networks of transition metals, bridged by TCNQ^{-1} radicals $M(TCNQ)_2$ (M = Mn, Fe, Co, Ni), among which was the compound Mn(TCNQ)$_2$ with T_c = 44 K [56,57].

In the crystal field approach, the structural characteristics of **1**·MeOH, **2** and **3** deny exchange coupling between the magnetic moments of Fe(III) in neighboring complexes. Moreover, the shortest distances d_{Fe-Fe} = 7.070 Å (350 K, **1**·MeOH) and 7.729 Å (325 K, **2**) are not sufficient for the effective spin-spin relaxation via dipole-dipole mechanism. Therefore, magnetic and resonance behavior of the Fe(III) magnetic moments, found in the current study, implies the key role of TCNQ molecules as a mediator. Schematic structural arrangements of the TCNQ molecules and [Fe(sal$_2$-trien)]$^+$ complexes in **1**·MeOH (A) and **2** (B) are shown in Figure 15. Dotted lines represent short contacts N...N, C...C, having distances less than the sum of their van der Waals radii including effect of a high spin Fe dilation. In **1**·MeOH at ambient conditions the distances N_{amine}...N_{CN} were 3.128 Å (N4, N8T), 3.226 Å (N4, N5T), N_{imine}...N_{CN} (N1, N5T)—3.292 Å. Short contacts allow the exchange interactions. In the Ligand Field Theory (LFT), π bonding between d-orbitals of transition metal in octahedral symmetry t_{2g}(d_{xy}, d_{yz}, d_{xz}) and p(π) orbitals of the ligand takes place and significantly diminishes a total energy. In [Fe(sal$_2$-trien)]$^+$complexes π bonding is expected along OY axes (N1–Fe1–N2): p(π)$_{imine}$-d_{yz}-p(π)$_{imine}$. Hybridization due to overlapping induces a local spin density on the ligand. In turn, non-bonding ligand orbitals of N_{amine} and N_{imine} overlap with valence π orbitals of TCNQ$_2^-$ (**1**·MeOH) or SOMO π^* orbitals of CN group in TCNQ$^{\bullet-}$ radical (**2**). Thus, an interplay between π bonding and intramolecular distortion could be the driving force for the exchange coupling. Note that that the acetonitrile molecule in **2** acts as an "anchor" due to short contact with N_{amine} (N3, Figure 15B). Therefore, the solvent removal may release the distortions. However, a key role belongs to the electrons in the TCNQ sublattice. Hopping electrons of semiconducting TCNQ layers in **1**·MeOH could serve as an efficient spin reservoir for cross relaxation between S = 1/2 and S = 5/2 local moments (Figure 15A). High conductivity along TCNQ stacks is capable to provide long-range spin-spin interactions. Localized spins of adjacent radicals TCNQ$^{\bullet-}$ in weakly conducting compounds **2** and **3** suggest intrinsic interactions in the form of singlet-triplet splitting or exchange coupling. Taking into consideration π-stacking between phenoxy groups of neighboring ligands in Figure 15B, one can consider an alternating spin chain with the units [–1/2–1/2–1/2–1/2–], [–1/2–1/2–1/2–5/2–], or [–5/2–1/2–1/2–5/2–] appearing upon passing the SCO transition. There were several drawbacks. We could not simulate the temperature dependences in Figure 14b by a Heisenberg model for the individual four spin unit with two exchange constant J_1, J_2 and one ZFS parameter D. We also did not find a coercivity in magnetization curves in Figures 11 and 12. Though the latter can be understood, as the measurements were performed before heating-cooling cycles.

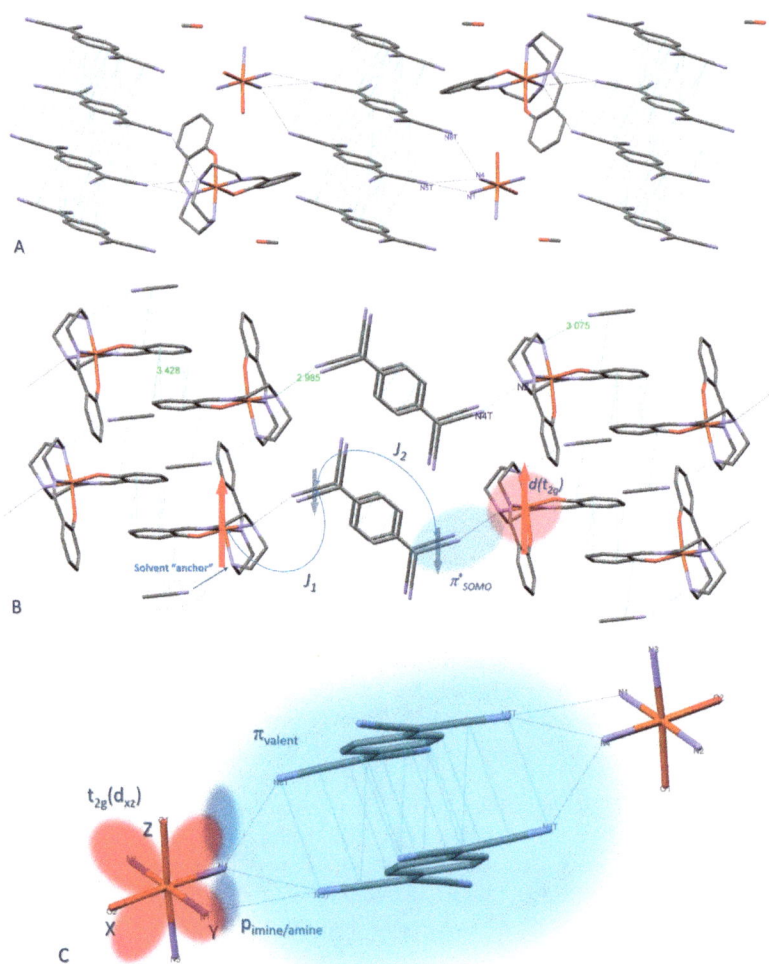

Figure 15. Schematic structural arrangements of the TCNQ molecules and [Fe(sal$_2$-trien)]$^+$ cation complexes in **1**·MeOH (**A**) and **2** (**B**). Dotted lines represent N...N, C...C short contacts of less than the sum of the van der Waals radii; (**C**) Mutual arrangement of coordination Fe(III) octahedra with neighbor TCNQ units **II** in **1**·MeOH. Orbitals are presented schematically.

Thus, the discussion of physical properties in this study was extended with somewhat hypothetical consideration of fundamental mechanisms. We believe, that holistically, this will draw attention to the important aspects of spin-spin interactions and electronic transport in [Fe(III)(sal$_2$-trien)](TCNQ)$_2$ (**1**·MeOH) system as well as an interplay between ferro/ferrimagnetic coupling and SCO in [Fe(III)(sal$_2$-trien)](TCNQ) (**2**, **3**). While the first compound has a record conductivity up-to-date and reversible SCO transition, the compound [Fe(III)(sal$_2$-trien)](TCNQ)·CH$_3$CN is the first system, where ferromagnetic coupling was triggered by the SCO transition. In other words, in the system [Fe(III)(sal$_2$-trien)](TCNQ)$_n$ switchable magnetic moments of isolated metallo-complexes coexist and interact with spin system of organic network in full or fractional reduction state. In turn, the organic network allows electronic/spin transport. The fact that the spin interactions are switchable, makes this conducting system a prospective candidate for molecular spintronics.

3. Materials and Methods

General remarks: LiTCNQ and [Fe(sal$_2$-trien)](NO$_3$)·1.5H$_2$O were obtained according to the literature procedures [19,58]. All other reagents and solvents were commercial products.

3.1. Synthesis

[Fe(sal$_2$-trien)](TCNQ)$_2$·CH$_3$OH (**1**·MeOH) was synthesized under argon atmosphere by mixing hot solutions of TCNQ (0.102 g, 0.5 mmol) in 10 mL of acetonitrile, LiTCNQ (0.106 g, 0.5mmol) and [Fe(sal$_2$-trien)](NO$_3$)·1.5H$_2$O (0.232 g, 0.5 mmol), each in a mixture of acetonitrile/methanol (10/5 mL). The resulting solution was placed in a refrigerator overnight. Black shiny plate crystals of **1**·MeOH were formed. They were collected and washed with cold methanol and dried on air. Yield: 0.253 g (59%). Elemental analysis calcd. (%) for **1**·MeOH (C$_{45}$H$_{36}$FeN$_{12}$O$_3$): C 63.68, H 4.27, N 19.8, O 5.66; found (%): C 63.21, H 4.50, N 19.54, O 5.85.

[Fe(sal$_2$-trien)]TCNQ·CH$_3$CN (**2**). Dark purple platelet-like crystals of **2** were obtained by recrystallization of **3** from acetonitrile. Yield: 70.0%. Elemental analysis (%): calc. for **2** (C$_{34}$H$_{31}$FeN$_9$O$_2$): C 62.49, H 4.78, N 19.29, O 4.9; found (%): C 62.24, H 4.50, N 18.96, O 5.27.

[Fe(sal$_2$-trien)](TCNQ)$_2$·H$_2$O (**3**) was obtained by mixing hot solutions [Fe(sal$_2$-trien)](NO$_3$) ·1.5H$_2$O (0.232 g, 0.5 mmol) and LiTCNQ (0.106 g, 0.5 mmol), each in 10 mL of methanol. The resulting solution was placed in the refrigerator overnight. Microcrystals were formed, which were collected, washed with ether and dried on air. Yield: 0.270 g (85%). Elemental analysis calcd. (%) for **3** (C$_{32}$H$_{30}$FeN$_8$O$_3$): C60.96, H4.80, N17.77, O7.60; found (%): C 60.94, H 4.54, N 17.51, O 7.97.

3.2. Thermogravimetric Analysis

The thermogravimetric analysis was performed in argon atmosphere with a heating rate 5.0 °C·min^{-1} using a NETZSCH STA 409 C /QMS 403 thermal analyzer (NETZSCH-Gerätebau GmbH, Selb, Germany), which allows simultaneous thermogravimetry (TG), differential scanning calorimetry (DSC) and mass-spectrometry measurements, which allows simultaneous thermogravimetry (TG), differential scanning calorimetry (DSC) and mass-spectrometry measurements.

3.3. X-ray Crystallography

Single crystal X-ray diffraction experiments were carried out on a Bruker SMART APEX2 CCD diffractometer (Bruker AXS Advanced X-ray Solutions GmbH, Karlsruhe, Germany) (for **1**·MeOH at 100, 220 K and **2** at 100 K) and an Oxford Diffraction Gemini-R CCD diffractometer (Oxford Diffraction, Oxford, Oxfordshire, United Kingdom) [for all other temperatures, λ(MoKα) = 0.71073 Å, graphite monochromator, ω-scan mode]. Multi temperature experiment procedure was used without control of cooling/warming rate; the average rate was roughly estimated as 2–4 K/min. Data collection for **1**·MeOH and **1** proceeded during 24 h at each temperature. Crystal of **1**·MeOH had begun to lose solvent at the end of the 340 K experiment; in this reason some latest X-ray frames were excluded from the data list for correct refinement of the solvate structure. Then the crystal was warmed to 350 K and kept at this temperature during one day before data collection to obtain the new phase **1** totally free of solvent. The structures were solved by the direct method and refined by the full-matrix least-squares technique against F^2 in an anisotropic approximation for all non-hydrogen atoms. Hydrogen atoms were localized from the Fourier synthesis of the electron density and refined in the isotropic approximation. MeOH sites in **1**·MeOH were finally refined as fully occupied at 295 and 340 K though occupancy refinement gave values of 0.988(4) and 0.956(6), respectively. All calculations were performed using SHELXTL PLUS 5.0 and SHELX-2016 program packages [59]. Selected crystallographic data and refinement parameters are given in Table 3. The full data of studies are available at the Cambridge Crystallographic Data Centre. Distortion of the lattice due to solvent loss transforms acute unit cell of **1**·MeOH into obtuse one for **1**, the latter is given in non-standard setting both in Table 3 and *cif*-files for direct comparison of the structure data.

Table 3. Crystal structure and refinement data for **1·MeOH**, **1** and **2**.

Parameter	1·MeOH 100 K	1·MeOH 220 K	1·MeOH 295 K	1·MeOH 340 K	1 350 K	1 385 K	1 325 K
Cell setting	triclinic	triclinic	triclinic	triclinic	triclinic	triclinic	triclinic
Formula	$C_{45}H_{36}FeN_{12}O_3$	$C_{45}H_{36}FeN_{12}O_3$	$C_{45}H_{36}FeN_{12}O_3$	$C_{45}H_{36}FeN_{12}O_3$	$C_{44}H_{32}FeN_{12}O_2$	$C_{44}H_{32}FeN_{12}O_2$	$C_{44}H_{32}FeN_{12}O_2$
Molecular weight	848.71	848.71	848.71	848.71	816.67	816.67	816.67
Crystal size (mm)	0.40 × 0.20 × 0.14	0.40 × 0.20 × 0.14	0.39 × 0.23 × 0.07	0.55 × 0.19 × 0.06	0.55 × 0.19 × 0.06	0.39 × 0.23 × 0.07	0.55 × 0.19 × 0.06
$\lambda(MoK\alpha)$ (Å)	0.71073	0.71073	0.71073	0.71073	0.71073	0.71073	0.71073
Space group, Z	$P\bar{1}$, 2	$P\bar{1}$, 2	$P\bar{1}$, 2	$P\bar{1}$, 2	$P\bar{1}$, 2	$P\bar{1}$, 2	$P\bar{1}$, 2
a (Å)	8.9571(5)	8.9675(3)	8.9616(2)	8.9838(3)	8.8932(5)	8.8948(3)	8.9218(2)
b (Å)	13.1036(7)	13.2483(5)	13.3489(2)	13.4223(5)	12.9755(8)	13.0374(7)	12.9832(4)
c (Å)	17.4234(9)	17.4794(7)	17.5163(3)	17.5869(5)	17.6089(5)	17.7142(7)	17.6134(4)
α (°)	89.338(1)	89.1589(8)	88.976(2)	88.985(3)	91.234(4)	91.106(4)	91.356(2)
β (°)	85.101(1)	85.24238(8)	85.3650(10)	85.425(3)	86.922(4)	87.062(3)	86.696(2)
γ (°)	78.027(1)	78.3151(8)	78.525(2)	78.782(3)	82.757(5)	82.846(4)	82.570(2)
Cell volume (Å³)	1993.2(2)	2026.58(13)	2046.83(7)	2073.54(12)	2012.01(18)	2034.87(15)	2018.73(9)
ρ (g/cm³)	1.414	1.391	1.377	1.359	1.348	1.333	1.344
μ, cm⁻¹	4.38	4.31	4.26	4.21	4.29	4.24	4.28
Refls collected/unique	24,684/10,591	21,590/10,764	22,180/10,116	20,254/10,910	16,091/9254	21,019/10,054	17,015/9203
R_{int}	0.0552	0.0371	0.0139	0.0515	0.0217	0.0353	0.0178
θ_{max} (°)	29.0	29.0	28.3	31.1	28.3	28.3	28.3
Refls with $[I > 2\sigma(I)]$	7166	7057	8542	7150	5480	4017	6038
Parameters refined	562	562	560	560	538	538	538
Final R_1, wR_2 $[I > 2\sigma(I)]$	0.0496, 0.1022	0.0426, 0.0920	0.0348, 0.0884	0.0597, 0.1327	0.0613/0.1386	0.0699, 0.1311	0.0569, 0.1244
Goodness-of-fit	1.011	1.000	1.002	1.000	1.006	1.003	1.003
Residual el. density (e·Å⁻³)	0.502/−0.633	0.349/−0.482	0.279/−0.219	1.109/−0.331	0.489/−0.621	0.495/−0.349	0.559/−0.705
CCDC reference	1527039	1527040	1527042	1527041	1527043	1527044	1527045

Table 3. *Cont.*

Parameter	1 295 K	1 260 K	1 220 K	2 100 K	2 220 K	2 295 K	2 325 K
Cell setting	triclinic	triclinic	triclinic	triclinic	triclinic	triclinic	triclinic
Formula	$C_{44}H_{32}FeN_{12}O_2$	$C_{44}H_{32}FeN_{12}O_2$	$C_{44}H_{32}FeN_{12}O_2$	$C_{34}H_{31}FeN_9O_2$	$C_{34}H_{31}FeN_9O_2$	$C_{34}H_{31}FeN_9O_2$	$C_{34}H_{31}FeN_9O_2$
Molecular weight	816.67	816.67	816.67	653.53	653.53	653.53	653.53
Crystal size (mm)	$0.55 \times 0.19 \times 0.06$	$0.55 \times 0.19 \times 0.06$	$0.55 \times 0.19 \times 0.06$	$0.13 \times 0.10 \times 0.08$	$0.52 \times 0.28 \times 0.05$	$0.52 \times 0.28 \times 0.05$	$0.52 \times 0.28 \times 0.05$
$\lambda(MoK\alpha)$ (Å)	0.71073	0.71073	0.71073	0.71073	0.71073	0.71073	0.71073
Space group, Z	$P\bar{1}$, 2	$P\bar{1}$, 2	$P\bar{1}$, 2	$P\bar{1}$, 2	$P\bar{1}$, 2	$P\bar{1}$, 2	$P\bar{1}$, 2
a (Å)	8.9230(2)	8.8950(2)	8.9120(2)	7.6085(8)	7.6695(4)	7.7072(4)	7.72945(18)
b (Å)	12.9439(3)	12.8594(3)	12.8379(3)	14.2710(16)	14.3596(5)	14.3879(6)	14.4016(4)
c (Å)	17.5498(3)	17.4386(3)	17.4233(4)	16.0300(18)	16.0693(5)	16.1428(6)	16.1913(6)
α (°)	91.432(2)	91.464(2)	91.468(2)	67.454(2)	66.939(3)	66.861(4)	66.872(3)
β (°)	86.497(2)	86.312(2)	86.152(2)	85.078(2)	84.870(3)	84.786(4)	84.831(3)
γ (°)	82.430(2)	82.310(2)	82.220(2)	76.296(2)	76.089(4)	75.980(4)	75.969(2)
Cell volume (Å³)	2004.44(7)	1971.49(7)	1969.41(8)	1561.7(3)	1580.51(11)	1597.03(13)	1608.02(9)
ρ (g/cm³)	1.353	1.376	1.377	1.390	1.373	1.359	1.350
μ, cm⁻¹	4.31	4.38	4.38	5.30	5.24	5.19	5.15
θ_{max} (°)	28.3	28.3	29.5	29.0	28.8	29.0	28.8
Refls collected/unique	16,900/9137	15,460/8994	16,836/9234	19,290/8302	16,642/8648	17,204/8850	16,515/8763
R_{int}	0.0165	0.0177	0.0171	0.0627	0.0179	0.0229	0.0220
Refls with [$I > 2\sigma(I)$]	6665	6990	7080	6003	7172	6739	6104
Parameters refined	538	538	538	423	426	426	426
Final R_1, wR_2 [$I > 2\sigma(I)$]	0.0479, 0.1080	0.0427, 0.0924	0.0387, 0.0877	0.0391, 0.0814	0.0348, 0.0776	0.0406, 0.0871	0.0459, 0.0953
Goodness-of-fit	1.003	1.004	1.001	1.007	1.006	1.005	1.001
Residual el. density (e·Å⁻³)	0.447/−0.652	0.387/−0.548	0.335/−0.349	0.487/−0.448	0.310/−0.437	0.312/−0.443	0.350/−0.202
CCDC reference	1527046	1527047	1527048	1527049	1527050	1527051	1527052

3.4. Transport and Magnetic Measurements

The dc resistivity measurements were performed on single crystals by a standard four-probe method with the current flow parallel to the TCNQ stacks (along the [1$\bar{1}$0] direction) in the temperature range 78–300 K. Four annealed platinum wires (0.02 mm in diameter) were attached to a crystal surface by a graphite paste (Figure S1). This geometry is convenient for the test measurements to reveal the features in the temperature dependences of the resistance including measurement time. In the strongly anisotropic sample the measured value contains the mixture of both in-plane and out-of-plane components of the resistivity tensor, due to the current distributed non-uniformly through the sample cross section. This explains why we measured the resistivity tensor components separately in the control experiments. To measure in-plane anisotropy we applied Montgomery method [60] for the samples in the shape of thin plates, elongated in the direction of TCNQ stacks (typical sample shape is shown in the Figure S1). Therefore, by using two pairs of contacts attached to the plate corners on the long sides of the plate, we could measure two components of the resistivity tensor along and perpendicular to the of TCNQ stacks. To measure the out-of-plane resistivity tensor we applied the modified Montgomery method [61] on the sample with two pairs of contacts attached to the opposite sample surfaces.

Magnetic measurements were performed by using a Quantum Design MPMS-5-XL SQUID magnetometer (Quantum Design, San Diego, CA, USA). The static magnetic susceptibility $\chi(T)$ of the polycrystalline sample was measured at the magnetic fields B = 0.1 T, 4.0 T at warming and cooling regimes in the temperature range of 2–400 K. Field dependence of the magnetization $M(B)$ were obtained at 2.0 K after several scans over the field range from −5.0 to +5.0 T. The sample had been cooled to 2.0 K in a magnetic field B = 4.0 T. Then the measurements were performed at the decreasing field with a sign reversal to −5.0 T and further increasing field to +5.0 T.

EPR spectra were recorded in the temperature range of 90–370 K on a standard homodyne X-band (9.4 GHz) Bruker ELEXSYS E580 FT/CW spectrometer (Bruker AXS GmbH, Karlsruhe, Germany). The temperature was set and stabilized at a rate of 0.5–5 K/min with an accuracy of 0.1 K using a liquid nitrogen gas-flow cryostat. The spin contribution to the magnetic susceptibility was determined by the double integration of the EPR signal (Schumacher-Slichter method) under conditions for the field sweep $\delta B_{sw} \geq 10 \Delta B$ (ΔB is the peak-to-peak EPR line width of the total spectrum). In this case, an error of the method for the Lorentz lineshape is ~10%. The pyrolytic coal product with $g = 2.00283$ was used as the standard of a spin concentration.

4. Conclusions

We reported synthesis and physical properties (structure, transport, magnetic susceptibility, magnetization and electron paramagnetic resonance characteristics) of the series of three compounds incorporating Fe(III) cation complexes [Fe(III)(sal$_2$-trien)]$^+$ in the TCNQ network in a fractional/full reduction state:[Fe(III)(sal$_2$-trien)](TCNQ)$_2$·CH$_3$OH (1·MeOH), [Fe(III)(sal$_2$-trien)](TCNQ)·CH$_3$CN (2), and [Fe(III)(sal$_2$-trien)](TCNQ)·H$_2$O (3). Spin-crossover transition was found in all three systems regardless solvent molecules.

Highly conducting system (1·MeOH) with $\sigma(300$ K$) = 5.4$ $\Omega^{-1} \cdot$cm^{-1} and narrow band gap $\Delta E = 0.05$ eV revealed a reversible SCO transition at 410 K. Resistivity and spin relaxation in the conducting tetradic TCNQ stacks were found sensitive to SCO but demonstrated qualitatively different behavior. Spin-spin relaxation between low-spin and high-spin moments of Fe(III) complexes took place via a spin reservoir of mobile electrons. Temperature evolution of the structural characteristics revealed the in-stack charge leveling and thermal hysteresis due to solvent loss.

Low conducting systems **2** and **3** demonstrated irreversible magnetic response and thermal hysteresis of SCO transition. Due to cooperative interactions, solvent loss led to the phase fractionation. The LS phase demonstrated SCO transition and revealed spin-spin interactions between low-spin and high-spin magnetic moments of Fe(III) ions. Residual HS phase discovered ferro(ferri)magnetic coupling at $T_c = 4.11$ K.

We did consider the arguments, promoting presence of superexchange coupling between sal$_2$-trien ligand and CN groups of TCNQ via N_{amine}/N_{imine} short contacts. Switchable magnetic moments, ferromagnetic coupling, and low-dimensional conductivity make the realization of spin-dependent electron transport prospective in such systems.

Supplementary Materials: The following are available online, Figure S1: A crystal of complex **1**·MeOH with the electrodes for measurement of conductivity, Figure S2: *TG-DSC* curves and mass spectra for **1**·MeOH, Figure S3: *TG-DSC* curves and mass spectra for **2**, Figure S4: The character of TCNQ overlap within the II-I-I-II tetrads (a) and between the tetrads (b) in **1**·MeOH at 100 K, Figure S5: View of the structure **2** along *a*, Figure S6: The $\pi \ldots \pi$ stacking in the pairs of cations in **2**, Figure S7: Temperature evolution of the EPR lineshape for **2** and **3**, Figure S8: EPR spectra for the compounds **1**·MeOH (1), **2** (2) and **3** (3), Figure S9: Evolution of the χT for **2** in the range of spin-crossover transition between the LS states, S = 1/2, and the HS states, S = 5/2, of Fe(III) ions, Figure S10: Best-fit curves for EPR linewidth broadening, $\Delta B(T)$, in **1**·MeOH, **2** and **3**, Table S1: The charges (δ) of TCNQ radical anions estimated from Kistenmacher's empirical formula, Table S2: Selected bond lengths (Å) and angles (°) in **1**·MeOH and **1**,Table S3: Selected bond lengths (Å) and angles (°) in **2**, Table S4: Parameters of exponential fitting curves for the EPR linewidth data in Figure S10.

Acknowledgments: This work was supported by the Russian Foundation for Basic Research, project No. 14-03-00119 and Presidium of the Russian Academy of Sciences, project No. 15-17-2-17. The authors would like to thank K.A. Lyssenko for the X-ray experiments at 100 K and 220 K for **1**·MeOH and at 100 K for **2**.

Author Contributions: A.I.K. synthesized the complexes; S.V.S. solved crystal structures; L.V.Z. analyzed the structure data; L.I.B. and V.N.Z. performed measurements of conductivity and anisotropy of conductivity; A.V.K. performed measurements on SQUID magnetometer; D.V.S. recorded EPR spectra and analyzed data; Y.N.S. supervised magnetic and resonance measurements, analyzed data and wrote the paper together with L.V.Z. and E.B.Y.; E.B.Y. supervised overall work and organized the project.

Conflicts of Interest: The authors declare no conflict of interest. The founding sponsors had no role in the design of the study; in the collection, analyses, or interpretation of data; in the writing of the manuscript, and in the decision to publish the results.

References

1. Sugawara, T. and Miyazaki, A. Magnetism and Conductivity. In *Multifunctional Molecular Materials*; Ouahab, L., Ed.; Pan Stanford Publishing, Pte. Ltd.: Singapore, 2013; pp. 1–60.

2. Coronado, E.; Day, P. Magnetic molecular conductors. *Chem. Rev.* **2004**, *104*, 5419–5449. [CrossRef] [PubMed]

3. Kobayashi, H.; Cui, H.; Kobayashi, A. Organic Metals and Superconductors Based on BETS (BETS = Bis(ethylenedithio)tetraselenafulvalene). *Chem. Rev.* **2004**, *104*, 5265–5288. [CrossRef] [PubMed]

4. Enoki, T.; Miyazaki, A. Magnetic TTF-based charge-transfer complexes. *Chem. Rev.* **2004**, *104*, 5449–5477. [CrossRef] [PubMed]

5. Kushch, N.D.; Yagubskii, E.B.; Kartsovnik, M.V.; Buravov, L.I.; Dubrovskii, A.D.; Biberacher, W. Pi-donor BETS based bifunctional superconductor with polymeric dicyanamidomanganate(II) anion layer: κ-(BETS)$_2$Mn[N(CN)$_2$]$_3$. *J. Am. Chem. Soc.* **2008**, *130*, 7238–7240. [CrossRef] [PubMed]

6. Vyaselev, O.M.; Kartsovnik, M.V.; Biberacher, W.; Zorina, L.V.; Kushch, N.D.; Yagubskii, E.B. Magnetic transformations in the organic conductor κ-(BETS)$_2$Mn[N(CN)$_2$]$_3$ at the metal-insulator transition. *Phys. Rev. B* **2011**, *83*, 094425. [CrossRef]

7. Kurmoo, M.; Graham, A.W.; Day, P.; Coles, S.J.; Hursthouse, M.B.; Caulfield, J.L.; Singleton, J.; Pratt, F.L.; Hayes, W.; Ducasse, L.; et al. Superconducting and semiconducting magnetic charge-transfer salts: (BEDT-TTF)$_4$AFe(C$_2$O$_4$)$_3$C$_6$H$_5$CN (A = H$_2$O, K, NH$_4$). *J. Am. Chem. Soc.* **1995**, *117*, 12209–12217. [CrossRef]

8. Rashid, S.; Turner, S.S.; Day, P.; Howard, J.A.K.; Guionneau, P.; McInnes, E.J.L.; Mabbs, F.E.; Clark, R.J.H.; Firth, S.; Biggs, T.J. New superconducting charge-transfer salts (BEDT-TTF)$_4$[A·M(C$_2$O$_4$)$_3$]·C$_6$H$_5$NO$_2$ (A = H$_3$O or NH$_4$, M = Cr or Fe, BEDT-TTF = bis(ethylenedithio)tetrathiafulvalene). *J. Mater. Chem.* **2001**, *11*, 2095–2101. [CrossRef]

9. Laukhin, V.N.; Audouard, A.; Fortin, J.-Y.; Vignolles, D.; Prokhorova, T.G.; Yagubskii, E.B.; Canadell, E. Quantum oscillations in coupled orbits networks of (BEDT-TTF) salts with tris(oxalato)metallate anions. *Low Temp. Phys. (Fiz. Nizk. Temp.)* **2017**, *43*, 33–40. [CrossRef]

10. Prokhorova, T.G.; Yagubskii, E.B. Organic conductors and superconductors based on bis(ethylenedithio) tetrathiafulvalene radical cation salts with supramolecular tris(oxalato)metallate anions. *Russ. Chem. Rev.* **2017**, *86*, 164–180. [CrossRef]

11. Otsuka, T.; Kobayashi, A.; Miyamoto, Y.; Kiuchi, J.; Nakamura, S.; Wada, N.; Fujiwara, E.; Fujiwara, H.; Kobayashi, H. Organic antiferromagnetic metals exhibiting superconducting transitions κ-(BETS)$_2$FeX$_4$ (*X* = Cl, Br): Drastic effect of halogen substitution on the successive phase transitions. *J. Solid State Chem.* **2001**, *159*, 407–412. [CrossRef]

12. Ojima, E.; Fujiwara, H.; Kato, K.; Kobayashi, H.; Tanaka, H.; Kobayashi, A.; Tokumoto, M.; Cassoux, P. Antiferromagnetic organic metal exhibiting superconducting transition, κ-(BETS)$_2$FeBr$_4$ [BETS = bis(ethylenedithio)tetraselenafulvalene]. *J. Am. Chem. Soc.* **1999**, *121*, 5581–5582. [CrossRef]

13. Coronado, E.; Galan-Mascaros, J.R.; Gomez-Garcia, C.J.; Laukhin, V.N. Coexistence of ferromagnetism and metallic conductivity in a molecule-based layered compound. *Nature* **2000**, *408*, 447–449. [CrossRef] [PubMed]

14. Uji, S.; Shinagawa, H.; Terashima, T.; Yakabe, T.; Terai, Y.; Tokumoto, M.; Kobayashi, A.; Tanaka, H.; Kobayashi, H. Magnetic-field-induced superconductivity in a two-dimensional organic conductor. *Nature* **2001**, *410*, 908–910. [CrossRef] [PubMed]

15. Fujiwara, H.; Kobayashi, H.; Fujiwara, E.; Kobayashi, A. An indication of magnetic-field-induced superconductivity in a bifunctional layered organic conductor, κ-(BETS)$_2$FeBr$_4$. *J. Am. Chem. Soc.* **2002**, *124*, 6816–6817. [CrossRef] [PubMed]

16. Valade, L.; Malfant, I.; Faulmann, C. Toward bifunctional materials with conducting, photochromic, and spin crossover properties. In *Multifunctional Molecular Materials*; Ouahab, L., Ed.; Pan Stanford Publishing, Pte. Ltd.: Singapore, 2013; pp. 149–184.

17. Sato, O.; Li, Z.-Y.; Yao, Z.-S.; Kang, S.; Kanegawa, S. Multifunctional materials combining spin-crossover with conductivity and magnetic ordering. In *Spin-Crossover Materials: Properties and Applications*; Halcrow, M.A., Ed.; John Wiley & Sons: Oxford, UK, 2013; pp. 304–319.

18. Kato, R. Conducting Metal Dithiolene Complexes: Structural and Electronic Properties. *Chem. Rev.* **2004**, *104*, 5319–5346. [CrossRef] [PubMed]

19. Melby, L.R.; Harder, R.J.; Hertler, W.R.; Mahler, W.; Benson, R.E.; Mochel, W.E. Substituted Quinodimethans. II. Anion-radical Derivatives and Complexes of 7,7,8,8-Tetracyano-quinodimethan. *J. Am. Chem. Soc.* **1962**, *84*, 3374–3387. [CrossRef]

20. Schegolev, I.F. Electric and magnetic properties of linear conducting chains. *Phys. Status Solidi (a)* **1972**, *12*, 9–45. [CrossRef]

21. Shibaeva, L.; Atovmyan, O. The structure of conducting 7,7,8,8-tetracyanoquinodimethane complexes. *J. Struct. Chem.* **1972**, *13*, 514–531. [CrossRef]

22. Yagubskii, E.B. From quasi-one-dimensional conductors based on TCNQ salts to the first quasi-two-dimensional superconductors at ambient pressure based on BEDT-TTF triiodides. In *Organic Conductor, Superconductors and Magnets: From Synthesis to Molecular Electronics*; Ouahab, L., Yagubskii, E.B., Eds.; Kluwer Academic Publishers: Dordrecht, the Netherlands, 2003; pp. 45–65.

23. Herbstein, F.H.; Kapon, M. Classification of closed shell TCNQ salts into structural families and comparison of diffraction and spectroscopic methods of assigning charge states to TCNQ moieties. *Crystallogr. Rev.* **2008**, *14*, 3–74. [CrossRef]

24. Halcrow, M.A. (Ed.) *Spin-Crossover Materials: Properties and Applications*; John Wiley & Sons: Oxford, UK, 2013.

25. Halcrow, M.A. Structure: Function relationships in molecular spin-crossover complexes. *Chem. Soc. Rev.* **2011**, *40*, 4119–4142. [CrossRef] [PubMed]

26. Nemec, I.; Herchel, R.; Salitros, I.; Travnicek, Z.; Moncol, J.; Fuess, H.; Ruben, M.; Linert, W. Anion driven modulation of magnetic intermolecular interactions and spin crossover properties in an isomorphous series of mononuclear iron(III) complexes with a hexadentate Schiff base ligand. *CrystEngComm* **2012**, *14*, 7015–7024. [CrossRef]

27. Murata, K.; Kagoshima, S.; Yasuzuka, S.; Yoshino, H.; Kondo, R. High-Pressure Research in Organic Conductors. *J. Phys. Soc. Jpn.* **2006**, *75*, 051015. [CrossRef]

28. Yasuzuka, S.; Murata, K. Recent progress in high-pressure studies on organic conductors. *Sci. Technol. Adv. Mater.* **2009**, *10*, 024307. [CrossRef] [PubMed]

29. Takahashi, K.; Cui, H.-B.; Okano, Y.; Kobayashi, H.; Einaga, Y.; Sato, O. Electrical Conductivity Modulation Coupled to a High-Spin–Low-Spin Conversion in the Molecular System [Fe^{III}(qsal)$_2$][Ni(dmit)$_2$]$_3$·CH$_3$CN·H$_2$O. *Inorg. Chem.* **2006**, *45*, 5739–5741. [CrossRef] [PubMed]

30. Takahashi, K.; Cui, H.-B.; Okano, Y.; Kobayashi, H.; Mori, H.; Tajima, H.; Einaga, Y.; Sato, O. Evidence of the Chemical Uniaxial Strain Effect on Electrical Conductivity in the Spin-Crossover Conducting Molecular System: [Fe^{III}(qnal)$_2$][Pd(dmit)$_2$]$_5$·Acetone. *J. Am. Chem. Soc.* **2008**, *130*, 6688–6689. [CrossRef] [PubMed]

31. Nihei, M.; Takahashi, N.; Nishikawa, H.; Oshio, H. Spin-crossover behavior and electrical conduction property in iron(II) complexes with tetrathiafulvalene moieties. *Dalton Trans.* **2011**, *40*, 2154–2156. [CrossRef] [PubMed]

32. Plan, H.; Benjamin, S.M.; Steven, E.; Brooks, J.S.; Shatruk, M. Photomagnetic Response in Highly Conductive Iron(II) Spin-Crossover Complexes with TCNQ Radicals. *Angew. Chem. Int. Ed.* **2015**, *54*, 823–827.

33. Zhang, X.; Wang, Z.-X.; Xie, H.; Li, M.-X.; Woods, T.J.; Dunbar, K.R. A cobalt(II) spin-crossover compound with partially charged TCNQ radicals and an anomalous conducting behavior. *Chem. Sci.* **2016**, *7*, 1569–1574. [CrossRef]

34. Dorbes, S.; Valade, L.; Real, J.A.; Faulmann, C. [Fe(sal$_2$-trien)][Ni(dmit)$_2$]: Towards switchable spin crossover molecular conductors. *Chem. Commun.* **2005**, 69–71. [CrossRef] [PubMed]

35. Faulmann, C.; Dorbes, S.; Real, J.A.; Valade, L. Electrical conductivity and spin crossover: Towards the first achievement with a metal bis dithiolene complex. *J. Low Temp. Phys.* **2006**, *142*, 261–266. [CrossRef]

36. Faulmann, C.; Dorbes, S.; Garreau de Bonneval, B.; Molnar, G.; Bousseksou, A.; Gomes-Garcia, C.J.; Coronado, E.; Valade, L. Towards molecular conductors with a spin-crossover phenomenon: Crystal structures, magnetic properties and Mossbauer spectra of [Fe(salten)Mepepy][M(dmit)$_2$] complexes. *Eur. J. Inorg. Chem.* **2005**, *2005*, 3261–3270. [CrossRef]

37. Faulmann, C.; Jacob, K.; Dorbes, S.; Lampert, S.; Malfant, I.; Doublet, M.-L.; Valade, L.; Real, J.A. Electrical conductivity and spin crossover: A new achievement with a metal bis dithiolene complex. *Inorg. Chem.* **2007**, *46*, 8548–8559. [CrossRef] [PubMed]

38. Fukuroi, K.; Takahashi, K.; Mochida, T.; Sakurai, T.; Ohta, H.; Yamamoto, T.; Einada, Y.; Mori, H. Synergistic Spin Transition between Spin Crossover and Spin-Peierls-like Singlet Formation in the Halogen-Bonded Molecular Hybrid System: [Fe(Iqsal)$_2$][Ni(dmit)$_2$]·CH$_3$CN·H$_2$O. *Angew. Chem. Int. Ed.* **2014**, *53*, 1983–1986. [CrossRef] [PubMed]

39. Nakano, M.; Fujita, N.; Matsubayashi, G.E.; Mori, W. Modified chesnut model for spin-crossover semiconductors [Fe(acpa)$_2$](TCNQ)n. *Mol. Cryst. Liq. Cryst.* **2002**, *379*, 365–370. [CrossRef]

40. Shvachko, Y.N.; Starichenko, D.V.; Korolyov, A.V.; Yagubskii, E.B.; Kotov, A.I.; Buravov, L.I.; Lyssenko, K.A.; Zverev, V.N.; Simonov, S.V.; Zorina, L.V.; et al. The Conducting Spin-Crossover Compound Combining Fe(II) Cation Complex with TCNQ in a Fractional Reduction State. *Inorg. Chem.* **2016**, *55*, 9121–9130. [CrossRef] [PubMed]

41. Kistenmacher, T.J.; Emge, T.J.; Bloch, A.N.; Cowan, D.O. Structure of the red, semiconducting form of 4,4',5,5'-tetramethyl-Δ2,2'-bi-1,3-diselenole-7,7,8,8-tetracyano-p-quinodimethane, TMTSF-TCNQ. *Acta Cryst. B* **1982**, *38*, 1193–1199. [CrossRef]

42. Radváková, A.; Kazheva, O.N.; Chekhlov, A.N.; Dyachenko, O.A.; Kucmin, M.; Kajňaková, M.; Feher, A.; Starodub, V.A. Two-gap magnetic structure of the two-stack anion-radical salt (N-Me-3,5-Di-Me-Py)(TCNQ)$_2$ (Py is pyridine). *J. Phys. Chem. Solids* **2010**, *71*, 752–757. [CrossRef]

43. Pukacki, W.; Graja, A. Electric and magnetic properties of organometallic TCNQ salts. *Synth. Met.* **1988**, *24*, 137–143. [CrossRef]

44. Petersen, R.L.; Symons, M.C.R.; Taiwo, F.A. Application of radiation and ESR spectroscopy to the study of ferryl haemoglobin. *J. Chem. Soc. Faraday Trans. 1* **1989**, *85*, 2435–2444. [CrossRef]

45. Pilbrow, J.R. *Transition Ion Electron Paramagnetic Resonance*; Clarendon Press: Oxford, UK, 1990.

46. Hagen, W.R. *Biomolecular EPR Spectroscopy*; CRC Press: Boca Raton, FL, USA, 2009.

47. Weil, J.A.; Wertz, J.E.; Bolton, J.R. *Electron Paramagnetic Resonance: Elementary Theory and Practical Applications*; John Wiley: New York, NY, USA, 1994.

48. Paradis, N.; Le Gac, F.; Guionneau, P.; Largeteau, A.; Yufit, D.S.; Rosa, P.; Létard, J.-F.; Chastanet, G. Effects of Internal and External Pressure on the [Fe(PM-PEA)$_2$(NCS)$_2$] Spin-Crossover Compound (with PM-PEA=N-(21-pyridylmethylene)-4(phenylethynyl)aniline). *Magnetochemistry* **2016**, *2*, 15–32. [CrossRef]

49. Barnes, S.E. Theory of electron spin resonance of magnetic ions in metals. *Adv. Phys.* **1981**, *30*, 801–938. [CrossRef]

50. Kürti, J.; Menczel, G. g-Factor anisotropy and charge transfer in three complex TCNQ salts. *Phys. Status Solidi B* **1980**, *102*, 639–645. [CrossRef]

51. Brooker, S. Spin crossover with thermal hysteresis: Practicalities and lessons learnt. *Chem. Soc. Rev.* **2014**, *44*, 2880–2892. [CrossRef] [PubMed]

52. Benelli, C.; Gatteschi, D. *Introduction to Molecular Magnetism: From Transition Metals to Lanthanides*; Wiley-VCH: Weinheim, Germany, 2015.

53. Létard, J.-F.; Guionneau, P.; Rabardel, L.; Howard, J.A.K.; Goeta, A.E.; Chasseau, D.; Kahn, O. Structural, Magnetic, and Photomagnetic Studies of a Mononuclear Iron(II) Derivative Exhibiting an Exceptionally Abrupt Spin Transition. Light-Induced Thermal Hysteresis Phenomenon. *Inorg. Chem.* **1998**, *37*, 4432–4441. [CrossRef] [PubMed]

54. Day, P.; Underhill, A.E. Metal-organic and Organic Molecular Magnets. *Phil. Trans. R. Soc. Lond. A* **1999**, *357*, 2849–3184.

55. Starodub, V.A.; Starodub, T.N. Radical anion salts and charge transfer complexes based on tetracyanoquinodimethane and other strong π-electron acceptors. *Russ. Chem. Rev.* **2014**, *83*, 391–438. [CrossRef]

56. Zhao, H.; Heintz, R.A.; Ouyang, X.; Grandinetti, G.; Cowen, J.; Dunbar, K.R. *Insight into the Behavior of M(TCNQ)n (n = 1, 2) Crystalline Solids and Films: X-ray, Magnetic and Conducting Properties. NATO ASI: Supramolecular Engineering of Synthetic Metallic Materials: Conductors and Magnets*; Veciana, J., Ed.; Kluwer: Dordrecht, The Netherlands, 1999; Volume 518, pp. 353–376.

57. Clerac, R.; O'Kane, S.; Cowen, J.; Ouyang, X.; Heintz, R.A.; Zhao, H.; Bazile, M.J., Jr.; Dunbar, K.R. Glassy Magnets Composed of Metals Coordinated to 7,7,8,8-tetracyanoquinodimethane: $M(TCNQ)_2$ (M = Mn, Fe, Co, Ni). *Chem. Mater.* **2003**, *15*, 1840–1850. [CrossRef]

58. Nweedle, M.F.; Wilson, L.J. Variable spin iron(III) chelates with hexadentate ligands derived from triethylenetetramine and various salicylaldehydes. Synthesis, characterization, and solution state studies of a new $^2T \leftrightarrow {}^6A$ spin equilibrium system. *J. Am. Chem. Soc.* **1976**, *98*, 4824–4834.

59. Sheldrick, G.M. A short history of SHELX. *Acta Cryst. A* **2008**, *64*, 112–122. [CrossRef] [PubMed]

60. Montgomery, H.C. Method for Measuring Electrical Resistivity of Anisotropic Materials. *J. Appl. Phys.* **1971**, *42*, 2971. [CrossRef]

61. Buravov, L.I. Calculation of resistance anisotropy with regard to model ends by conformal-transformation. *Zhurnal Tekh. Fiz.* **1989**, *59*, 138–142.

magnetochemistry

MDPI

Article

Dye-Sensitized Molecular Charge Transfer Complexes: Magnetic and Conduction Properties in the Photoexcited States of Ni(dmit)$_2$ Salts Containing Photosensitive Dyes

Ryoma Yamamoto [1], Takashi Yamamoto [1], Keishi Ohara [1] and Toshio Naito [1,2,*]

[1]   Graduate School of Science and Engineering, Ehime University, 2-5, Bunkyo-cho, Matsuyama 790-8577, Japan; c851018x@mails.cc.ehime-u.ac.jp (R.Y.); yamataka@ehime-u.ac.jp (T.Y.); ohara.keishi.mg@ehime-u.ac.jp (K.O.)
[2]   Division of Material Science, Advanced Research Support Center (ADRES), Ehime University, 2-5, Bunkyo-cho, Matsuyama 790-8577, Japan
*   Correspondence: tnaito@ehime-u.ac.jp; Tel.: +81-(0)89-927-9604

Academic Editor: Manuel Almeida
Received: 3 March 2017; Accepted: 12 May 2017; Published: 19 May 2017

Abstract: Photosensitive dyes often induce charge transfer (CT) between adjacent chemical species and themselves under irradiation of appropriate wavelengths. Because of the reversibility and selectivity of such CT, it is considered to be interesting to utilize such dyes as optically controllable trigger components for conduction and magnetism in the photoexcited states of organic materials. Based on this idea, such a type of new salts, i.e., γ- and δ-DiCC[Ni(dmit)$_2$] in addition to DiCC$_2$[Ni(dmit)$_2$]$_3$ have been prepared, characterized and their physical and structural properties have been examined both under dark and irradiated conditions (dmit^{2-} = 1,3-dithiole-2-thione-4,5-dithiolate, DiCC$^+$ = 3,3'-Dihexyloxacarbocyanine monocation). Among them, under UV (254–450 nm) irradiation, δ-DiCC[Ni(dmit)$_2$] exhibited photoconductivity being six times as high as its dark conductivity at room temperature. The electron spin resonance (ESR) spectra have demonstrated that there are photoexcited spins on both DiCC and [Ni(dmit)$_2$] species as a result of the CT transition between them, serving as localized spins (DiCC) and carriers ([Ni(dmit)$_2$]), respectively. The results obtained in this work have indicated that the strategy mentioned above is effective in developing organic photoresponsive semiconductors with paramagnetism.

Keywords: Ni(II)-dithiolene complex; cyanine dye; charge transfer complex; photoconduction; molecular crystal

1. Introduction

Some kind of aromatic amines and bipyridyl derivatives have attracted attention in various research fields for a long time as photosensitive dyes, which exhibit strong reducing or oxidizing abilities under irradiation [1–8]. Accordingly, they induce charge transfer (CT) between adjacent chemical species and the dyes in the photoexcited states in both solution and solid. This property often triggers various series of redox-type photochemical reactions. Some of them were originally synthesized as photosensitizers for color photographic films, and are now utilized for simplified model systems of photosynthesis [6], dye sensitized solar cells [7], and photocatalysts for clean energy [8].

Different but also interesting utilization of them includes combining them with molecular building blocks for conduction and magnetism [9–33]. Irradiation of light with appropriate wavelengths would bring about CT transition to produce unpaired electrons/holes on both of the dyes and the building blocks in a transient way. As long as the irradiation continues, the excitation and relaxation happen one after another in a continual, rapid and repetitive manner, which practically retains the conduction and magnetism during irradiation. This would enable optical/remote control of appearance and disappearance of conduction and magnetism in an immediate and reversible way.

In fact, such examples have been recently reported [10–12,16]. It has been demonstrated that metallic and paramagnetic properties can be realized in some CT complexes containing bipyridyl derivatives and π-acceptor molecules under UV irradiation. They have mixed-stacking structures with almost fully ionized cations and anions, and are practically diamagnetic insulators under dark conditions. These physical properties are natural judging from their crystal structures and formal charges. Since this structure–property correlation is convincing and generally the case, these kinds of solids has been avoided in the field of molecular conductors. This assumption is true under dark conditions. However, in the photoexcited states, there is metallic conduction coexisting with semiconducting contribution from thermal carriers even in such solids. In addition, one requires low-conducting materials under the dark condition for high photoresponse. Thus, the finding above suggests that there should be a different guideline for development of conducting and magnetic materials under irradiation from the established one for thermodynamically stable metals.

Based on this idea, various kinds of organic CT salts have been prepared. Particular attention has been paid to well-known photosensitizers such as methyl viologen (Scheme 1) and cyanine dyes. By utilizing them as counter cations, a series of [Ni(dmit)$_2$] radical anion salts (dmit^{2-} = 1,3-dithiole-2-thione-4,5-dithiolate; Scheme 1) has been found to form a group of interesting candidates. Among them, there are four kinds of DiCC$^+$ salts (DiCC$^+$ = 3,3′-Dihexyloxacarbocyanine monocation; Scheme 1) with the same stoichiometry but with different crystal structures; α-, β-, γ-, δ-DiCC[Ni(dmit)$_2$]. The α-salt is known [13], yet because of its poor crystal quality and because it is seldom obtained, details still remain elusive. The structural and physical properties of β-salt under dark conditions have been recently reported by our group [13], while the two salts (γ- and δ-salts) are newly obtained. In addition, single crystals with a different stoichiometry, DiCC$_2$[Ni(dmit)$_2$]$_3$ (2:3-salt), have been also newly obtained. In this article, the physical and structural properties of the new salts (γ-, δ- and 2:3-salts) are presented, and are compared with each other in order to discuss the factors and conditions for efficient photoresponsive conducting and magnetic properties.

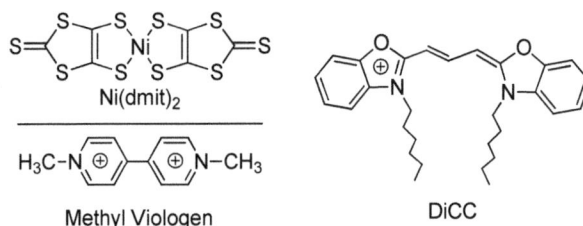

Scheme 1. Chemical structures of compounds.

2. Results

2.1. Crystal Structures, Similarities and Differences among the DiCC Salts

The crystal structures of γ-, δ- and 2:3-salts are shown in Figures 1–3. The crystallographic data are summarized in Table 1. As for the molecular structures, all of the interatomic distances and angles were normal for both anions and cations, and their arrangements were also standard for these types of compounds. In order to make it clear what the important features are for high photoresponse in conduction and magnetism, we should now pay more attention to comparison of their structures than details of each structure. The single crystals of these salts were all (elongated) black platelets. Considering the color of the starting materials (dark green and reddish orange for [Ni(dmit)$_2$]$^-$ and

DiCC$^+$, respectively) of their synthesis, the black color suggests some intermolecular interaction involving CT.

Table 1. Summary of crystallographic data.

Salt	γ	δ	2:3
Cation:Anion	1:1	1:1	2:3
Formula	$C_{35}H_{37}N_2NiO_2S_{10}$	$C_{35}H_{37}N_2NiO_2S_{10}$	$C_{76}H_{74}N_4Ni_3O_4S_{30}$
M (g mol^{-1})	896.99	896.99	2245.26
Temperature (K)	296	296	296
Crystal system	Triclinic	Monoclinic	Triclinic
Space Group	$P\bar{1}$(#2)	$P2_1/n$ (#14)	$P\bar{1}$(#2)
a (Å)	8.36861(17)	10.8163(5)	7.7933(4)
b (Å)	13.7285(3)	30.5876(12)	12.7677(6)
c (Å)	18.2519(4)	12.4468(5)	25.2676(10)
$α$ (°)	79.6321(11)	–	76.977(2)
$β$ (°)	80.7564(11)	95.6803(17)	88.978(3)
$γ$ (°)	83.0074(11)	–	76.505(3)
V (Å^3)	2026.57(7)	4097.7(3)	2380.29(19)
Z	2	4	1
D_{calc} (g cm^{-3})	1.470	1.454	1.566
$μ$ (Cu $Kα$) (cm^{-1})	57.784	57.155	72.288
CCDC deposit #	1526722	1526720	1526726
Reflection/Parameter	15.82	16.49	15.87
Max peak (e$^-$/Å^3)	0.66	0.63	0.89
Min peak (e$^-$/Å^3)	−0.48	−0.71	−0.60
R_1, wR_2	0.0563 [1], 0.1950 [2]	0.0731 [1], 0.2486 [2]	0.0733 [1], 0.2194 [2]
GOF	1.124	1.038	0.929
Max Shift/Error	0.007	0.000	0.066

[1] $I > 2.00σ$ (*I*). [2] All reflections.

The three salts were basically comprised of mixed-stacking structures or alternate arrangement of cations and anions like ionic crystals. This structural common feature suggests that Coulombic attraction between anions and cations should overwhelm the π–π interaction between ions with the same charge in these salts. Closer examination clarified that both cations and anions were almost completely planar except for the hexyl groups of the DiCC cations, and neighboring molecules were parallel and/or close to each other, suggesting π–π interaction between them. This feature manifested itself in 2:3-salt, where there were two types of stacking columns, i.e., anion-only columns in addition to anion-cation mixed columns. Thus, it appears that overall molecular arrangement in these salts should be governed by the Coulombic and π–π interactions. It is the unique feature common to these salts that the crystals most developed along the cation–anion interactions, in contrast to other Ni(dmit)$_2$ semiconducting salts. It is often the case with the Ni(dmit)$_2$ salts that their crystals mostly develop in parallel with the Ni(dmit)$_2$ sheets or columns in which the Ni(dmit)$_2$ anions closely interact with each other.

The differences in the conformation of hexyl groups in the cations affected the resultant molecular arrangements only in a quantitative way such as intermolecular distances. The following subsections briefly describe selected details of the structure of each salt, which will provide us information necessary for understanding the physical properties discussed below.

2.1.1. γ-Salt

The single crystals were elongated hexagonal platelets and most developed along the *a*-axis. The unit cell consisted of an asymmetric unit containing a cation and an anion (Figure 1a), and possessed a mixed-stacking structure along the *a*-axis.

The two hexyl groups of the cation extended approximately in the plane of the π-conjugated part of the molecule. In other words, the entire cation was nearly planar (Figure 1b).

For conduction and magnetism, the interaction between the Ni(dmit)$_2$ anions is important. In γ-salt, the anions formed a loosely-assembled sheet in the *bc*-plane (Figure 1b,c), and there were some short distances between them shorter than or comparable to the van der Waals distance between two sulfur atoms (3.70 Å; Figure 1d).

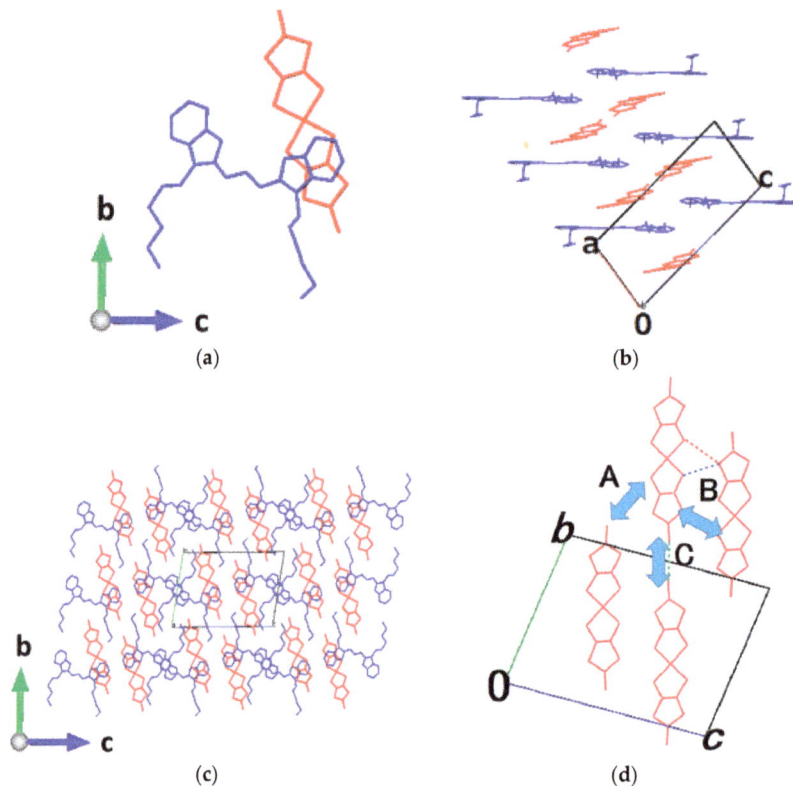

Figure 1. Crystal structure of γ-DiCC[Ni(dmit)$_2$] (DiCC = 3,3′-Dihexyloxacarbocyanine monocation). Red; Ni(dmit)$_2$ anon, blue; DiCC cation, hydrogen atoms are omitted for clarity: (**a**) asymmetric unit; (**b**) view down along *b*-axis; (**c**) view down along *a*-axis; (**d**) interaction between anions in the *bc*-plane. In (**d**), the shortest sulfur–sulfur distances between the two anions designated by broken lines are 3.280(2) (green, in relation C), 3.572(2) (blue, B), 3.705(2) Å (red, B), respectively.

2.1.2. δ-Salt

The single crystals were elongated thick platelets and most developed along the *a*-axis. The unit cell of δ-salt consisted of an asymmetric unit containing a cation and an anion, and can be regarded as a mixed-stacking structure along the *a*-axis.

The direction in which the two hexyl groups extended (Figure 2a) was different from that in γ-salt (Figure 1a). The two hexyl groups of the cation extended almost vertically to the plane of the π-conjugated part of the molecule, and they extended in a parallel way to each other but in an opposite way to those of the neighboring cation (Figure 2b).

Still, in γ-salt, the neighboring mixed-stacking columns are solid-crossing to each other (Figure 2c,d), and, at the crossing points, there were short distances between the adjacent anions in different columns shorter than or comparable to the van der Waals distance between two sulfur atoms.

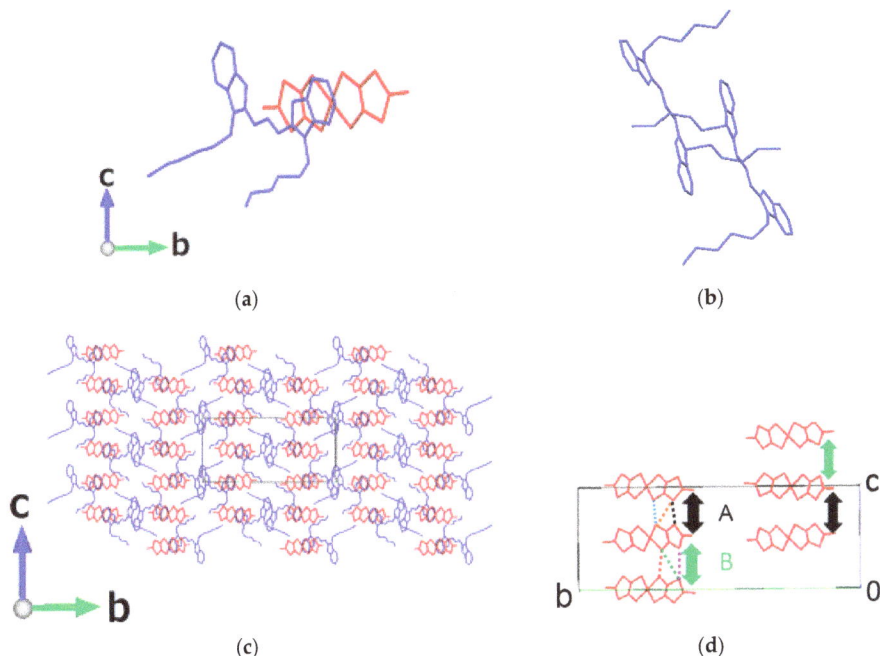

(a)

(b)

(c)

(d)

Figure 2. Crystal structure of δ-DiCC[Ni(dmit)$_2$]. Red; Ni(dmit)$_2$ anon, blue; DiCC cation, hydrogen atoms are omitted for clarity: (**a**) asymmetric unit; (**b**) neighboring cation arrangement; (**c**) view down along *a*-axis; (**d**) interaction between anions in the *bc*-plane. In (**d**), the sulfur–sulfur distances between the two anions designated by broken lines are 3.747(2) (blue), 3.519(2) (orange) and 3.656(2) Å (black) in the intermolecular relation A and 5.312(2) (red), 4.474(2) (green) and 3.819(2) Å (violet) in the relation B, respectively.

2.1.3. 2:3-Salt

The single crystals were elongated thick platelets (parallelepipeds) and most developed along the [210]-direction, which was nearly parallel (~15°) with the *a*-axis. The unit cell of 2:3-salt consisted of an asymmetric unit containing a whole cation and one and a half of anions (Figure 3a). There were two kinds of stacking columns along the *a*-axis: cation–anion mixed-stacking columns and anion-only columns. Accordingly, the structure of 2:3-salt possessed both features of insulating and conducting molecular CT complexes; the former was common to the structures of the 1:1-salts (β- [13], γ-, and δ-) and those often observed in the fully ionized (ionic) molecular CT complexes. The latter structural feature is considered to originate from two factors; one is the non-integer charges on cations and anions due to the small amount of CT between cations and anions in the ground state, and the other is the molecular shapes and sizes to allow them to share the similar intermolecular distances between the two different types of columns (anion-only, and anion–cation mixed types).

The direction in which the two hexyl groups extended was different either from that in γ-salt or that in δ-salt. The two hexyl groups of the cation extended as if they surround the anion-only columns (Figure 3b). There were some short distances (<3.70 Å) between sulfur atoms on neighboring anions, which could form conduction pathways (Figure 3c). Considering the sulfur–sulfur interatomic contacts, the most conductive direction should be in the *a*-axis.

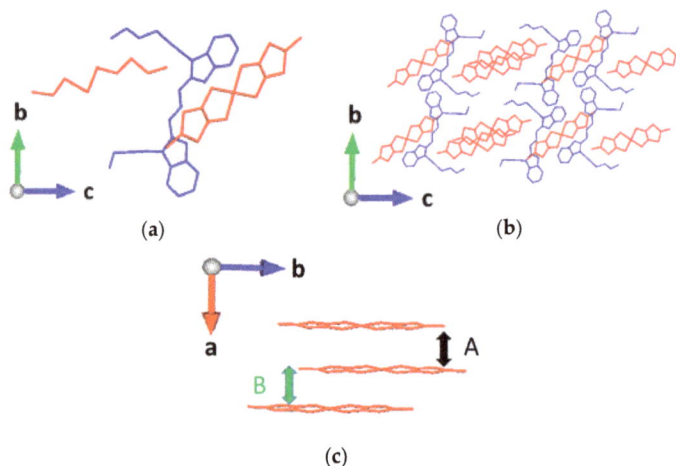

Figure 3. Crystal structure of DiCC$_2$[Ni(dmit)$_2$]$_3$. Red; Ni(dmit)$_2$ anon, blue; DiCC cation, hydrogen atoms are omitted for clarity: (**a**) asymmetric unit containing one and a half of the anions and a whole cation; (**b**) view down along *a*-axis; (**c**) interaction between anions in stacking columns along the *a*-axis. In (**c**), the shortest sulfur–sulfur distances between the two anions designated as A and B are 3.792(3) and 3.688(3) Å, respectively.

2.2. Electrical Behavior

As the conducting and magnetic properties of these salts are expected to be governed by the Ni(dmit)$_2$ anions, resistivity should be measured in the direction where the anions aggregate themselves to form conduction pathways of columns or sheets. However, in these salts, the dimensions of the single crystals in such directions were too small to carry out electrical measurements, suggesting that the interactions between the anions are not so large compared with other intermolecular interactions. Based on the structural analyses above, the common feature of these three salts is that cations and anions alternate with each other along the longest axis of each crystal. In these types of crystals, the electrical resistivity along the longest axis is considered to be dominated by the hopping of carriers between cations and anions. Additionally, the photoresponse should be largest when the CT bands between cations and anions are excited, since the CT transitions produce holes and electrons, corresponding to net carrier doping to both species. Such CT requires cation–anion π–π interactions, which can be estimated by resistivity measurements in the penetrating direction through cations and anions. Therefore, the electrical resistivity measurements were carried out on the single crystals of these salts along the longest axes concerning both dark conductivity and photoresponse.

The γ-salt was highly insulating with room temperature conductivity of 7.9×10^{-9} Scm^{-1} (// *a*-axis) both under dark and irradiated conditions irrespective of incident wavelength in UV-Vis region (Figure 4a). The photoresponse is corroborated by its diffuse reflectance spectra (Figure A1 in Appendix), where no CT bands were observed in 200–2500 nm. This result is also consistent with the electron spin resonance (ESR) spectra under irradiation, where no response (spectral change) was observed either.

The δ-salt exhibited 100 times as high dark conductivity as the remaining salts, 8.6×10^{-7} Scm^{-1} (// *a*-axis) at room temperature, and even higher conductivity under UV-irradiation by six times, 5.2×10^{-6} Scm^{-1} (// *a*-axis) at room temperature, compared with its dark conductivity (Figure 4b).

The 2:3-salt was also highly insulating with room temperature conductivity of 8.3×10^{-9} Scm^{-1} (// *a*-axis) both under dark and irradiated conditions irrespective of incident wavelength in UV-Vis region (Figure 4c). Such negligible photoresponse in conduction is consistent with the diffuse reflectance spectra of UV-Vis-NIR (Figure A1 in Appendix), where CT bands were hardly observed like the case of γ-salt.

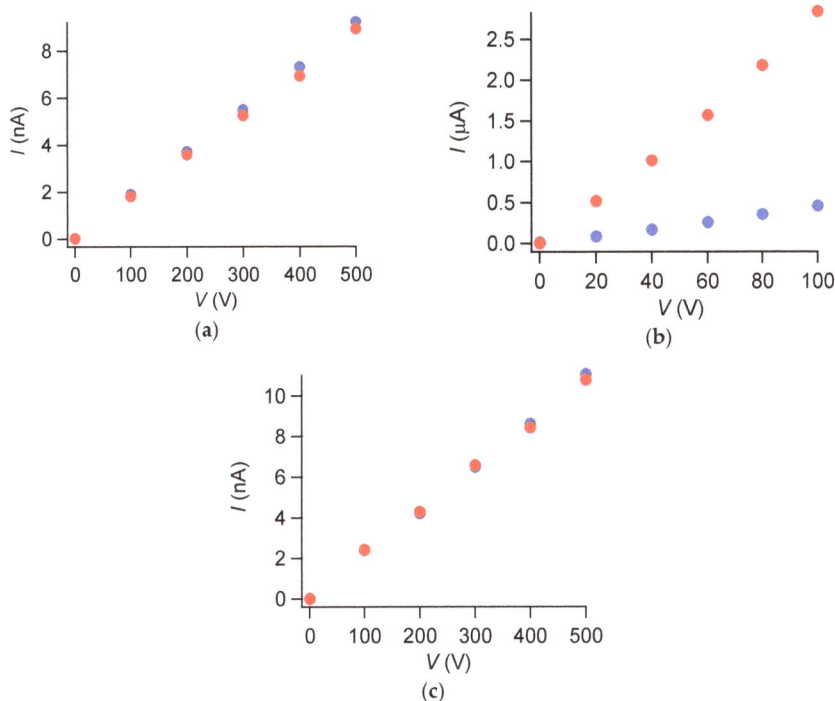

Figure 4. Observed current under dark and UV-irradiated conditions of γ-, δ- and 2:3-salts. Red; under UV-irradiation (254–450 nm), blue; under dark condition: (**a**) γ-salt; (**b**) δ-salt; (**c**) 2:3-salt.

2.3. Electron Spin Resonance (ESR)

The ESR spectra under dark and irradiated conditions for γ- and δ-salts are shown in Figures 5 and 6, together with simulated spectra considering hyperfine interactions with nuclear spins, linewidths, and *g*-values in an anisotropic way. As a result, the obtained parameters are rather isotropic or two-dimensional, as is often the case with π-spins. The best fit parameters corresponding to the simulated spectra are summarized in Tables A1–A3 in the Appendix. The irradiated conditions were identical with those in the electrical resistivity measurements. The ESR was not measured for 2:3-salt because it was highly insulating and did not exhibit any photoresponse in conduction.

Based on the ESR of β-salt in our previous work [13] as well as the spectral simulation in this work, all of the spectra consisted of two sets of peaks originating from the spins on the cations (~325–330 mT) and those on the anions (~315–320 mT), respectively. However, their relative intensities, overall lineshapes, and photoresponses depended on the salts. Some of the peaks could not be reproduced in a quantitative way, i.e., in all of the peak positions, the lineshapes, the linewidths and the intensities. Still, overall spectral features are well reproduced by the simulation for all the three salts, enabling semi-quantitative discussion for us.

The spectra of γ-salt did not change at all between dark and irradiated conditions (Figure 5), which agreed with the photoresponse in electrical behavior. This is consistent with the crystal structure.

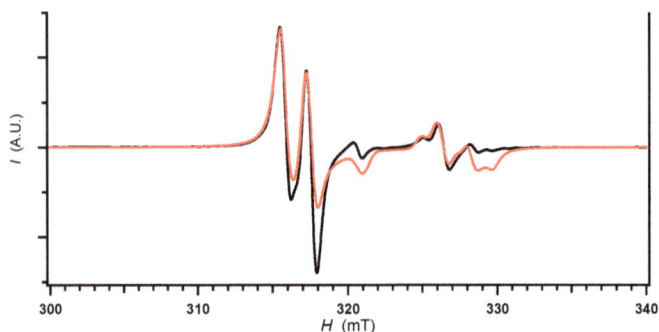

Figure 5. Electron spin resonance (ESR) spectra of γ-salt measured under dark and irradiated (254–450 nm) conditions. Black; observed spectra, red; simulated spectra. The observed spectra were completely identical and overlapped with each other under dark and irradiated conditions.

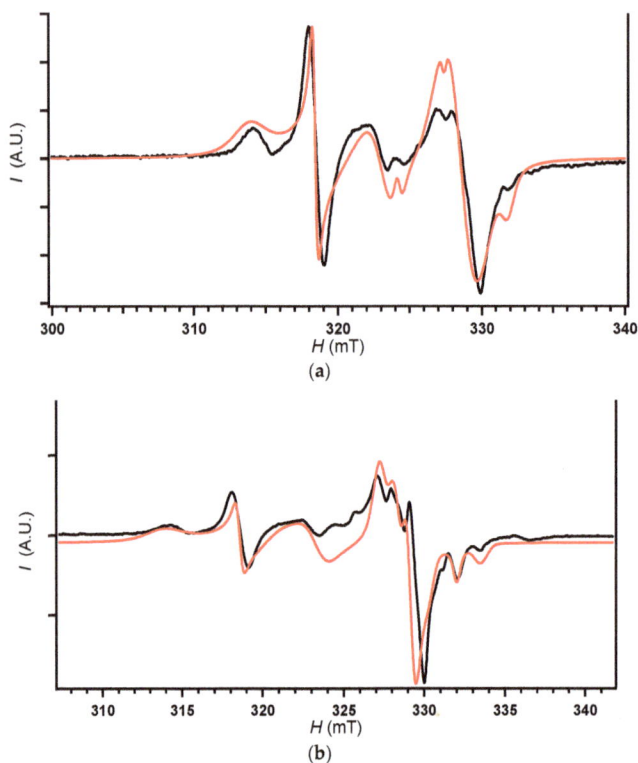

Figure 6. ESR spectra of δ-salt measured under dark and irradiated (254–450 nm) conditions. Black; observed, red; simulated spectra: (**a**) spectrum under dark condition and its simulated spectrum; (**b**) spectrum under irradiated condition and its simulated spectrum.

The spectra of δ-salt under UV-irradiation were clearly different from its spectra under dark conditions (Figure 6). Under dark conditions, the total intensity of peaks due to the spins on the anions (anions' peaks) was comparable to or slightly larger than that due to the spins on the cations (cations' peaks). In contrast, under UV-irradiation, the relative intensity of the cations' peaks reversibly increased to become much more intense than those of anions' peaks, almost retaining the lineshapes and *g*-values. This implies that CT transition occurs between cations and anions under UV-irradiation. As the CT transitions between two different components in solids produce net photocarriers and photoexcited spins at the same time [10–12,16], the observed photoresponse in ESR spectra accounts for the observed photoresponse in electrical behavior mentioned above. The *g*-values of a part of the spins on the N atoms (#3 (N) in the dark conditon in Table A2 were markedly enhanced compared with those of isolated spins on the bipyridine derivatives (*g* ~2.00 for N atoms), indicating strong interaction with the spins having larger *g*-values such as those in heavy atoms and transition metals, i.e., indicating strong interaction between cations and anions. This is consistent with the crystal structure. By comparison between Tables A2 and A3, the simulation of the ESR spectra indicates that the spin densities on the cations relative to the anions increases in the photoexcited state than that in the dark state. This indicates that the photoexcitation corresponds to the CT from the anions to the cations producing localized spins on the cations and doping holes in the anions' bands at the same time.

3. Discussion

3.1. Structure–Property Relations

The obtained results thus far are summarized in Table 2 for comparison. The CT interactions between cations and anions and those between anions are important for production of charge carriers and forming conduction pathways, respectively. However, if they are too strong between a particular pair of molecules, other interactions would inevitably become small. Accordingly, an occurrence of a strong interaction will lead to strong dimerization as is the case in γ- and 2:3-salts. Such a situation corresponds to uneven potential with deep valleys and high barriers for carriers to go through, which is evidently unfavorable to electrical conduction and favorable for localization of unpaired electrons in deep valleys.

As regards carriers, because the photoexcitation can produce/increase them to some extent, the lack of carriers in the ground state does not matter so seriously compared to dark conductivity. In principle, a small (large) cation–anion CT interaction generally involves a large (small) amount of charge transfer between them with a small (large) transition probability. In addition, the actual conduction depends also on the produced carriers' mobility and the mean free path in the excited states/bands as well as those in the originally fully-occupied states/bands. The mobility and relaxation times depend on the characteristics of the bands, and is practically independent of the number of carriers produced unless their densities are high. Therefore, the interrelation between photoresponse and the strength of cation–anion interaction is complicated. The cation–anion CT interaction is necessary for high photoresponse ($\sigma_{photo}/\sigma_{dark}$), yet interaction that is too strong often leads to unfavorable situations for photoconduction, as is the case in the dark conduction.

Table 2. Comparison of properties of DiCC salts (DiCC = 3,3′-Dihexyloxacarbocyanine monocation).

Salts	Dark Conductivity σ_{dark} (Scm^{-1})	Photoconductivity [1] σ_{photo} (Scm^{-1})	$\sigma_{photo}/\sigma_{dark}$	Anion–Anion Interaction [2]	Cation–Anion Interaction [2]
γ	7.9×10^{-9}	7.9×10^{-9}	1	Uneven	Weak
δ	8.6×10^{-7}	5.2×10^{-6}	6	Very weak	Moderate
2:3	8.3×10^{-9}	8.3×10^{-9}	1	Strongly dimerized columns	Strong

[1] Here, "photoconductivity" means the conductivity under irradiation, and the observed values may include contributions of dark conductivity and thermal carriers. [2] Estimated by the structural features.

Thus far, we have not explicitly discussed the possible structural differences between dark and photoexcited states. We have not obtained any data indicating or excluding such a structural change on photoexcitation in these salts, and nothing can be said until some evidence will be obtained in future work.

As regards the relaxation times of photoexcited carriers and localized spins, the ESR spectra of the "photosilent" salts (γ- and 2:3-salts) left us an important message. Whether they exhibit photoresponse in conduction or not, there should be photoexcited spins/carriers in them during the ESR measurements under photoirradiation as long as they absorb the light. However, irradiation did not change the ESR spectra of γ-salt or conduction behavior of γ- and 2:3-salts at all. This can be explained by considering the relaxation times. In the ESR spectra under UV-irradiation, there are at least two kinds of relaxation times, that of UV-excitation (τ_{UV}) and that of ESR, i.e., microwave-excitation (τ_{ESR}). For the observation of ESR on the photoexcited unpaired electrons, the condition $\tau_{UV} \gg \tau_{ESR}$ is required. In addition, if τ_{ESR} in these salts are too short or too long compared with 10^{-10} s, they are not observed in ESR spectra measured using X-band microwaves (9–10 GHz). General substances do not satisfy these requirements, and do not exhibit photoresponses in ESR. Similarly, the increase in conductivity would not be observed under irradiation, when τ_{UV} is far shorter than the time scale of resistivity measurement (~1–10 ms), even if there are a sufficient number of photocarriers with sufficient mobilities. This situation is clearly different from dark conductivity, where carriers can be discussed based on a single kind of relaxation time characteristic of their mean free path. Since all of these salts absorb UV-Vis-NIR light (~10^{15} Hz), τ_{UV} should be on the order of 10^{-15} s, and succeeding processes are considered to differ from each other to result in different relaxation times of photoexcited carriers on the anions and localized spins on the cations. Based on the discussion thus far, the overall relaxation time of photoexcited spins on the cations and anions is unusually prolonged to produce the observed photoresponse in ESR and conduction. Such prolonged relaxation times have been observed only in the CT complexes containing bipyridine derivatives with apparently mixed-stacking structures [10–12]. The mechanism can be related to the characteristic or advantage in this kind of molecular CT complexes such as CT interactions with photosensitive dyes, which would stabilize the photoexcited states. It can be also the key feature for mechanism that they exhibit rather high conduction along the mixed-stacking direction. The detailed experiments are now under way to clarify the mechanism.

3.2. Material Design for and beyond Photoconductors

The Ni(dmit)$_2$ salts with photosensitizer cations were obtained and one of them exhibited semiconducting behavior with a moderately high ratio between photoconduction relative to its dark conduction and unusually long relaxation times of photoexcited carriers and localized spins. This photoexcited behavior is considered to be closely related to the interaction between DiCC$^+$ and [Ni(dmit)$_2$]$^-$ species, which produces carriers and localized spins under UV-irradiation. This structural feature originates from the intermediate molecular packing between ionic and molecular crystal structures with retaining anion–anion molecular orbital overlaps narrowly.

The molecular design and combination strategy here is different from that for molecular metals and superconductors. In the design of highly conducting materials, smaller counter ions are favorable for close packing and self-aggregation of [Ni(dmit)$_2$]$^-$. Additionally, the partial oxidation of [Ni(dmit)$_2$]$^-$ is required for production of carriers. However, highly conducting materials generally do not exhibit high photoconductivity; semiconducting or insulating behavior is required under the dark condition. Here in this work, a rather bulky π-conjugated counter species (DiCC$^+$) and fully ionized [Ni(dmit)$_2$]$^-$ were utilized, which produced a semiconductor with low conductivity.

For efficient photoresponsive conductors and magnets, the relaxation times of photoexcited electrons are one of the most important factors. The long relaxation times observed here have never been reported in known materials for photoelectric conversion, whether they are organic or inorganic, and thus can be a suitable or superior new material for solar cells/photovoltaics, for example. Accordingly, δ-salt proves that the combination of dye sensitizers and π-conjugated radical anions/cations is a promising strategy to photoresponsive semiconductors and paramagnets as next-generation molecular functional materials.

4. Materials and Methods

4.1. Sample Preparation

3,3'-*Dihexyloxacarbocyanine iodide* (DiCC·I) (Sigma-Aldrich, St. Louis, MO, USA, 98%) and acetonitrile (Wako Pure Chemicals, Osaka, Japan, Super Dehydrated Grade) were purchased and used as received. [n-$(C_4H_9)_4$N][Ni(dmit)$_2$] (abbreviated as TBA[Ni]) was synthesized by following the reported procedure [34]. The single crystals of all kinds of salts DiCC$_n$[Ni(dmit)$_2$]$_m$ {(n, m) = (1,2) or (2,3)} were obtained from the salt metathesis of DiCC·I and $(C_4H_9)_4$N[Ni(dmit)$_2$] in CH$_3$CN. The polycrystalline solid of DiCC·I (3 mg) and TBA[Ni] (3 mg) were separately dissolved in 10 mL of acetonitrile and filtered. The former solution was added to the latter solution, and the resulting mixed solution stood still being loosely sealed with a sheet of plastic (polyvinylidene chloride) film at room temperature (RT) under dark conditions to allow gradual evaporation of the solvent until precipitation of the crystals were observed. This procedure yielded γ-, δ- and 2:3-salts as a mixture after a week. Attempts to find synthetic conditions for them in a selective way remain unsuccessful. The single crystals were filtered and washed thoroughly with acetonitrile and acetone, and dried in vacuo. The obtained crystals were identified based on the X-ray oscillation photographs (cell parameters), and subjected to electrical resistivity and ESR measurements.

4.2. X-ray Structural Analysis

The data collection was carried out at 296 K for γ-, δ-, and 2:3-salts using R-AXIS RAPID (Rigaku, Tokyo, Japan). Radiation used was graphite monochromated Cu Kα (λ = 1.54187 Å) for all the salts. The structural analysis was carried out using CrystalStructure 4.1 (Rigaku, Tokyo, Japan). The details are summarized in Supplementary Materials. Crystallographic data have been deposited with Cambridge Crystallographic Data Centre (CCDC): the deposition numbers are summarized in Table 1.

4.3. Physical Measurements

4.3.1. Electrical Resistivity Measurements

The single crystals were freshly prepared and checked in advance by X-ray oscillation photographs for the quality, morphology (to confirm the direction of the crystallographic axes for identifying most conductive direction) and phase (α-, β-, γ-, δ- or 2:3-salts). Both ends of the needle or platelet crystal (~0.5–1 mm) were freshly cut immediately before every measurement, and gold wires were attached upon the fresh cross sections.

As the electrical resistivity of these compounds were found to be generally high (bulk resistance $R > 10^7$ ohm) based on the preliminary measurements, the resistivity measurements were carried out by a direct-current two-probe method. All of the measurements were carried out at 296 K under ambient pressure, and the temperature raise due to Joule heating effects was checked by time-resolved resistivity measurements with the time resolution of 18 ms by the same equipment [13] and on the spot, and temperature and/or resistivity data were taken at the equilibrium/constant states. Their resistivity was too high to measure at low temperature whether under dark or irradiated conditions, and thus the temperature dependence was not measured. A constant voltage (Table 3) was applied and the resultant current was measured. The voltage was varied in a range to confirm the linearity between the voltage and the current to exclude any artifact. The constant voltage was applied along the largest dimensions (longest axes, edges or diagonals) of the crystals. The equipment was homemade cryostat consisting of a picoammeter/voltage source (Keithley 6487, Tektronix, Inc., Beaverton, OR, USA) or a sourcemeter (Keithley 2400, Tektronix, Inc., Beaverton, OR, USA), a digital temperature controller (Model 331, Lake Shore Cryotronics, Inc., Westerville, OH, USA), and a rotary/diffusion packaged pumping system (DS-A412N, Diavac Limited, Yachiyo, Japan). Gold paste (No. 8560, Tokuriki Honten Co., Ltd., Tokyo, Japan) and gold wires (25 μm in diameter; Nilaco 171086, The Nilaco Corporation,

Tokyo, Japan) were used as the electrical contacts. All the measurements of dark conductivity were carried out in a double-shielded copper sample room with a helium atmosphere (~2–5 Pa at RT) under complete darkness. After dark conductivity, the photoconductivity was measured on the same sample in an open atmosphere. The light was guided through an optical fiber to be normal on the sample surface and the distance between the end of the fiber and sample was 1 cm. The light source used was Hg/Xe-Lamp (200 W, Supercure-203S, San-Ei Electric, Osaka, Japan) equipped with adjustable power gain and filters. The irradiation conditions are summarized in Table 4.

Table 3. Conditions in electrical resistivity measurements. [1]

Salt	Direction	Start (V)	End (V)	Interval (V)
γ	// a-axis	100	500	100
δ	// a-axis	20	100	20
2:3	~// a-axis [2]	100	500	100

[1] Common to dark and photoconductivity measurements. The current and voltage were applied and measured in the direction indicated. In order to check linearity between voltage and current, which is required for electrical resistivity measurements, the measurement was carried out by applying six different voltages in series by increasing the voltage from the value of Start [V] to that of End [V] with the interval [V] indicated in the table. [2] The exact direction was the [210]-direction, which was nearly parallel (~15°) with the a-axis.

Table 4. Irradiation conditions in photoconductivity measurements. [1]

Salt	Wavelength Range (1) (nm)	Intensity (1) (Wcm^{-2})	Wavelength Range (2) (nm)	Intensity (2) (Wcm^{-2})
γ	254–450	6.7	254–1100	12.3
δ	254–450	6.7	NA	NA
2:3	254–450	6.7	254–1100	12.3

[1] Some samples were examined under two different irradiation conditions.

4.3.2. Electron Spin Resonance (ESR)

All of the ESR measurements were carried out using single crystals. The single crystal, which was briefly checked by X-ray oscillation photographs, was set on a piece of silicone sheet with Apiezon N grease (Apiezon, Manchester, UK). This was set on a piece of Teflon with Apiezon N in a quartz ESR tube (5 mm in diameter) with the proper alignment relative to the magnetic field, and the tube was sealed in a helium atmosphere (~10–20 kPa). The measurements were carried out using a JES-FA100 (X-band spectrometer; 9.3 GHz, JEOL Ltd., Tokyo, Japan) at a constant temperature. The temperature was controlled so as not to allow the temperature variation to exceed ±0.5 K during the field sweep. The magnetic field was corrected by a Gaussian meter (NMR Field Meter ES-FC5, JEOL Ltd., Tokyo, Japan) at the end of every measurement. The accuracy of magnetic fields was further confirmed by a single crystal of DPPH (1,1-Diphenyl-2-picrylhydrazyl; $g = 2.00366 \pm 0.00004$) [35–37], which was used as a standard of g-values. The temperature (77 K), sweep time (30 s), modulation (100 kHz), microwave power (0.998 mW), and time constant (0.01 s) were identical for each salt and common to the measurements under dark and irradiated conditions. The time constants were varied and finally determined to be the best values considering signal:noise ratio and observation of details of the lineshapes (hyperfine structures). For every salt, the ESR measurements under dark conditions were carried out first. Then, the sample in the ESR tube was irradiated for 10 min through an optical fiber and a focusing lens in situ through an optical window of the ESR cavity. For all of the salts, the ESR spectra gradually changed during the first several minutes of irradiation. After confirming that the spectra did not change any more, the spectra were recorded under continuous irradiation by integrating the spectra by 10–20 times. The spectra simulation was carried out using Anisotropic Simulation software AniSim/FA ver. 2.2.0 (JEOL Ltd., Tokyo, Japan). In the simulation, the parameters for g-value, hyperfine coupling constants A (mT) of ^{14}N ($I = 1$) and ^{1}H ($I = 1/2$), linewidth Γ (mT), and the ratio between Lorentzian and Gaussian were considered in an anisotropic way.

4.3.3. Diffuse Reflectance Spectra

Diffuse reflectance spectra in the UV-vis-NIR (200–2500 nm) region were measured for the samples dispersed in a KBr pellet and sandwiched with a pair of quartz glasses with a U-4000 spectrophotometer (Hitachi High-Technologies Corporation, Tokyo, Japan) at RT with a resolution of 2 nm.

Supplementary Materials: The following are available online at www.mdpi.com/2312-7481/3/2/20/s1

Acknowledgments: This work was partially supported by Ehime University GP, and an Ehime University. Grant for Interdisciplinary Research.

Author Contributions: T.N. and R.Y. conceived and designed the experiments; R.Y. performed the experiments; T.N. and R.Y. analyzed the data; T.Y. and K.O. helped R.Y. in the measurements of resistivity and ESR, respectively; and T.N. wrote the paper.

Conflicts of Interest: The authors declare no conflict of interest. The founding sponsors had no role in the design of the study; in the collection, analyses, or interpretation of data; in the writing of the manuscript, and in the decision to publish the results.

Appendix A. Diffuse Reflectance Spectra

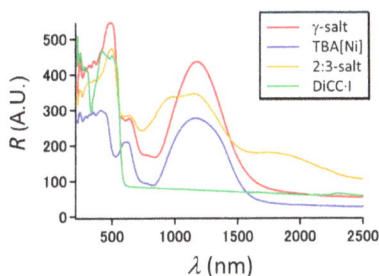

Figure A1. Diffuse reflectance spectra of γ- and 2:3-salts (powder). Spectra of [n-(C$_4$H$_9$)$_4$N][Ni(dmit)$_2$] (TBA[Ni]) and DiCC·I (both in powder) are also shown for comparison. Red; γ-salt, blue; [n-(C$_4$H$_9$)$_4$N][Ni(dmit)$_2$], yellow; 2:3-salt, green; DiCC·I.

Appendix B. Electron Spin Resonance (ESR) Spectra

Table A1. Parameters for simulated spectrum of γ-salt [a,b].

Spin # [c]	#1	#2	#3	#4
I_{rel} (%)	82.8	13.9	1.65	1.65
I	0 (^{32}S)	0 (^{32}S)	1 (^{14}N)	1 (^{14}N)
g_x	2.0855	2.0743	2.0060	2.0180
g_y	2.0855	2.0743	2.0060	2.0175
g_z	1.9970	2.0510	2.0058	2.0160
A_x (mT)	NA	NA	0.100	1.300
A_y (mT)	NA	NA	0.100	0.100
A_z (mT)	NA	NA	0.100	0.100
Γ_x (mT)	0.660	0.700	0.800	0.700
Γ_y (mT)	0.660	0.700	0.800	0.800
Γ_z (mT)	0.650	0.800	1.000	1.000
Lorentzian/Gaussian	100/0	0/100	100/0	100/0

[a] The observed spectra were completely identical and overlapped with each other under dark and irradiated conditions. Thus, the parameters shown above are common to both spectra under dark and irradiated conditions. [b] I_{rel}, I, g_i, A_i and Γ_i ($i = x, y, z$) designate relative intensity, nuclear spin, g-value, hyperfine coupling constant and linewidth in the i-direction, respectively. [c] Spin # designate the serial numbers of the oscillators required for the reproduction of the observed spectra. It does not mean that there are such numbers of independent spins in the sample.

Table A2. Parameters for simulated spectrum of δ-salt under dark conditions [a].

Spin # [b]	#1	#2	#3	#4
I_{rel} (%)	82.5	6.19	1.03	10.3
I	0 (^{32}S)	0 (^{32}S)	1 (^{14}N)	1 (^{14}N)
g_x	2.0980	2.0500	2.0310	2.0040
g_y	2.0680	2.0400	2.0308	2.0040
g_z	1.9843	2.0328	2.0112	2.0040
A_x (mT)	NA	NA	0.100	0.100
A_y (mT)	NA	NA	0.100	0.100
A_z (mT)	NA	NA	0.100	0.100
Γ_x (mT)	2.000	2.000	0.800	2.200
Γ_y (mT)	0.300	1.300	0.300	2.200
Γ_z (mT)	0.800	1.000	0.150	2.200
Lorentzian/Gaussian	80/20	100/0	100/0	50/50

[a] I_{rel}, I, g_i, A_i and Γ_i ($i = x, y, z$) designate relative intensity, nuclear spin, g-value, hyperfine coupling constant and linewidth in the i-direction, respectively. [b] Spin # designate the serial numbers of the oscillators required for the reproduction of the observed spectra. It does not mean that there are such numbers of independent spins in the sample.

Table A3. Parameters for simulated spectrum of δ-salt under irradiated condition [a].

Spin # [b]	#1	#2	#3	#4
I_{rel} (%)	63.8	26.6	7.98	1.60
I	0 (^{32}S)	0 (^{32}S)	1 (^{14}N)	1 (^{14}N)
g_x	2.0980	2.0500	2.0048	2.0008
g_y	2.0670	2.0370	2.0048	2.0006
g_z	1.9832	1.9745	2.0043	2.0004
A_x (mT)	NA	NA	1.000	1.200
A_y (mT)	NA	NA	1.000	0.100
A_z (mT)	NA	NA	0.100	0.100
Γ_x (mT)	2.000	2.000	0.800	0.600
Γ_y (mT)	0.400	1.300	0.800	0.600
Γ_z (mT)	0.400	0.800	0.800	0.600
Lorentzian/Gaussian	80/20	100/0	80/20	100/0

[a] I_{rel}, I, g_i, A_i and Γ_i ($i = x, y, z$) designate relative intensity, nuclear spin, g-value, hyperfine coupling constant and linewidth in the i-direction, respectively. [b] Spin # designate the serial numbers of the oscillators required for the reproduction of the observed spectra. It does not mean that there are such numbers of independent spins in the sample.

References

1. Watanabe, T.; Honda, K. Measurement of the excitation coefficient of the methyl viologen cation radical and the efficiency of its formation by semiconductor photocatalysis. *J. Phys. Chem.* **1982**, 86, 2617–2619. [CrossRef]
2. Mohammad, M. Methyl viologen neutral MV: 1. Preparation and some properties. *J. Org. Chem.* **1987**, 52, 2779–2782. [CrossRef]
3. Yoon, K.B.; Kochi, J.K. Direct observation of superoxide electron transfer with viologens by immobilization in zeolite. *J. Am. Chem. Soc.* **1988**, 110, 6586–6588. [CrossRef]
4. Yoon, K.B.; Kochi, J.K. Shape-selective access to zeolite supercages. Arene charge-transfer complexes with viologens as visible probes. *J. Am. Chem. Soc.* **1989**, 111, 1128–1130. [CrossRef]
5. Bockman, T.M.; Kochi, J.K. Isolation and oxidation-reduction of methylviologen cation radicals. Novel disproportionation in charge-transfer salts by X-ray crystallography. *J. Org. Chem.* **1990**, 55, 4127–4135. [CrossRef]
6. Akins, D.L.; Guo, C. Photoinduced electron transfer in synthetic model systems. *Adv. Mater.* **1994**, 6, 512–516. [CrossRef]

7. Guo, J.; Nie, J.; Lv, Z. Synthesis and properties of bipyridyl-based dye-sensitizers. *Res. Chem. Intermed.* **2013**, *39*, 4247–4257. [CrossRef]

8. Coe, B.J.; Sanchez, S. Synthesis and properties of new mononuclear Ru(II)-based photocatalysts containing 4,4′-diphenyl-2,2′-bipyridyl ligands. *Dalton Trans.* **2016**, *45*, 5210–5222. [CrossRef] [PubMed]

9. Naito, T.; Inabe, T. Molecular conductors containing photoreactive species. *J. Phys. IV* **2004**, *114*, 553–555. [CrossRef]

10. Naito, T.; Karasudani, T.; Mori, S.; Ohara, K.; Konishi, K.; Takano, T.; Takahashi, Y.; Inabe, T.; Nishihara, S.; Inoue, K. Molecular photoconductor with simultaneously photocontrollable localized spins. *J. Am. Chem. Soc.* **2012**, *134*, 18656–18666. [CrossRef] [PubMed]

11. Naito, T.; Karasudani, T.; Ohara, K.; Takano, T.; Takahashi, Y.; Inabe, T.; Furukawa, K.; Nakamura, T. Simultaneous control of carriers and localized spins with light in organic materials. *Adv. Mater.* **2012**, *24*, 6153–6157. [CrossRef] [PubMed]

12. Naito, T.; Karasudani, T.; Nagayama, N.; Ohara, K.; Konishi, K.; Mori, S.; Takano, T.; Takahashi, Y.; Inabe, T.; Kinose, S.; et al. Giant photoconductivity in NMQ[Ni(dmit)$_2$]. *Eur. J. Inorg. Chem.* **2014**, *2014*, 4000–4009. [CrossRef]

13. Saiki, T.; Mori, S.; Ohara, K.; Naito, T. Capacitor-like behavior of molecular crystal β-DiCC[Ni(dmit)$_2$]. *Chem. Lett.* **2014**, *43*, 1119–1121. [CrossRef]

14. Noma, H.; Ohara, K.; Naito, T. [Cu(dmit)$_2$]$^{2−}$ Building block for molecular conductors and magnets with photocontrollable spin distribution. *Chem. Lett.* **2014**, *43*, 1230–1232. [CrossRef]

15. Noma, H.; Ohara, K.; Naito, T. Direct control of spin distribution and anisotropy in Cu-dithiolene complex anions by light. *Inorganics* **2016**, *4*. [CrossRef]

16. Naito, T. Development of control method of conduction and magnetism in molecular crystals. *Bull. Chem. Soc. Jpn.* **2017**, *90*, 89–136. [CrossRef]

17. Kisch, H.; Fernández, A.; Wakatsuki, Y.; Yamazaki, H. Charge transfer complexes of nickel dithiolenes with methyl viologen. *Z. Naturforsch.* **1985**, *40B*, 292–297. [CrossRef]

18. Fernández, A.; Görner, H.; Kisch, H. Photoinduzierte elektronenübertragung mit metalldithiolenen. *Chem. Berichte* **1985**, *118*, 1936–1948. [CrossRef]

19. Lahner, S.; Wakatsuki, Y.; Kisch, H. Charge-transfer-komplexe von metalldithiolenen mit viologenen. *Chem. Berichte* **1987**, *120*, 1011–1016. [CrossRef]

20. Nüßlein, F.; Peter, R.; Kisch, H. Viologene als redoxaktive akzeptoren–synthese und electrische leifähigkeit. *Chem. Berichte* **1989**, *122*, 1023–1030. [CrossRef]

21. Kisch, H.; Nüsslein, F.; Zenn, I. Molekulare steuerung von festkörpereigenschaften: Die elecktrische leitfähigkeit von metal-organischen charge-transfer komplexen. *Z. Anorg. Allg. Chem.* **1991**, *600*, 67–71. [CrossRef]

22. Kisch, H.; Dümler, W.; Nüßlein, F.; Zenn, I.; Chiorboli, C.; Scandola, F.; Albrecht, W.; Meier, H. Consequences of thermal and photochemical electron transfer on solution and solid state properties of metal organic ion pairs. *Z. Phys. Chem.* **1991**, *170*, 117–127.

23. Meier, H.; Albrecht, W.; Kisch, H.; Nunn, I.; Nüsslein, F. Photoconductivity of metal dithiolate ion-pair complexes. *Synth. Met.* **1992**, *48*, 111–127. [CrossRef]

24. Kisch, H. Charge-transfer in ion pairs: Design of photoconductivity in solution and electrical dark-and photoconductivity in the solid. *Coord. Chem. Rev.* **1993**, *125*, 155–172. [CrossRef]

25. Lemke, M.; Knoch, F.; Kisch, H. Structure of the ion-pair charge-transfer complex (methylviologen)$^{2+}$[Pd(mnt)$_2$]$^{2−}$. *Acta Cryst.* **1993**, *C49*, 1630–1632. [CrossRef]

26. Nunn, I.; Eisen, B.; Benedix, R.; Kisch, H. Control of electrical conductivity by supramolecular charge-transfer interactions in (dithiolene)metalate-viologen ion pairs. *Inorg. Chem.* **1994**, *33*, 5079–5085. [CrossRef]

27. Kisch, H. Electron transfer modelling of electrical dark- and photoconductivity of redoxactive ion pairs. *Comments Inorg. Chem.* **1994**, *16*, 113–132. [CrossRef]

28. Knoch, F.; Ammon, U.; Kisch, H. Crystal structure of *N*,*N*′-dimethyl-2,2′-bipyridinium bis(*cis*-1,2-dicyanoethane-1,2-dithiolato)palatinate(II), ((CH$_3$)C$_5$NH$_4$C$_5$NH$_4$(CH$_3$))(Pt(S$_2$C$_2$(CN)$_2$)$_2$). *Z. Kristallographie* **1995**, *210*, 77–78.

29. Hofbauer, M.; Möbius, M.; Knoch, F.; Benedix, R. Ion-pair charge-transfer complexes of a dithiooxalate zinc donor component with viologens. Synthesis, structural and electronic characterization. *Inorg. Chim. Acta* **1996**, *247*, 147–154. [CrossRef]

30. Götz, B.; Knoch, F.; Kisch, H. Bis(maleonitriledithiolato)oxomolybdate(IV)-bipyridinium ion pairs. *Chem. Berichte* **1996**, *129*, 33–37. [CrossRef]

31. Kisch, H. Tailoring of solid state electrical conductivity and optical electron transfer activation of dioxygen in solution through supramolecular charge-transfer interaction in ion pairs. *Coord. Chem. Rev.* **1997**, *159*, 385–396. [CrossRef]

32. Handrosch, C.; Dinnebier, R.; Bondarenko, G.; Bothe, E.; Heinemann, F.; Kisch, H. Charge-transfer complexes of metal dithiolenes XXVI Azobipyridinium dications and radical monocations as acceptors. *Eur. J. Inorg. Chem.* **1999**, *1999*, 1259–1269. [CrossRef]

33. Kisch, H.; Eisen, B.; Dinnebier, R.; Shankland, K.; David, W.I.F.; Knoch, F. Chiral metal-dithiolene/viologen ion pairs: Synthesis and electrical conductivity. *Chem. Eur. J.* **2001**, *7*, 738–748. [CrossRef]

34. Steimecke, G.; Sieler, H.-J.; Kirmse, R.; Hoyer, E. 1,3-dithiol-2-thion-4,5-dithiolat aus schwefelkohlenstoff und alkalimetall. *Phosphorus Sulfur* **1979**, *7*, 49–55. [CrossRef]

35. Yordanov, N.D. Quantitative EPR spectrometry—"State of the art". *Appl. Magn. Reson.* **1994**, *6*, 241–257. [CrossRef]

36. Kai, A.; Miki, T. Electron spin resonance of sulfite radicals in irradiated calcite and aragonite. *Radiat. Phys. Chem.* **1992**, *40*, 469–476. [CrossRef]

37. Inokuchi, H.; Kinoshita, M. The oxygen effect on electronic properties of α,α'-diphenyl-β-picrylhydrazyl. *Bull. Chem. Soc. Jpn.* **1960**, *33*, 1627–1629. [CrossRef]

MDPI

St. Alban-Anlage 66

4052 Basel

Switzerland

Tel. +41 61 683 77 34

Fax +41 61 302 89 18

www.mdpi.com

Magnetochemistry Editorial Office

E-mail: magnetochemistry@mdpi.com

www.mdpi.com/journal/magnetochemistry

www.ingramcontent.com/pod-product-compliance
Lightning Source LLC
Chambersburg PA
CBHW051837210326
41597CB00033B/5689